Algebra and Computer Science

CONTEMPORARY MATHEMATICS

677

Algebra and Computer Science

Joint AMS-EMS-SPM Meeting
Algebra and Computer Science
June 10–13, 2015: Porto, Portugal

Joint Mathematics Meetings
Groups, Algorithms, and Cryptography
January 10–13, 2015: San Antonio, TX

Joint AMS-Israel Mathematical Union Meeting
Applications of Algebra to Cryptography
June 16–19, 2014: Tel-Aviv, Israel

Delaram Kahrobaei
Bren Cavallo
David Garber
Editors

American Mathematical Society
Providence, Rhode Island

EDITORIAL COMMITTEE

Dennis DeTurck, Managing Editor

Michael Loss   Kailash Misra   Catherine Yan

2010 *Mathematics Subject Classification.* Primary 20-XX, 68-XX.

Library of Congress Cataloging-in-Publication Data

Library of Congress Cataloging-in-Publication Data

Names: Kahrobaei, Delaram, 1975– editor. | Cavallo, Bren, 1989– editor. | Garber, David, 1973– editor.

Title: Algebra and computer science / Delaram Kahrobaei, Bren Cavallo, David Garber, editors.

Description: Providence, Rhode Island : American Mathematical Society, [2016] | Series: Contemporary mathematics ; volume 677 | Special Session at the Joint AMS-EMS-SPM Meeting, Algebra and Computer Science, June 10–13, 2015, Porto, Portugal. | Special Session at the Joint Mathematics Meetings, Groups, Algorithms, and Cryptography, January 10–13, 2015, San Antonio, TX. | Special Session at the Joint AMS-Israel Mathematical Union Meeting, Applications of Algebra to Cryptography, June 16–19, 2014, Tel-Aviv, Israel. | Includes bibliographical references.

Identifiers: LCCN 2016019097 | ISBN 9781470423032 (alk. paper)

Subjects: LCSH: Logic, Symbolic and mathematical–Congresses. | Algebra–Congresses. | Computer science–Mathematics–Congresses. | AMS: Group theory and generalizations. msc | Computer science. msc

Classification: LCC QA9.A1 A44 2016 | DDC 512.0285–dc23 LC record available at https://lccn.loc.gov/2016019097

Contemporary Mathematics ISSN: 0271-4132 (print); ISSN: 1098-3627 (online)

DOI: http://dx.doi.org/10.1090/conm/677

Color graphic policy. Any graphics created in color will be rendered in grayscale for the printed version unless color printing is authorized by the Publisher. In general, color graphics will appear in color in the online version.

Copying and reprinting. Individual readers of this publication, and nonprofit libraries acting for them, are permitted to make fair use of the material, such as to copy select pages for use in teaching or research. Permission is granted to quote brief passages from this publication in reviews, provided the customary acknowledgment of the source is given.

Republication, systematic copying, or multiple reproduction of any material in this publication is permitted only under license from the American Mathematical Society. Permissions to reuse portions of AMS publication content are handled by Copyright Clearance Center's RightsLink® service. For more information, please visit: http://www.ams.org/rightslink.

Send requests for translation rights and licensed reprints to reprint-permission@ams.org.

Excluded from these provisions is material for which the author holds copyright. In such cases, requests for permission to reuse or reprint material should be addressed directly to the author(s). Copyright ownership is indicated on the copyright page, or on the lower right-hand corner of the first page of each article within proceedings volumes.

© 2016 by the American Mathematical Society. All rights reserved.
The American Mathematical Society retains all rights
except those granted to the United States Government.
Copyright of individual articles may revert to the public domain 28 years
after publication. Contact the AMS for copyright status of individual articles.
Printed in the United States of America.

∞ The paper used in this book is acid-free and falls within the guidelines
established to ensure permanence and durability.
Visit the AMS home page at http://www.ams.org/

10 9 8 7 6 5 4 3 2 1   21 20 19 18 17 16

Contents

Preface

This volume consists of contributions by participants and speakers in special sessions at three AMS meetings. These special sessions concerned algorithmic problems in algebra and applications to computer science and cryptography. One of the special sessions was at Tel Aviv University, Israel in June 2014, another at the University of Porto, Portugal in June 2015, and the other one at the San Antonio Convention Center, Texas in January 2015.

Over the past few years the field of interaction between computer science and algebra has attracted the attention of both algebraists and computer scientists.

This volume contains both survey and research papers on algorithmic algebra and applications in computer science, particularly cryptography and complexity theory.

We are grateful to the American Mathematical Society for their help in the publication of this volume. In particular we thank Christine Thivierge for her patience and assistance in putting this volume together.

Delaram Kahrobaei
Bren Cavallo
David Garber

Contemporary Mathematics
Volume **677**, 2016
http://dx.doi.org/10.1090/conm/677/13619

Generic properties of subgroups of free groups and finite presentations

Frédérique Bassino, Cyril Nicaud, and Pascal Weil

Abstract. Asymptotic properties of finitely generated subgroups of free groups, and of finite group presentations, can be considered in several fashions, depending on the way these objects are represented and on the distribution assumed on these representations: here we assume that they are represented by tuples of reduced words (generators of a subgroup) or of cyclically reduced words (relators). Classical models consider fixed size tuples of words (e.g. the few-generator model) or exponential size tuples (e.g. Gromov's density model), and they usually consider that equal length words are equally likely. We generalize both the few-generator and the density models with probabilistic schemes that also allow variability in the size of tuples and non-uniform distributions on words of a given length.

Our first results rely on a relatively mild prefix-heaviness hypothesis on the distributions, which states essentially that the probability of a word decreases exponentially fast as its length grows. Under this hypothesis, we generalize several classical results: exponentially generically a randomly chosen tuple is a basis of the subgroup it generates, this subgroup is malnormal and the tuple satisfies a small cancellation property, even for exponential size tuples. In the special case of the uniform distribution on words of a given length, we give a phase transition theorem for the central tree property, a combinatorial property closely linked to the fact that a tuple freely generates a subgroup. We then further refine our results when the distribution is specified by a Markovian scheme, and in particular we give a phase transition theorem which generalizes the classical results on the densities up to which a tuple of cyclically reduced words chosen uniformly at random exponentially generically satisfies a small cancellation property, and beyond which it presents a trivial group.

This paper is part of the growing body of literature on asymptotic properties of subgroups of free groups and of finite group presentations, which goes back at least to the work of Gromov [**10**] and Arzhantseva and Ol'shanskii [**1**]. As in much of the recent literature, the accent is on so-called generic properties, that is, properties whose probability tends to 1 when the size of instances grows to infinity. A theory

2010 *Mathematics Subject Classification.* Primary 20E05, 60J10; Secondary 20E07, 05A16, 68Q17.

Key words and phrases. Asymptotic properties, generic properties, random subgroups, random presentations, Markovian automata, malnormality, small cancellation.

The authors acknowledge partial support from ANR projects ANR 2010 Blan 0202_01 Frec, ANR 2012 JCJC JS02-012-0 MealyM and ANR 2010 Blan 0204_07 Magnum, as well as from ERC grant PCG-336983 and the Programme IdEx Bordeaux - CPU (ANR-10-IDEX-03-02).

©2016 American Mathematical Society

of genericity and its applications to complexity theory was initiated by Kapovich, Myasnikov, Schupp and Shpilrain [**14**], and developed in a number of papers, see Kapovich for a recent discussion [**13**].

Genericity, and more generally asymptotic properties, depends on the fashion in which input is represented: finitely presented groups are usually given by finite presentations, i.e. tuples of cyclically reduced words; finitely generated subgroups of free groups can be represented by tuples of words (generators) or Stallings graphs. The representation by Stallings graphs is investigated by the authors, along with Martino and Ventura in [**3**–**5**] but we will not discuss it in this paper: we are dealing, like most of the literature, with tuples of words.

There are, classically, two main models (see Section 2.2): the few words model, where an integer k is fixed and one considers k-tuples of words of length at most n, when n tends to infinity, see e.g. [**1**, **3**, **5**, **12**]; and the density model, where we consider tuples of cyclically reduced words of length n, whose size grows exponentially with n, see e.g. [**7**, **10**, **23**, **25**].

Typical properties investigated include the following (see in particular Sections 1.2 and 1.3): whether a random tuple $\vec{h}$ freely generates the subgroup $H = \langle\vec{h}\rangle$ [**1**, **12**], whether H is malnormal [**3**, **12**] or Whitehead minimal [**5**, **27**], whether the finite presentation with relators $\vec{h}$ has a small cancellation property, or whether the group it presents is infinite or trivial [**23**].

All these models implicitly assume the uniform distribution on the set of reduced words of equal length (Ollivier also considers non-uniform distributions in [**23**]).

We introduce (Section 3) a model for probability distributions on tuples of reduced words that is sufficiently general to extend the few words model and Gromov's density model mentioned above, and to leave space for non uniform distributions. Like these two models, ours assumes that a tuple $\vec{h}$ of words is generated by independently drawing words of given lengths, but it also handles independently the size of $\vec{h}$ and the lengths of the words in $\vec{h}$.

Our first set of results assumes a *prefix-heaviness* hypothesis on the probability distribution on words: the probability of drawing a word decreases exponentially fast as its length grows (precise definitions are given in Section 3). It is a natural hypothesis if we imagine that our probabilistic source generates words one letter at a time, from left to right. This relatively mild hypothesis suffices to obtain general results on the exponential genericity of a certain geometric property of the Stallings graph of the subgroup H generated by a randomly chosen tuple $\vec{h}$ (the *central tree property*, implicitly considered in [**1**, **12**] and explicitly in [**5**]), of the fact that $\vec{h}$ freely generates H, and of the malnormality of H, see Section 3.5.

In Section 3.6, we apply these general results to the uniform distribution and generalize known results in two directions. Firstly we consider random exponential size tuples, for which we give a phase transition theorem for the central tree property: it holds exponentially generically up to density $\frac{1}{4}$, and fails exponentially generically at densities greater than $\frac{1}{4}$ (Proposition 3.21). In particular, a random tuple is exponentially generically a basis of the subgroup it generates up to density $\frac{1}{4}$, but we cannot say anything of that property at higher densities.

We also extend Jitsukawa's result on malnormality [**12**], from fixed size to exponential size tuples under uniform distribution up to density $\frac{1}{16}$ (Proposition 3.22).

In view of the methods used to establish this result, it is likely that the value $\frac{1}{16}$ is not optimal.

Secondly, we show that the height of the central tree of a random fixed size tupe $\vec{h}$, which measures the amount of initial cancellation between the elements of $\vec{h}$ and $\vec{h}^{-1}$, is generically less than any prescribed unbounded non-decreasing function (Proposition 3.24). Earlier results only showed that this height was exponentially generically bounded by any linear function.

We then introduce *Markovian automata*, a probabilistic automata-theoretic model, to define explicit instances of prefix-heavy distributions (Section 4). Additional assumptions like irreducibility or ergodicity lead to the computation of precise bounds for the parameters of prefix-heaviness. In particular, we prove a phase transition theorem for ergodic Markovian automata (Section 4.4), showing that small cancellation properties generically hold up to a certain density, and generically do not hold at higher densities. More precisely, if $\alpha_{[2]}$ is the coincidence probability of the Markovian automaton, Property $C'(\lambda)$ holds exponentially generically at $\alpha_{[2]}$-density less than $\frac{\lambda}{2}$ (that is: for random tuples of size $\alpha_{[2]}^{-dn}$ for some $d < \frac{\lambda}{2}$), and fails exponentially generically at $\alpha_{[2]}$-densities greater than $\frac{\lambda}{2}$. We also show that at $\alpha_{[2]}$-densities greater than $\frac{1}{2}$, a random tuple of cyclically reduced words generically presents a degenerate group (see Proposition 4.23 for a precise definition). These results generalize the classical results on uniform distribution in Ollivier [**23, 24**]. It remains to be seen whether our methods can be applied to fill the gap, say, between $\alpha_{[2]}$-density $\frac{1}{12}$ and $\frac{1}{2}$, where small cancellation property $C'(\frac{1}{6})$ generically does not hold yet the presented group might be hyperbolic, see [**23, 24**].

Some of the definitions in this paper, notably that of Markovian automata, were introduced by the authors in [**2**], and some of the results were announced there as well. The results in the present paper are more precise, and subsume those of [**2**].

1. Free groups, subgroups and presentations

In this section, we set the notation and basic definitions of the properties of subgroups of free groups and finite presentations which we will consider.

1.1. Free groups and reduced words. Let A be a finite non-empty set, which will remain fixed throughout the paper, with $|A| = r$, and let $\tilde{A}$ be the symmetrized alphabet, namely the disjoint union of A and a set of formal inverses $A^{-1} = \{a^{-1} \in A \mid a \in A\}$. By convention, the formal inverse operation is extended to $\tilde{A}$ by letting $(a^{-1})^{-1} = a$ for each $a \in A$. A word in $\tilde{A}^*$ (that is: a word written on the alphabet $\tilde{A}$) is *reduced* if it does not contain length 2 factors of the form aa^{-1} ($a \in \tilde{A}$). If a word is not reduced, one can *reduce* it by iteratively deleting every factor of the form aa^{-1}. The resulting reduced word is uniquely determined: it does not depend on the order of the cancellations. For instance, $u = aabb^{-1}a^{-1}$ reduces to aaa^{-1}, and thence to a.

The set F of reduced words is naturally equipped with a group structure, where the product $u \cdot v$ is the (reduced) word obtained by reducing the concatenation uv. This group is called the *free group* on A. More generally, every group isomorphic to F, say, $G = \varphi(F)$ where φ is an isomorphism, is said to be a free group, *freely generated by* $\varphi(A)$. The set $\varphi(A)$ is called a *basis* of G. Note that if $r \geq 2$, then F has infinitely many bases: if, for instance, $a \neq b$ are elements of A, then replacing a by b^nab^m (for some integers n, m) yields a basis. The *rank* of F (or of any isomorphic free group) is the cardinality $|A|$ of A, and one shows that this notion is well-defined in the following sense: every basis of F has the same cardinality.

Let x, y be elements of a group G. We say that y is a *conjugate* of x if there exists an element $g \in G$ such that $y = g^{-1}xg$, which we write $y = x^g$. The notation is extended to subsets of G: if $H \subseteq G$, then $H^g = \{x^g \mid x \in H\}$. Conjugacy of elements of the free group F is characterized as follows. Say that a word u is *cyclically reduced word* if it is non-empty, reduced and its first and last letters are not mutually inverse (or equivalently, if u^2 is non-empty and reduced). For instance, $ab^{-1}a^{-1}bbb$ is cyclically reduced, but $ab^{-1}a^{-1}bba^{-1}$ is not.

For every reduced word u, let $\kappa(u)$ denote its *cyclic reduction*, which is the shortest word v such that $u = wvw^{-1}$ for some word w. For instance, $\kappa(ab^{-1}a^{-1}bba^{-1}) = a^{-1}b$. It is easily verified that two reduced words u and v are conjugates if and only if $\kappa(u)$ and $\kappa(v)$ are *cyclic conjugates* (that is: there exist words x and y such that $\kappa(u) = xy$ and $\kappa(v) = yx$).

Let $\mathcal{R}_n$ (resp. $\mathcal{C}_n$) denote the set of all reduced (resp. cyclically reduced) words of length $n \geq 1$, and let $\mathcal{R} = \bigcup_{n\geq 1} \mathcal{R}_n$ and $\mathcal{C} = \bigcup_{n\geq 1} \mathcal{C}_n$ be the set of all reduced words, and all cyclically reduced words, respectively.

Every word of length 1 is cyclically reduced, so $|\mathcal{R}_1| = |\mathcal{C}_1| = 2r$. A reduced word of length $n \geq 2$ is of the form ua, where u is reduced and a is not the inverse of the last letter of u. An easy induction shows that there are $|\mathcal{R}_n| = 2r(2r-1)^{n-1} = \frac{2r}{2r-1}(2r-1)^n$ reduced words of length $n \geq 2$.

Similarly, if $n \geq 2$, then $\mathcal{C}_n$ is the set of words of the form ua, where u is a reduced word and $a \in \tilde{A}$ is neither the inverse of the first letter of u, nor the inverse of its last letter: for a given u, there are either $2r-1$ or $2r-2$ such words, depending whether the first and last letter of u are equal. In particular, the number of words in $\mathcal{C}_n$ satisfies $\frac{2r}{2r-1}(2r-1)^{n-1}(2r-2) \leq |\mathcal{C}_n| \leq \frac{2r}{2r-1}(2r-1)^n$, and in particular, $|\mathcal{C}_n| = \Theta((2r-1)^n)$.

1.2. Subgroups and presentations. Given a tuple $\vec{h} = (h_1, \ldots, h_k)$ of elements of F, let $\vec{h}^{\pm} = (h_1, h_1^{-1}, \ldots, h_k, h_k^{-1})$ and let $\langle \vec{h} \rangle$ denote the subgroup of F generated by the elements of $\vec{h}$, that is, the set of all the elements of F which can be written as a product of elements of $\vec{h}^{\pm}$. It is a classical result of Nielsen that every such subgroup is free [**22**].

An important property of subgroups is malnormality, which is related to geometric considerations (*e.g.* [**9, 17**]): a subgroup H of a group G is *malnormal* if $H \cap H^x$ is trivial for every $x \notin H$. It is decidable whether a finitely generated subgroup $\langle \vec{h} \rangle$ is malnormal ([**12, 15**], see Section 1.3), whereas malnormality is not decidable in general hyperbolic groups [**6**].

A tuple $\vec{h}$ of elements of $F(A)$ can also be considered as a set of relators in a group presentation. More precisely, we denote by $\langle A \mid \vec{h} \rangle$ the group with generator set A and relators the elements of $\vec{h}$, namely the quotient of $F(A)$ by the normal subgroup generated by $\vec{h}$. It is customary to consider such a group presentation only when $\vec{h}$ consists only of cyclically reduced words, since $\langle A \mid \vec{h} \rangle = \langle A \mid \kappa(\vec{h}) \rangle$.

The small cancellation property is a combinatorial property of a group presentation, with far-reaching consequences on the quotient group. Let $\vec{h}$ be a tuple of cyclically reduced words. A *piece* in $\vec{h}$ is a word u with at least two occurrences as a prefix of a cyclic conjugate of a word in $\vec{h}^{\pm}$. Let $0 < \lambda < 1$. The tuple $\vec{h}$ (or the group presentation $\langle A \mid \vec{h}\rangle$) has the *small cancellation property* $C'(\lambda)$ if whenever a piece u occurs as a prefix of a cyclic conjugate w of a word in $\vec{h}^{\pm}$, then $|u| < \lambda|w|$.

The following properties are well-known. We do not give the definition of the group-theoretic properties in this statement and refer the reader to [**19**] or to the comprehensive survey [**24**].

PROPOSITION 1.1. *If $\vec{h}$ is a tuple of cyclically reduced words satisfying $C'(\frac{1}{6})$, then $G = \langle A \mid \vec{h}\rangle$ is infinite, torsion-free and word-hyperbolic. In addition, it has solvable word problem (by Dehn's algorithm) and solvable conjugacy problem.*

Moreover, if $\vec{h}$ and $\vec{g}$ both have property $C'(\frac{1}{6})$ and if they present the same group, then $\vec{h}^{\pm} = \vec{g}^{\pm}$ up to the order of the elements in the tuples.

1.3. Graphical representation of subgroups and the central tree property. A privileged tool for the study of subgroups of free groups is provided by *Stallings graphs*: if H is a finitely generated subgroup of F, its Stallings graph $\Gamma(H)$ is a finite graph of a particular type, uniquely representing H, whose computation was first made explicit by Stallings [**31**]. The mathematical object itself is already described by Serre [**29**]. The description we give below differs slightly from Serre's and Stallings', it follows [**15, 21, 30, 33, 35**] and it emphasizes the combinatorial, graph-theoretical aspect, which is more conducive to the discussion of algorithmic properties.

A *finite A-graph* is a pair $\Gamma = (V, E)$ with V finite and $E \subseteq V \times A \times V$, such that if both (u, a, v) and (u, a, v') are in E then $v = v'$, and if both (u, a, v) and (u', a, v) are in E then $u = u'$. Let $v \in V$. The pair (Γ, v) is said to be *admissible* if the underlying graph of Γ is connected (that is: the undirected graph obtained from Γ by forgetting the letter labels and the orientation of edges), and if every vertex $w \in V$, except possibly v, occurs in at least two edges in E.

Every admissible pair $(\Gamma, 1)$ represents a unique subgroup H of $F(A)$ in the following sense: if u is a reduced word, then $u \in H$ if and only if u labels a loop at 1 in Γ (by convention, an edge (u, a, v) can be read from u to v with label a, or from v to u with label a^{-1}). One can show that H is finitely generated. More precisely, the following procedure yields a basis of H: choose a spanning tree T of Γ; for each edge $e = (u, a, v)$ of Γ not in T, let $b_e = x_u a x_v^{-1}$, where x_u (resp. x_v) is the only reduced word labeling a path in T from 1 to u (resp. v); then the b_e freely generate H and as a result, the rank of H is exactly $|E| - |V| + 1$.

Conversely, if $\vec{h} = (h_1, \ldots, h_k)$ is a tuple of reduced words, then the subgroup $H = \langle\vec{h}\rangle$ admits a Stallings graph, written $(\Gamma(H), 1)$, which can be computed effectively and efficiently. A quick description of the algorithm is as follows. We first build a graph with edges labeled by letters in $\tilde{A}$, and then reduce it to an A-graph using *foldings*. First build a vertex 1. Then, for every $1 \leq i \leq k$, build a loop with label h_i from 1 to 1, adding $|h_i| - 1$ new vertices. Change every edge (u, a^{-1}, v) labeled by a letter of A^{-1} into an edge (v, a, u). At this point, we have constructed the so-called *bouquet of loops* labeled by the h_i.

Then iteratively identify the vertices v and w whenever there exists a vertex u and a letter $a \in A$ such that either both (u, a, v) and (u, a, w) or both (v, a, u) and (w, a, u) are edges in the graph (the corresponding two edges are *folded*, in Stallings' terminology).

The resulting graph Γ is such that $(\Gamma, 1)$ is admissible, the reduced words labeling a loop at 1 are exactly the elements of H and, very much like in the (1-dimensional) reduction of words, that graph does not depend on the order used to perform the foldings. The graph $(\Gamma(H), 1)$ can be computed in time almost linear (precisely: in time $\mathcal{O}(n \log^* n)$ [**33**]).

Some algebraic properties of H can be directly seen on its Stallings graph $(\Gamma(H), 1)$. For instance, one can show that H is malnormal if and only if there exists no non-empty reduced word u which labels a loop in two distinct vertices of $\Gamma(H)$ [**12, 15**]. This property leads to an easy decision procedure of malnormality for subgroups of a free group. We refer the reader to [**15, 21, 31, 35**] for more information about Stallings graphs.

If $\vec{h}$ is a tuple of elements of F, let $\mathsf{min}(\vec{h})$ be the minimum length of an element of $\vec{h}$ and let $\mathsf{lcp}(\vec{h})$ be the length of the longest common prefix between two words in $\vec{h}^{\pm}$, see Figure 1[1]. We say that $\vec{h}$ has the *central tree property* if $2\,\mathsf{lcp}(\vec{h}) < \mathsf{min}(\vec{h})$.

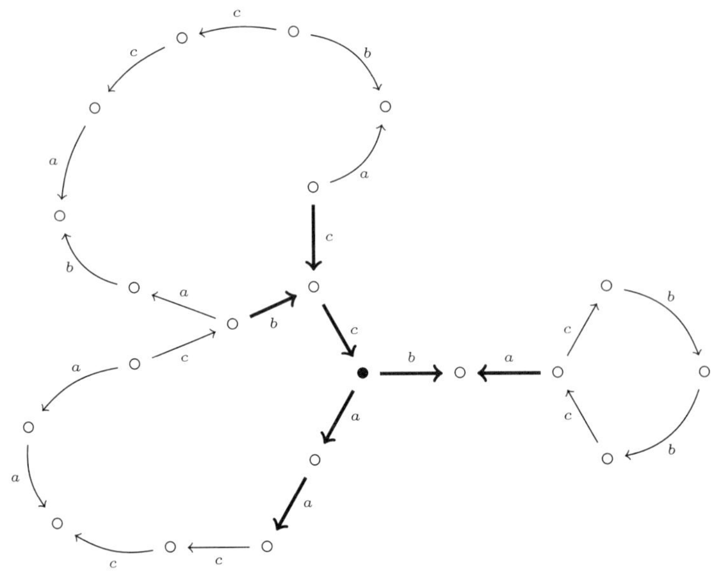

FIGURE 1. The Stallings graph of the subgroup generated by $\vec{h} = (ba^{-1}cb^2a^2b^{-1}, a^2c^2a^{-2}cbc, c^{-1}b^{-1}aba^{-1}c^{-2}ba^{-1}c^2)$, has the central tree property and satisfies $\mathsf{lcp}(\vec{h}) = 2$. The origin is denoted by $\bullet$ and the central tree is depicted in bold arrows.

PROPOSITION 1.2. *Let $\vec{h} = (h_1, \ldots, h_k)$ be a tuple of elements of $F(A)$ with the central tree property and let $H = \langle \vec{h} \rangle$. Then the Stallings graph $\Gamma(H)$ consists of a* central tree *of height $t = \mathsf{lcp}(\vec{h})$ and of k* outer loops, *one for each h_i, connecting*

[1]This definition is closely related with the notion of *trie* of $\vec{h}^{\pm}$. The height of the trie of $\vec{h}^{\pm}$ is $1 + \mathsf{lcp}(\vec{h})$.

the length t prefix and the length t suffix of h_i (two leaves of the central tree), of length $|h_i| - 2t$ respectively. The set of vertices of the central tree can be identified with the set of prefixes of length at most t of the words of $\vec{h}^{\pm}$.

In particular, $\vec{h}$ is a basis of H. Moreover, if $\vec{g}$ is a basis of H also with the central tree property, then $\vec{h}^{\pm}$ and $\vec{g}^{\pm}$ coincide up to the order of their elements.

PROOF. The central tree property shows that the cancellation (folding) that occurs when one considers the bouquet of h_i-labeled loops around the origin, stops before canceling entirely any one of the h_i. The result follows immediately. □

Under the central tree property, we record an interesting sufficient condition for malnormality.

PROPOSITION 1.3. *Let $\vec{h} = (h_1, \ldots, h_k)$ be a tuple of elements of $F(A)$ with the central tree property and let $H = \langle\vec{h}\rangle$. Let us assume additionally that $3\,\mathsf{lcp}(\vec{h}) < \mathsf{min}(\vec{h})$ and that no word of length at least $\frac{1}{2}(\mathsf{min}(\vec{h}) - 3\,\mathsf{lcp}(\vec{h}))$ has several occurrences as a factor of an element of $\vec{h}^{\pm}$, then H is malnormal.*

REMARK 1.4. In the proof below, and in several other statements and proofs later in the paper, we consider words whose length is specified by an algebraic expression which does not always compute to an integer (*e.g.*, $\frac{1}{2}(\mathsf{min}(\vec{h}) - 3\,\mathsf{lcp}(\vec{h}))$). To be rigorous, we should consider only the integer part of these expressions. For the sake of simplicity, we dispense with this extra notation, and implicitly consider that if a word of length ℓ is considered, then we mean that its length is $\lfloor \ell \rfloor$.

PROOF. Let $m = \mathsf{min}(\vec{h})$ and $t = \mathsf{lcp}(\vec{h})$. Proposition 1.2 shows that $\Gamma(H)$ consists of a central tree of height t and of outer loops, one for each h_i, of length $|h_i| - 2t \geq m - 2t$.

If H is not malnormal, then a word u labels a loop at two distinct vertices of $\Gamma(H)$. Without loss of generality, u is cyclically reduced. Moreover, given the particular geometry of $\Gamma(H)$, both loops visit the central tree. Without loss of generality, we may assume that one of the u-labeled loops starts in the central tree, at distance exactly t from the base vertex 1, and travels away from 1. In particular, $|u| \geq m - 2t$, and if v is the prefix of u of length $m - 2t$, then v is a factor of some $h_i^{\pm 1}$.

Let s be the start state of the second u-labeled loop: reading this loop starts with reading the word v. Suppose that s is in the central tree: either reading u (and v) from s takes us away from 1 towards a leaf of the central tree and into an outer loop, and v is a factor of some $h_j^{\pm 1}$; or reading v from s moves us towards 1 for a distance at most t, after which the path travels away from 1, along a path labeled by a factor of some $h_j^{\pm 1}$, for a distance at least $m - 3t$. In either case, a factor of u of length $m - 3t > \frac{1}{2}(m - 3t)$ has two occurrences in $\vec{h}^{\pm}$.

Suppose now that s is on an outer loop (say, associated to $h_j^{\pm 1}$) and that s' is the first vertex of the central tree reached along the loop. If s' is reached after reading a prefix of u of length greater than $\frac{1}{2}(m - 3t)$, then the prefix of v of length $\frac{1}{2}(m - 3t)$ is a factor of $h_j^{\pm 1}$. Otherwise v labels a path from s which first reaches s', then travels towards 1 in the central tree for a distance at most t, and thence away from 1, along a path labeled by some $h_\ell^{\pm 1}$, which it follows over a length at least equal to $(m - 2t) - \frac{1}{2}(m - 3t) - t = \frac{1}{2}(m - 3t)$.

Thus, in every case, u contains a factor of length $\frac{1}{2}(m-3t)$ with two distinct occurrences as a factor of an element of $\vec{h}^{\pm}$ and this concludes the proof. □

To conclude this section, we note that the properties discussed above are preserved when going from a tuple $\vec{h}$ to a sub-tuple: say that a tuple $\vec{g}$ is *contained in* a tuple $\vec{h}$, written $\vec{g} \leq \vec{h}$, if every element of $\vec{g}$ is an element of $\vec{h}$.

PROPOSITION 1.5. *Let $\vec{g}, \vec{h}$ be tuples of reduced words such that $\vec{g} \leq \vec{h}$.*

- *If $\vec{h}$ has the central tree property, so does $\vec{g}$.*
- *If $\vec{h}$ consists of cyclically reduced words and $\vec{h}$ has Property $C'(\lambda)$, then so does $\vec{g}$.*
- *If $\vec{h}$ has the central tree property, then $\langle\vec{g}\rangle$ is a free factor of $\langle\vec{h}\rangle$, and $\langle\vec{g}\rangle$ is malnormal if $\langle\vec{h}\rangle$ is.*

PROOF. The first two properties are immediate from the definition. Supose now that $\vec{h}$ has the central tree property. Then by Proposition 1.2, $\vec{h}$ is a basis of $\langle\vec{h}\rangle$, and by the first statement of the current proposition, $\vec{g}$ is a basis of $\langle\vec{g}\rangle$. Since $\vec{g} \leq \vec{h}$, $\langle\vec{g}\rangle$ is a free factor of $\langle\vec{h}\rangle$.

In particular, $\langle\vec{g}\rangle$ is malnormal in $\langle\vec{h}\rangle$ (a free factor always is, by elementary reasons). It is immediate from the definition that malnormality is transitive, so if $\langle\vec{h}\rangle$ is malnormal in F, then so is $\langle\vec{g}\rangle$. □

2. Random models and generic properties

We will discuss several models of randomness for finitely presented groups and finitely generated subgroups, or rather, for finite tuples of cyclically reduced words (group presentations) and finite tuples of reduced words. In this section, we fix a general framework for these models of randomness and we survey some of the known results.

2.1. Generic properties and negligible properties. Let us say that a function f, defined on $\mathbb{N}$ and such that $\lim f(n) = 0$, is *exponentially* (resp. *super-polynomially, polynomially*) *small* if $f(n) = o(e^{-dn})$ for some $d > 0$ (resp. $f(n) = o(n^{-d})$ for every positive integer d, $f(n) = o(n^{-d})$ for some positive integer d).

Given a sequence of probability laws $(\mathbb{P}_n)_n$ on a set S, we say that a subset $X \subseteq S$ is *negligible* if $\lim_n \mathbb{P}_n(X) = 0$, and *generic* if its complement is negligible.[2]

We also say that X is *exponentially* (resp. *super-polynomially, polynomially*) *negligible* if $\mathbb{P}_n(X)$ tends to 0 and is exponentially (resp. super-polynomially, polynomially) small. And it is *exponentially* (resp. *super-polynomially, polynomially*) *generic* if its complement is exponentially (resp. super-polynomially, polynomially) negligible.

In this paper, the set S will be the set of all finite tuples of reduced words, or cyclically reduced words, and the probability laws $\mathbb{P}_n$ will be such that every subset is measurable: we will therefore not specify in the statements that we consider only measurable sets.

The notions of genericity and negligibility have elementary closure properties that we will use freely in the sequel. For instance, a superset of a generic set is

[2]This is the same notion as *with high probability* or *with overwhelming probability*, which are used in the discrete probability literature.

generic, as well as the intersection of finitely many generic sets. Dual properties hold for negligible sets.

2.2. The few-generator model and the density model. In this section, we review the results known on two random models, originally introduced to discuss finite presentations. We discuss more general models in Section 3 below.

2.2.1. *The few-generator model.* In the *few-generator model*, an integer $k \geq 1$ is fixed, and we let $\mathbb{P}_n$ be the uniform probability on the set of k-tuples of words of F of length at most n. Proposition 2.1 is established by elementary counting arguments, see Gromov [**10**, Prop. 0.2.A] or Arzhantseva and Ol'shanskii [**1**, Lemma 3].

PROPOSITION 2.1. *Let $k \geq 1$, $0 < \alpha < \frac{1}{2}$, $2\alpha < \beta < 1$ and $0 < \lambda < 1$. Then a k-tuple $\vec{h}$ of elements of F of length at most n picked uniformly at random, exponentially generically satisfies the following properties:*

- *$\textsf{min}(\vec{h}) > \beta n$,*
- *$\textsf{lcp}(\vec{h}) < \alpha n$,*
- *no word of length λn has two occurrences as a factor of an element of $\vec{h}^{\pm}$.*

In view of Propositions 1.2 and 1.3, this yields the following corollary ([**3**], and [**12**] for the malnormality statement).

COROLLARY 2.2. *Let $k \geq 1$. If $\vec{h}$ is a k-tuple of elements of F of length at most n picked uniformly at random and $H = \langle \vec{h} \rangle$, then*

- *exponentially generically, $\vec{h}$ has the central tree property, and in particular, $\Gamma(H)$ can be constructed in linear time (in $k \cdot n$), simply by computing the initial cancellation of the elements of $\vec{h}^{\pm}$; H is freely generated by the elements of $\vec{h}$, and H has rank k;*
- *exponentially generically, H is malnormal.*

Moreover, if $\vec{h}$ and $\vec{g}$ generate the same subgroup, then exponentially generically, $\vec{h}^{\pm} = \vec{g}^{\pm}$ up to the order of the elements in the tuples.

The following statement follows from Proposition 1.5, and from Theorem 2.4 below (which is independent).

COROLLARY 2.3. *In the few-generator model, if $\vec{h}$ is a k-tuple of cyclically reduced words of length at most n, then*

- *for any $0 < \lambda < \frac{1}{2}$, $\vec{h}$ exponentially generically satisfies the small cancellation property $C'(\lambda)$;*
- *exponentially generically, the group $\langle A \mid \vec{h} \rangle$ is infinite, torsion-free, word-hyperbolic, it has solvable word problem (by Dehn's algorithm) and solvable conjugacy problem.*

2.2.2. *The density model.* In the *density model*, a density $0 < d < 1$ is fixed, and a tuple of cyclically reduced elements of the n-sphere of density d is picked uniformly at random: that is, the tuple $\vec{h}$ consists of $|\mathcal{C}_n|^d$ cyclically reduced words of length n. This model was introduced by Gromov [**11**] and complete proofs were given by Ol'shanskii [**25**], Champetier [**7**] and Ollivier [**23**].

THEOREM 2.4. *Let $0 < \alpha < d < \beta < 1$. In the density model, the following properties hold:*

(1) *exponentially generically, every word of length αn occurs as a factor of a word in $\vec{h}$, and some word of length βn fails to occur as a factor of a word in $\vec{h}^{\pm}$;*
(2) *if $d < \frac{1}{2}$, then exponentially generically, $\vec{h}$ satisfies property $C'(\lambda)$ for $\lambda > 2d$ but $\vec{h}$ does not satisfy $C'(\lambda)$ for $\lambda < 2d$; in particular, at density $d < \frac{1}{12}$, $\vec{h}$ satisfies exponentially generically property $C'(\frac{1}{6})$ and the group $\langle A \mid \vec{h}\rangle$ is infinite and hyperbolic; and at density $d > \frac{1}{12}$, exponentially generically, $\vec{h}$ does not satisfy $C'(\frac{1}{6})$;*
(3) *at density $d > \frac{1}{2}$, exponentially generically, $\langle \vec{h}\rangle$ is equal to $F(A)$, or has index 2. In particular, the group $\langle A \mid \vec{h}\rangle$ is either trivial or $\mathbb{Z}/2\mathbb{Z}$;*
(4) *at density $d < \frac{1}{2}$, the group $\langle A \mid \vec{h}\rangle$ is generically infinite and hyperbolic.*

Properties (1)-(3) in Theorem 2.4 are obtained by counting arguments. Property (4) is the "hard part" of the theorem, where hyperbolicity does not follow from a small cancellation property.

As pointed out by Ollivier [**24**, Sec. I.2.c], the statement of Theorem 2.4 still holds if a tuple of cyclically reduced elements is chosen uniformly at random at density d in the n-ball rather than in the n-sphere (that is, it consists of words of length at most n). We will actually verify this fact again in Section 3.6.

3. A general probabilistic model

We introduce a fairly general probabilistic model, which generalizes both the few-generator and the density models.

3.1. Prefix-heavy sequences of measures on reduced words. For every reduced word $u \in \mathcal{R}$, let $\mathcal{P}(u)$ be the set of all reduced words v of which u is a prefix (that is: $\mathcal{P}(u) = u\tilde{A}^* \cap \mathcal{R}$). Let also $\mathcal{P}_n(u)$ be the set $\mathcal{R}_n \cap \mathcal{P}(u)$. The notation $\mathcal{P}$ can also be extended to a set U of reduced words: $\mathcal{P}(U) = \bigcup_{u\in U}\mathcal{P}(u)$.

Let $(\mathbb{R}_n)_{n\geq 0}$ be a sequence of probability measures on $\mathcal{R}$ and let $C \geq 1$ and $\alpha \in (0,1)$. We say that the sequence $(\mathbb{R}_n)_{n\geq 0}$ is a *prefix-heavy sequence of measures on $\mathcal{R}$* of parameters (C,α) if:

(1) for every $n \geq 0$, the support of the measure $\mathbb{R}_n$ is included in $\mathcal{R}_n$;
(2) for every $n \geq 0$ and for every $u \in \mathcal{R}$, if $\mathbb{R}_n(\mathcal{P}(u)) \neq 0$ then for every $v \in \mathcal{R}$

$$\mathbb{R}_n\big(\mathcal{P}(uv) \mid \mathcal{P}(u)\big) \leq C\alpha^{|v|}.$$

This prefix-oriented definition is rather natural if one thinks of a source as generating reduced words from left to right, as is usual in information theory.

REMARK 3.1. Taking $u = \varepsilon$ in the definition yields $\mathbb{R}_n\big(\mathcal{P}(v)\big) \leq C\alpha^{|v|}$. For $n = |v|$, we have $\mathcal{P}(v) \cap \mathcal{R}_n = \{v\}$, so the probability of v decreases exponentially with the length of v.

EXAMPLE 3.2. The sequence of uniform distributions on $\mathcal{R}_n$ is a prefix-heavy sequence of measures with parameters $C = 1$ and $\alpha = \frac{1}{2r-1}$. Indeed, if u is a reduced word of length at most $n \geq 0$ (for a longer u, $\mathbb{R}_n(\mathcal{P}(u)) = 0$), and if uv is reduced, we have

$$\mathbb{R}_n\big(\mathcal{P}(uv) \mid \mathcal{P}(u)\big) = \begin{cases} \frac{1}{(2r-1)^{|v|}} & \text{if } |u|+|v| \leq n \text{ and } u \neq \varepsilon, \\ \frac{1}{2r(2r-1)^{|v|-1}} & \text{if } |v| \leq n \text{ and } u = \varepsilon, \\ 0 & \text{otherwise.} \end{cases}$$

EXAMPLE 3.3. By a similar computation, one verifies that the sequence of uniform distributions on $\mathcal{C}_n$, the cyclically reduced words, is also a prefix-heavy sequence of measures, with parameters $C = \frac{2r-1}{2r-2}$ and $\alpha = \frac{1}{2r-1}$ (see Section 1.1).

For the rest of this section, we fix a sequence of measures $(\mathbb{R}_n)_{n \geq 0}$ on $\mathcal{R}$, which is prefix-heavy with parameters (C, α). All probabilities refer to this sequence, that is: the probability of a subset of $\mathcal{R}_n$ is computed according to $\mathbb{R}_n$.

REMARK 3.4. If X and Y are subsets of $\mathcal{R}$, the notation $\mathbb{R}_n(X \mid Y)$ is technically defined only if $\mathbb{R}_n(Y) \neq 0$. To avoid stating cumbersome hypotheses, we adopt the convention that $\mathbb{R}_n(X \mid Y)\, \mathbb{R}_n(Y) = 0$ whenever $\mathbb{R}_n(Y) = 0$.

3.2. Repeated factors in random reduced words. Let us first evaluate the probability of occurrence of prescribed, non-overlapping factors in a reduced word. Let $m \geq 0$, $\vec{v} = (v_1, \ldots, v_m)$ be a vector of non-empty reduced words and $\vec{\imath} = (i_1, \ldots, i_m)$ be a vector of integers. We denote by $E(\vec{v}, \vec{\imath})$ denote the set of reduced words of length n, admitting v_j as a factor at position i_j for every $1 \leq j \leq m$ (if $m = 0$, then $E(\vec{v}, \vec{\imath}) = \mathcal{R}$). If $n \geq 1$, we also write $E_n(\vec{v}, \vec{\imath})$ for $E(\vec{v}, \vec{\imath}) \cap \mathcal{R}_n$.

LEMMA 3.5. *Let $\vec{v} = (v_1, \ldots, v_m)$ be a sequence of non-empty reduced words and $\vec{\imath} = (i_1, \ldots, i_m)$ be a sequence of integers satisfying*

$$1 \leq i_1 < i_1 + |v_1| \leq i_2 < i_2 + |v_2| \leq \ldots \leq i_m + |v_m| \leq n.$$

Then the following inequality holds:

$$\mathbb{R}_n\left(E(\vec{v}, \vec{\imath})\right) \leq C^m \alpha^{|v_1 v_2 \cdots v_m|}.$$

In addition, if $m \geq 1$ and $\vec{x} = (v_1, \ldots, v_{m-1})$ and $\vec{\jmath} = (i_1, \ldots, i_{m-1})$, then

$$\mathbb{R}_n(E(\vec{v}, \vec{\imath})) \leq C\alpha^{|v_m|}\mathbb{R}_n(E(\vec{x}, \vec{\jmath})).$$

PROOF. The proof is by induction on m and the case $m = 0$ is trivial. We now assume that $m \geq 1$ and that the inequality holds for vectors of length $m-1$. Since $(\mathbb{R}_n)_n$ is prefix-heavy, we have

$$\mathbb{R}_n(\mathcal{P}(uv_m)) = \mathbb{R}_n(\mathcal{P}(uv_m) \mid \mathcal{P}(u))\, \mathbb{R}_n(\mathcal{P}(u)) \leq C\alpha^{|v_m|}\mathbb{R}_n(\mathcal{P}(u))$$

for each u. Since $E(\vec{v}, \vec{\imath}) = \mathcal{P}(E_{i_m-1}(\vec{x}, \vec{\jmath})v_m)$, summing the previous inequality over all $u \in E_{i_m-1}(\vec{x}, \vec{\jmath})$ yields

$$\mathbb{R}_n(E(\vec{v}, \vec{\imath})) \leq C\alpha^{|v_m|}\mathbb{R}_n(\mathcal{P}(E_{i_m-1}(\vec{x}, \vec{\jmath}))) = C\alpha^{|v_m|}\mathbb{R}_n(E(\vec{x}, \vec{\jmath}))$$

since $n \geq i_m + |v_m|$. This concludes the proof. □

COROLLARY 3.6. *Let $v_1, \ldots, v_m$ be non-empty reduced words. The probability that a word of length n admits $v_1, \ldots, v_m$ in that order as non-overlapping factors, is at most $C^m n^m \alpha^{|v_1 \cdots v_m|}$.*

PROOF. This is a direct consequence of Lemma 3.5, summing over all possible position vectors. □

We now consider repeated non-overlapping occurrences of factors of a prescribed length.

LEMMA 3.7. *Let $1 \leq i, j, t \leq n$ be such that $i + t \leq j$. The probability that a word of length t occurs (resp. a word of length t and its inverse occur) at positions i and j in a reduced word of length n is at most equal to $C\alpha^t$.*

The probability that a reduced word of length n has two non-overlapping occurrences of a factor of length t (resp. occurrences of a factor of length t and its inverse) is at most equal to $Cn^2\alpha^t$.

PROOF. Let $E_n(t, i, j)$ be the set of reduced words of length n in which the same factor of length t occurs at positions i and j. Then $E_n(t, i, j)$ is the disjoint union of the sets $E_n((v, v), (i, j))$, where v runs over $\mathcal{R}_t$. By Lemma 3.5, we have

$$\mathbb{R}_n(E_n(t, i, j)) = \sum_{v \in \mathcal{R}_t} \mathbb{R}_n(E((v, v), (i, j))) \leq C\alpha^t \sum_{v \in \mathcal{R}_t} \mathbb{R}_n(E((v), (i))) = C\alpha^t,$$

where the last equality is due to the fact that the $E_n((v), (i))$ form a partition of $\mathcal{R}_n$ when v runs over $\mathcal{R}_t$.

The same reasoning applied to the vectors (v, v^{-1}) yields the analogous inequality for words containing non-overlapping occurrences of a word and its inverse.

The last part of the statement follows by summing over all possible values of i and j. □

Applying Lemma 3.7 with $i = 1$ and $j = n - t + 1$, we get the following useful statement.

COROLLARY 3.8. *For every positive integers n, t such that $n > 2t$, the probability that a reduced word $u \in \mathcal{R}_n$ is of the form vwv^{-1}, for some word v of length t, is at most $C\alpha^t$.*

Finally, we also estimate the probability that a word has two overlapping occurrences of a factor. Note that we do not need to consider overlapping occurrences of a word v and its inverse, since a reduced word cannot overlap with its inverse.

LEMMA 3.9. *Let $1 \leq t < n$. The probability that a reduced word of length n has overlapping occurrences of a factor of length t is at most $Cnt\alpha^t$.*

PROOF. If a word v overlaps with itself, more precisely, if $xv = vz$ for some words x, z such that $0 < |x| = |z| < |v|$, then it is a classical result from combinatorics on words that $v = x^s y$ where $s = \left\lfloor \frac{|v|}{|x|} \right\rfloor \geq 1$ and y is the prefix of x of length $|v| - s|x|$ (see Figure 2).

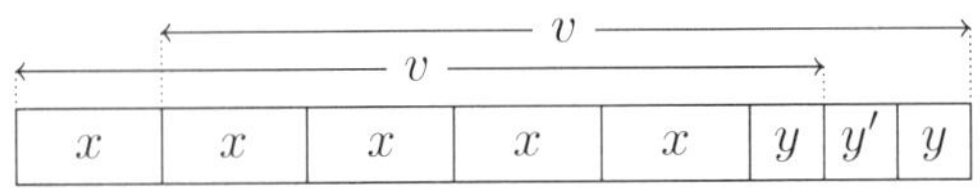

FIGURE 2. A classical result from combinatorics of words: if $xv = vz$ with $0 < |x| < |v|$, then v is of the form $v = x^s y$ for some positive integer s and some prefix y of x.

It follows that, if a reduced word u has (overlapping) occurrences of a factor v of length t at positions i and j ($j < i+t$), then u admits a factor of the form xv at position i, where x is the prefix of v of length $j-i$. Note that, once t and $j-i$ are fixed, v is entirely determined by x. Therefore this occurs with probability

$$P \leq \sum_{i=1}^{n} \sum_{j=i+1}^{i+t-1} \sum_{x\in\mathcal{R}_{j-i}} \mathbb{R}_n(E((xv),(i))) = \sum_{i=1}^{n} \sum_{j=i+1}^{i+t-1} \sum_{x\in\mathcal{R}_{j-i}} \mathbb{R}_n(E((x,v),(i,j))).$$

It follows that

$$P \leq \sum_{i=1}^{n} \sum_{j=i+1}^{i+t-1} C\alpha^t \sum_{x\in\mathcal{R}_{j-i}} \mathbb{R}_n(E((x),(i))) = \sum_{i=1}^{n} \sum_{j=i+1}^{i+t-1} C\alpha^t \leq Cnt\alpha^t.$$

by Lemma 3.5 and using the fact that the $E_n((x),(i))$ form a partition of $\mathcal{R}_n$ when x runs over $\mathcal{R}_{j-i}$. □

3.3. Repeated cyclic factors in random reduced words. A word v is a *cyclic factor* of a word u if either $u \in \tilde{A}^*v\tilde{A}^*$, or $v = v_1v_2$ and $u \in v_2\tilde{A}^*v_1$ – in which case we say that v is a *straddling factor*. For now, we only assume that u is reduced, but we will be ultimately interested in the cyclically reduced case, see Corollary 3.14.

LEMMA 3.10. *Let $1 \leq i, t \leq n$ such that $i+t \leq n$ and let v be reduced word v of length t. Then the probability that v is a cyclic factor at position i of an element of $\mathcal{R}_n$, is at most $(Cn + C^2t)\alpha^t \leq 2C^2n\alpha^t$.*

PROOF. The probability that v occurs as a (regular) factor of an element of $\mathcal{R}_n$ is at most $Cn\alpha^t$ by Corollary 3.6.

On the other hand, v occurs as a straddling factor of $u \in \mathcal{R}_n$ if $v = v_2v_1$, with $1 \leq \ell = |v_2| < t$ and $u \in v_1\tilde{A}^*v_2$, that is, $u \in E((v_1,v_2),(1,n-\ell+1))$. By Lemma 3.5, this happens with probability at most $C^2\alpha^t$. Summing over the possible values of ℓ, we find that that v occurs as a straddling factor of an element of $\mathcal{R}_n$ with probability at most $C^2t\alpha^t$.

Therefore the probability that v occurs in u as a cyclic factor is at most $(Cn + C^2t)\alpha^t$, as announced. □

We now consider multiple occurrences of cyclic factors of a given length.

LEMMA 3.11. *Let $1 \leq t < n$. The probability that a reduced word of length n has two non-overlapping occurrences of a cyclic factor of length t (resp. an occurrence of a cyclic factor of length t and its inverse), is at most $(Cn^2+C^2nt)\alpha^t \leq 2C^2n^2\alpha^t$.*

PROOF. Again there are several cases, depending whether the occurrences of the word (or the word and its inverse) are both standard factors, or one of them is straddling.

The probability that a reduced word $u \in \mathcal{R}_n$ admits two non-overlapping occurrences of a (standard) factor of length t (resp. occurrences of a factor of length t and its inverse), is at most $Cn^2\alpha^t$ by Lemma 3.7.

We now consider the situation where u has two occurrences of the same word of length t, one as a standard factor and one straddling: there exist integers ℓ, i and reduced words v_1, v_2 such that $0 < \ell < t$, $\ell \leq i \leq n-2t+\ell$, $|v_2| = \ell$, $|v_1v_2| = t$ and

$$u \in E((v_2, v_1v_2, v_1),(1,i,n-t+\ell+1)) = E((v_2,v_1,v_2,v_1),(1,i,i+\ell,n-t+\ell+1)).$$

Applying Lemma 3.5 twice, we find that the probability of this event according to $\mathbb{R}_n$ is at most equal to $C^2\alpha^t\mathbb{R}_n(E((v_2,v_1),(1,i)))$.

Then the probability P that a word in $\mathcal{R}_n$ admits two non-overlapping occurrences of a factor of length t, one standard and one straddling, is bounded above by the sum of these values when ℓ, i, v_1, v_2 run over all possible values:

$$P \leq \sum_{\ell=0}^{t} \sum_{i=\ell}^{n-2t+\ell} \sum_{v_2\in\mathcal{R}_\ell} \sum_{v_1\in\mathcal{R}_{t-\ell}} C^2\alpha^t\mathbb{R}_n(E((v_2,v_1),(1,i))).$$

For fixed values of ℓ and i, $\mathcal{R}_n$ is the disjoint union of the $E((v_2,v_1),(1,i))$ when v_2 runs over $\mathcal{R}_\ell$ and v_1 runs over $\mathcal{R}_{t-\ell}$. So we get

$$P \leq \sum_{\ell=0}^{t} \sum_{i=\ell}^{n-2t+\ell} C^2\alpha^t \leq C^2nt\alpha^t.$$

Thus the probability that a reduced word of length n has two non-overlapping occurrences of a word of length t as cyclic factors is at most equal to $(Cn^2 + C^2nt)\alpha^t \leq 2C^2n^2\alpha^t$, as announced.

Finally, we consider the situation where a factor of length t and its inverse occur in u, with one of the occurrences straddling: that is, there exist integers ℓ, i and reduced words v_1, v_2 such that $0 < \ell < t$, $\ell \leq i \leq n-2t+\ell$, $|v_2| = \ell$, $|v_1v_2| = t$ and u lies in

$$E((v_2, v_2^{-1}v_1^{-1}, v_1),(1,i,n-t+\ell+1)) = E((v_2,v_2^{-1},v_1^{-1},v_1),(1,i,i+\ell,n-t+\ell+1)).$$

As above, the probability of this event according to $\mathbb{R}_n$ is at most

$$C\alpha^{t-\ell}\mathbb{R}_n(E((v_2,v_2^{-1},v_1^{-1}),(1,i,i+\ell)))$$

and the probability P' that a reduced word of length n has two non-overlapping occurrences of a word of length t as cyclic factors, with one of them straddling, satisfies

$$P' \leq \sum_{\ell=1}^{t-1} \sum_{i=\ell}^{n-2t+\ell} \sum_{v_2\in\mathcal{R}_\ell} \sum_{v_1\in\mathcal{R}_{t-\ell}} C\alpha^{t-\ell}\mathbb{R}_n(E((v_2,v_2^{-1},v_1^{-1}),(1,i,i+\ell))).$$

For fixed values of ℓ, i and v_2, $E_n((v_2,v_2^{-1}),(1,i))$ is the disjoint union of the $E_n((v_2,v_2^{-1},v_1^{-1}),(1,i,i+\ell))$ when v_1 runs over $\mathcal{R}_{t-\ell}$. Therefore we have

$$P' \leq \sum_{\ell=1}^{t-1} \sum_{i=\ell}^{n-2t+\ell} \sum_{v_2\in\mathcal{R}_\ell} C\alpha^{t-\ell}\mathbb{R}_n(E((v_2,v_2^{-1}),(1,i))).$$

By Lemma 3.5 again, $\mathbb{R}_n(E((v_2,v_2^{-1}),(1,i))) \leq C\alpha^\ell\mathbb{R}_n(E((v_2)(1)))$ and we get, by the same reasoning as above,

$$P' \leq \sum_{\ell=1}^{t-1} \sum_{i=\ell}^{n-2t+\ell} \sum_{v_2\in\mathcal{R}_\ell} C^2\alpha^t\mathbb{R}_n(E((v_2),(1))) = \sum_{\ell=1}^{t-1} \sum_{i=\ell}^{n-2t+\ell} C^2\alpha^t \leq C^2nt\alpha^t.$$

Thus the probability that a reduced word of length n has an occurrence of a word of length t and its inverse as a cyclic factor is, again, at most equal to $(Cn^2+C^2nt)\alpha^t \leq 2C^2n^2\alpha^t$, as announced. □

Finally, we give an upper bound to the probability that a reduced word has overlapping occurrences of a cyclic factor of length t (observing again that a reduced word cannot have overlapping occurrences of a (cyclic) factor and its inverse).

LEMMA 3.12. *Let $1 \leq t < n$. The probability that a reduced word of length n has overlapping occurrences of a cyclic factor of length t is at most equal to $\left(Cnt + 2C^2t^2\right)\alpha^t \leq 3\, C^2nt\alpha^t$.*

PROOF. The probability that a reduced word of length n has overlapping occurrences of a non-straddling factor of length t is at most $Cnt\alpha^t$ by Lemma 3.9.

Let us now assume that the reduced word $u \in \mathcal{R}_n$ has overlapping occurrences of a cyclic factor v of length t, with one at least of these occurrences straddling. Note that any cyclic factor of u is a factor of u^2. Therefore, using the same arguments as for Lemma 3.9, u has a straddling cyclic factor of the form $xv = x^{s+1}y$, where $|x| > 0$, y is a prefix of x and $s \geq 1$. In particular, $v = x^sy$ and $t = s|x| + |y|$.

It follows that u is in $v_2\tilde{A}^*v_1$, for some v_1, v_2 such that $v_1v_2 = x^{s+1}y$. Denote by $\mathsf{pref}_\ell(z)$ and $\mathsf{suff}_\ell(z)$ the prefix and the suffix of length ℓ of a word z. Then there exist a cyclic conjugate z of x and integers $0 \leq h, \ell < |z| = |x|$ and $m, m' \geq 0$ such that $v_1 = \mathsf{suff}_h(z)z^{m'}$ and $v_2 = z^m\mathsf{pref}_\ell(z)$. Note that $x^{s+1}y = \mathsf{suff}_h(z)z^{m+m'}\mathsf{pref}_\ell(z)$ and

$$h + \ell = |y| \pmod{|z|}$$
$$m + m' = \begin{cases} s+1 & \text{if } h+\ell = |y| \\ s & \text{if } h+\ell = |z| + |y| \end{cases}$$
$$t + |z| = (m+m')|z| + h + \ell.$$

Observe also that $|y|$ is determined by $|z|$ ($|y| = t \pmod{|z|}$), that h is determined by ℓ and $|z|$, and that m' is determined by m, ℓ and $|z|$. Then

$$u \in \bigcup_{k=1}^{t-1}\bigcup_{\ell=0}^{k-1}\bigcup_{m=0}^{1+\lfloor\frac{t}{k}\rfloor}\bigcup_{z\in\mathcal{R}_k} X_{z,m,\ell}, \text{ where}$$
$$X_{z,\ell,m} = E((z^m\mathsf{pref}_\ell(z), \mathsf{suff}_h(z)z^{m'}), (1, n - m'|z| - h + 1))$$

and h and m' take the values imposed by those of $k = |z|$, ℓ and m. In particular, the probability P that a reduced word in $\mathcal{R}_n$ has overlapping occurrences of a cyclic factor of length t, with at least one of these occurrences straddling, satisfies

$$P \;\leq\; \sum_{k=1}^{t-1}\sum_{\ell=0}^{k-1}\sum_{m=0}^{1+\lfloor\frac{t}{k}\rfloor}\sum_{z\in\mathcal{R}_k}\mathbb{R}_n(X_{z,\ell,m}),$$

If $m \geq 1$, then

$$X_{z,\ell,m} = E((z, z^{m-1}\mathsf{pref}_\ell(z), \mathsf{suff}_h(z)z^{m'}), (1, |z|+1, n - m'|z| - h + 1))$$

and a double application of Lemma 3.5 shows that

$$\mathbb{R}_n(X_{z,\ell,m}) \leq C^2\alpha^{m'|z|+h}\alpha^{(m-1)|z|+\ell}\mathbb{R}_n(E((z),(1))) = C^2\alpha^t\mathbb{R}_n(E((z),(1))).$$

Summing these over $z \in \mathcal{R}_k$ (with k, ℓ and m fixed, $m \geq 1$), we get

$$\sum_{z\in\mathcal{R}_k}\mathbb{R}_n(X_{z,\ell,m}) \leq \sum_{z\in\mathcal{R}_k} C^2\alpha^t\mathbb{R}_n(E((z),(1))) \leq C^2\alpha^t,$$

since $\mathcal{R}_n$ is partitioned by the $\mathbb{R}_n(E((z),(1)))$ ($z \in \mathcal{R}_k$).

If $m = 0$ and $h + \ell = |y|$, then $m'|z| = t + |z| - |y|$ and we note that

$$\begin{aligned} X_{z,\ell,0} &= E((\mathsf{pref}_\ell(z), \mathsf{suff}_h(z)z^{m'}), (1, n-t-|z|+\ell+1)) \\ &\subseteq E((\mathsf{pref}_\ell(z), \mathsf{suff}_h(z), \mathsf{suff}_{|y|}(z)z^{m'-1}), (1, n-t-|z|+\ell+1, n-t+1)). \end{aligned}$$

By Lemma 3.5, we get

$$\mathbb{R}_n(X_{z,\ell,0}) \leq C\alpha^t \mathbb{R}_n(E((\mathsf{pref}_\ell(z), \mathsf{suff}_h(z)), (1, n-t-|z|+\ell+1))).$$

Summing over all $z \in \mathcal{R}_k$ (k and ℓ fixed), we get

$$\begin{aligned} \sum_{z\in\mathcal{R}_k} \mathbb{R}_n(X_{z,\ell,0}) &\leq \sum_{z\in\mathcal{R}_k} C\alpha^t \mathbb{R}_n(E((\mathsf{pref}_\ell(z), \mathsf{suff}_h(z)), (1, n-t-k+\ell+1))) \\ &\leq \sum_{z_1\in\mathcal{R}_\ell}\sum_{z_2\in\mathcal{R}_h} C\alpha^t \mathbb{R}_n(E((z_1, z_2), (1, n-t-k+\ell+1))) \\ &\leq C\alpha^t, \end{aligned}$$

since $\mathcal{R}_n$ is partitioned by the $\mathbb{R}_n(E((z_1, z_2), (1, n-t-k+\ell+1)))$ ($z_1 \in \mathcal{R}_\ell$, $z_2 \in \mathcal{R}_h$).

Finally, if $m = 0$ and $h + \ell = |z| + |y|$, then $m'|z| = t - |y|$. Therefore

$$\begin{aligned} X_{z,\ell,0} &= E((\mathsf{pref}_\ell(z), \mathsf{suff}_h(z)z^{m'}), (1, n-t-|z|+\ell+1)) \\ &= E((\mathsf{pref}_\ell(z), \mathsf{pref}_{|z|-\ell}(\mathsf{suff}_h(z)), \mathsf{suff}_{|y|}(z)z^{m'}), (1, n-t-|z|+\ell+1, n-t+1)). \end{aligned}$$

By Lemma 3.5, this yields

$$\mathbb{R}_n(X_{z,\ell,0}) \leq C\alpha^t \mathbb{R}_n(E((\mathsf{pref}_\ell(z), \mathsf{pref}_{|z|-\ell}(\mathsf{suff}_h(z))), (1, n-t+|z|+\ell+1))).$$

As in the previous case, summing over all $z \in \mathcal{R}_k$ (k and ℓ fixed) yields

$$\sum_{z\in\mathcal{R}_k} \mathbb{R}_n(X_{z,\ell,0}) \leq C\alpha^t.$$

Then we get the following upper bound for the probability P:

$$\begin{aligned} P &\leq \sum_{k=1}^{t-1}\sum_{\ell=0}^{k-1}\sum_{m=1}^{1+\lfloor \frac{t}{k} \rfloor} C^2\alpha^t + \sum_{k=1}^{t-1}\sum_{\ell=0}^{k-1} C\alpha^t \\ &\leq C^2\frac{3}{2}t(t-1)\alpha^t + C\frac{1}{2}t(t-1)\alpha^t \\ &\leq\ 2C^2 t(t-1)\alpha^t. \end{aligned}$$

This concludes the proof. □

In order to extend the results of this section to cyclically reduced words, we need an additional hypothesis, essentially stating that the probability of cyclically reduced words does not vanish. In fact, we have the following general result.

LEMMA 3.13. *Let $(\mathbb{R}_n)_{n\geq 0}$ be a sequence of measures satisfying* $\liminf \mathbb{R}_n(\mathcal{C}_n) = p > 0$. *Let X be a subset of $\mathcal{R}$. Then for each $\delta > 1$ and for every large enough n, the probability $\mathbb{R}_n(X \mid \mathcal{C})$ that a cyclically reduced word of length n is in X is at most equal to $\frac{\delta}{p}\mathbb{R}_n(X)$. In particular, if X is exponentially (resp. super-polynomially, polynomially, simply) negligible, then so is $X \cap \mathcal{C}$ in $\mathcal{C}$.*

PROOF. By definition, $\mathbb{R}_n(X \mid \mathcal{C}) = \mathbb{R}_n(X \cap \mathcal{C} \mid \mathcal{C}) = \frac{\mathbb{R}_n(X\cap\mathcal{C})}{\mathbb{R}_n(\mathcal{C}_n)} \leq \frac{\delta}{p}\mathbb{R}_n(X)$, which concludes the proof. □

The following statement is an immediate consequence.

COROLLARY 3.14. *Let $(\mathbb{R}_n)_{n\geq 0}$ be a prefix-heavy sequence of parameters (C,α), with the property that* $\liminf_n \mathbb{R}_n(\mathcal{C}_n) = p > 0$. *Then for every $\delta > 1$ and every large enough n, the probability that a cyclically reduced word of length n has two non-overlapping occurrences of a cyclic factor of length t (resp. an occurrence of a cyclic factor of length t and its inverse, two overlapping occurrences of a cyclic factor of length t) is at most $\frac{\delta}{p}(Cn^2+C^2nt)\alpha^t$ (resp. $\frac{\delta}{p}(Cn^2+C^2nt)\alpha^t$, $\frac{3\delta}{p}C^2nt\alpha^t$).*

PROOF. Let X be the set of reduced words of length n with two non-overlapping occurrences of a cyclic factor of length t (resp. an occurrence of a cyclic factor of length t and its inverse, two overlapping occurrences of a cyclic factor of length t). It suffices to apply Lemma 3.13 to the set X, and to use the results of Lemmas 3.11 and 3.12. $\square$

3.4. Measures on tuples of lengths and on tuples of words. For every positive integer k, let $\mathcal{T}_k$ denote the set of k-tuples of non-negative integers and $\mathcal{TW}_k$ denote the set of k-tuples of reduced words. Let also $\mathcal{T} = \bigcup_k \mathcal{T}_k$ and $\mathcal{TW} = \bigcup_k \mathcal{TW}_k$ be the sets of all tuples of non-negative integers, and of reduced words respectively.

For a given $\vec{h} = (h_1, \ldots, h_k)$ of $\mathcal{TW}_k$, let $\|\vec{h}\|$ be the element of $\mathcal{T}_k$ given by

$$\|\vec{h}\| = (|h_1|, \ldots, |h_k|)\,.$$

A *prefix-heavy sequence of measures on tuples of reduced words* is a sequence $(\mathbb{P}_n)_{n\geq 0}$ of measures on $\mathcal{TW}$ such that for every $\vec{h} = (h_1, \ldots, h_k)$ of $\mathcal{TW}$,

$$\mathbb{P}_n(\vec{h}) = \mathbb{T}_n(\|\vec{h}\|)\prod_{i=1}^{k}\mathbb{R}_{|h_i|}(h_i),$$

where $(\mathbb{T}_n)_{n\geq 0}$ is a sequence of measures on $\mathcal{T}$ and $(\mathbb{R}_n)_{n\geq 0}$ is a prefix-heavy sequence of measures on $\mathcal{R}$. If $(\mathbb{R}_n)_{n\geq 0}$ is prefix-heavy with parameters (C,α), then we say that $(\mathbb{T}_n)_{n\geq 0}$ is prefix-heavy with parameters (C,α).

REMARK 3.15. In the definition above, to draw a tuple of words according to $\mathbb{P}_n$, one can first draw a tuple of lengths $(\ell_1, \ldots, \ell_k)$ following $\mathbb{T}_n$, and then draw, independently for each coordinate, an element of $\mathcal{R}_{\ell_i}$ following $\mathbb{R}_{\ell_i}$.

EXAMPLE 3.16. Let $\nu(n)$ be an integer-valued function. The uniform distribution on the $\nu(n)$-tuples of reduced words of length exactly n is a prefix-heavy sequence of measures: one needs to take $\mathbb{T}_n$ to be the measure whose weight is entirely concentrated on the $\nu(n)$-tuple $(n, \ldots, n)$ and $\mathbb{R}_n$ to be the uniform distribution on $\mathcal{R}_n$ (see Example 3.2).

The uniform distribution on the $\nu(n)$-tuples of reduced words of length at most n is also a prefix-heavy sequence of measures. Here the support of $\mathbb{T}_n$ must be restricted to the tuples $(x_1, \ldots, x_{\nu(n)})$ such that $x_i \leq n$ for each i, with $\mathbb{T}_n(x_1, \ldots, x_{\nu(n)}) = \prod_i \frac{|\mathcal{R}_{x_i}|}{|\mathcal{R}_{\leq n}|}$.

Both can be naturally adapted to handle the uniform distribution on the $\nu(n)$-tuples of cyclically reduced words of length exactly (resp. at most) n.

For appropriate functions $\nu(n)$, we retrieve the few-generator and the density models discussed in Section 2.2. We will see a more general class of examples in Section 4.

3.5. General statements. If $\vec{x} \in \mathcal{T}$, we denote by $\mathsf{max}(\vec{x})$ and $\mathsf{min}(\vec{x})$ the maximum and minimum element of $\vec{x}$. We also denote by $\mathsf{size}(\vec{x})$ the integer k such that $\vec{x} \in \mathcal{T}_k$.

The statistics min, max, and size are extended to tuples of words by setting $\mathsf{min}(\vec{h}) = \mathsf{min}(\|\vec{h}\|)$, $\mathsf{max}(\vec{h}) = \mathsf{max}(\|\vec{h}\|)$ and $\mathsf{size}(\vec{h}) = \mathsf{size}(\|\vec{h}\|)$. In the sequel we consider sequences of probability spaces on $\mathcal{TW}$ and min, max, and size are seen as random variables.

The following statements give general sufficient conditions for a tuple to generically have the central tree property, generate a malnormal subgroup, or satisfy a small cancellation property.

PROPOSITION 3.17. *Let $(\mathbb{P}_n)_{n\geq 0}$ be a prefix-heavy sequence of measures on tuples of reduced words of parameters (C, α). Let $f\colon \mathbb{N} \to \mathbb{N}$ such that $f(\ell) \leq \frac{\ell}{2}$ for each ℓ. If there exists a sequence $(\eta_n)_{n\geq 0}$ of positive real numbers such that*

$$\lim_{n\to\infty} \mathbb{P}_n\left(\mathsf{size}^2\,\alpha^{f(\mathsf{min})} > \eta_n\right) = 0 \quad \text{and} \quad \lim_{n\to\infty} \eta_n = 0, \tag{1}$$

then a random tuple of words generically satisfies $\mathsf{lcp}(\vec{h}) < f(\mathsf{min}(\vec{h}))$.

If the limits in Equation (1) *converge polynomially (resp. super-polynomially, exponentially) fast, then $\mathsf{lcp}(\vec{h}) < f(\mathsf{min}(\vec{h}))$ polynomially (resp. super-polynomially, exponentially) generically.*

PROOF. The set of all tuples $\vec{h}$ that fail to satisfy the inequality $\mathsf{lcp}(\vec{h}) < f(\mathsf{min}(\vec{h}))$ is the union $\mathcal{G}_1 \cup \mathcal{G}_2$ of the two following sets:

- the set $\mathcal{G}_1$ of all tuples $\vec{h} = (h_1, \ldots, h_k)$ such that for some $1 \leq i < j \leq k$, a word of length $f(\mathsf{min}(\vec{h}))$ occurs as a prefix of h_i or h_i^{-1}, and also of h_j or h_j^{-1},
- the set $\mathcal{G}_2$ of all tuples $\vec{h} = (h_1, \ldots, h_k)$ such that for some $1 \leq i \leq k$, h_i and h_i^{-1} have a common prefix of length $f(\mathsf{min}(\vec{h}))$,

and we only need to prove that $\lim_n \mathbb{P}_n(\mathcal{G}_1) = \lim_n \mathbb{P}_n(\mathcal{G}_2) = 0$.

Let k, ℓ be positive integers and let $X_{k,\ell}$ be the set of tuples $\vec{h} \in \mathcal{TW}_k$ such that $\mathsf{min}(\vec{h}) = \ell$. If $\vec{h} \in X_{k,\ell}$ and $1 \leq i < j \leq k$, then the probability that h_i and h_j have the same prefix of length $t = f(\ell)$ is

$$\sum_{w\in\mathcal{R}_t} \mathbb{R}_{|h_i|}(\mathcal{P}(w))\mathbb{R}_{|h_j|}(\mathcal{P}(w)) \;\leq\; C\alpha^t \sum_{w\in\mathcal{R}_t} \mathbb{R}_{|h_j|}(\mathcal{P}(w)) \;\leq\; C\alpha^t.$$

Then we have $\mathbb{P}_n(\mathcal{G}_1 \mid X_{k,\ell}) \leq 4k^2C\alpha^{f(\ell)}$, or rather

$$\mathbb{P}_n(\mathcal{G}_1 \mid X_{k,\ell}) \leq \min(1, 4k^2C\alpha^{f(\ell)}),$$

where the factor k^2 corresponds to the choice of i and j and the factor 4 corresponds to the possibilities that h_i or h_i^{-1}, and h_j or h_j^{-1} have a common prefix of length $f(\ell)$. Therefore we have $\mathbb{P}_n(\mathcal{G}_1 \cap X_{k,\ell}) \leq \min(1, 4k^2C\alpha^{f(\ell)})\,\mathbb{P}_n(X_{k,\ell})$

We can split the set of pairs (k, ℓ) into those pairs such that $k^2\alpha^{f(\ell)} > \eta_n$ and the others, for which $k^2\alpha^{f(\ell)} \leq \eta_n$. Then we have

$$\mathbb{P}_n(\mathcal{G}_1) \;=\; \sum_{k,\ell} \mathbb{P}_n(\mathcal{G}_1 \cap X_{k,\ell}) \;\leq\; \mathbb{P}_n(\mathsf{size}^2\,\alpha^{f(\mathsf{min})} > \eta_n) + 4C\,\eta_n,$$

which tends to 0 under the hypothesis in Equation (1).

Similarly, if $\vec{h} \in X_{k,\ell}$ and $i \le k$, the probability that h_i and h_i^{-1} have a common prefix of length $f(\ell)$ is at most $C\alpha^{f(\ell)}$ by Corollary 3.8. It follows that $\mathbb{P}_n(\mathcal{G}_2 \mid X_{k,\ell}) \le \min(1, kC\alpha^{f(\ell)})$, and $\mathbb{P}_n(\mathcal{G}_2 \cap X_{k,\ell}) \le \min(1, kC\alpha^{f(\ell)})\, \mathbb{P}_n(X_{k,\ell})$.

Splitting the set of pairs (k,ℓ) into those pairs such that $k\alpha^{f(\ell)} > \eta_n$ and those for which $k\alpha^{f(\ell)} \le \eta_n$, yields

$$\mathbb{P}_n(\mathcal{G}_2) \;=\; \sum_{k,\ell} \mathbb{P}_n(\mathcal{G}_2 \cap X_{k,\ell}) \;\le\; \mathbb{P}_n(\mathsf{size}\,\alpha^{f(\mathsf{min})} > \eta_n) + C\,\eta_n.$$

Now $\mathsf{size}\,\alpha^{f(\mathsf{min})} < \mathsf{size}^2\,\alpha^{f(\mathsf{min})}$, so $\mathbb{P}_n(\mathsf{size}\,\alpha^{f(\mathsf{min})} > \eta_n) \le \mathbb{P}_n(\mathsf{size}^2\,\alpha^{f(\mathsf{min})} > \eta_n)$. It follows that $\lim_n \mathbb{P}_n(\mathsf{size}\,\alpha^{f(\mathsf{min})} > \eta_n) = 0$, and hence $\lim_n \mathbb{P}_n(\mathcal{G}_2) = 0$, which concludes the proof. □

THEOREM 3.18 (Central tree property). *Let $(\mathbb{P}_n)_{n\ge 0}$ be a prefix-heavy sequence of measures on tuples of reduced words of parameters (C,α). If there exists a sequence $(\eta_n)_{n\ge 0}$ of positive real numbers such that*

$$\lim_{n\to\infty} \mathbb{P}_n\left(\mathit{size}^2\,\alpha^{\frac{\mathit{min}}{2}} > \eta_n\right) = 0 \quad \textit{and} \quad \lim_{n\to\infty} \eta_n = 0, \tag{2}$$

then a random tuple of words generically has the central tree property. In particular, such a tuple is a basis of the subgroup it generates.

If the limits in Equation (2) converge polynomially (resp. super-polynomially, exponentially) fast, then the central tree property holds polynomially (resp. super-polynomially, exponentially) generically.

PROOF. By definition, a tuple $\vec{h} \in \mathcal{TW}$ satisfies the central tree property if $\mathsf{lcp}(\vec{h}) < \frac{\mathsf{min}(\vec{h})}{2}$, so the theorem is a direct application of Proposition 3.17 to the function $f(\ell) = \frac{\ell}{2}$, and of Proposition 1.2. □

THEOREM 3.19 (Malnormality). *Let $(\mathbb{P}_n)_{n\ge 0}$ be a prefix-heavy sequence of measures on tuples of reduced words of parameters (C,α). If there exists a sequence $(\eta_n)_{n\ge 0}$ of positive real numbers such that*

$$\lim_{n\to\infty} \mathbb{P}_n\left(\mathit{size}^2\,\mathit{max}^2\,\alpha^{\frac{\mathit{min}}{8}} > \eta_n\right) = 0 \quad \textit{and} \quad \lim_{n\to\infty} \eta_n = 0, \tag{3}$$

then a random tuple of words generically generates a malnormal subgroup.

If the limits in Equation (3) converge polynomially (resp. super-polynomially, exponentially) fast, then malnormality holds polynomially (resp. super-polynomially, exponentially) generically.

PROOF. By Proposition 1.3, a sufficient condition for a tuple $\vec{h} \in \mathcal{TW}$ to generate a malnormal subgroup is to have $\mathsf{lcp}(\vec{h}) < \frac{1}{3}\mathsf{min}(\vec{h})$, and to not have two occurrences of a word of length $\frac{1}{2}(\mathsf{min}(\vec{h}) - 3\,\mathsf{lcp}(\vec{h}))$ as a factor of a word in $\vec{h}^{\pm}$. This condition is satisfied in particular if $\mathsf{lcp}(\vec{h}) < \frac{1}{4}\mathsf{min}(\vec{h})$ and no word of length $\frac{1}{8}\mathsf{min}(\vec{h})$ has two occurrences as a factor of a word in $\vec{h}^{\pm}$.

Therefore the set of all tuples $\vec{h}$ that generate a non malnormal subgroup is contained in the union $\mathcal{G}_1 \cup \mathcal{G}_2 \cup \mathcal{G}_3 \cup \mathcal{G}_4$ of the following sets:

- the set $\mathcal{G}_1$ of all tuples $\vec{h} = (h_1, \dots, h_k)$ such that $\mathsf{lcp}(\vec{h}) \ge \frac{1}{4}\mathsf{min}(\vec{h})$,
- the set $\mathcal{G}_2$ of all tuples $\vec{h} = (h_1, \dots, h_k)$ such that for some $1 \le i < j \le k$, a word of length $\frac{1}{8}\mathsf{min}(\vec{h})$ occurs as a factor of h_i, and also of h_j or h_j^{-1},

- the set $\mathcal{G}_3$ of all tuples $\vec{h} = (h_1, \dots, h_k)$ such that for some $1 \leq i \leq k$, h_i and h_i^{-1} have a common factor of length $\frac{1}{8}\,\mathsf{min}(\vec{h})$,
- the set $\mathcal{G}_4$ of all tuples $\vec{h} = (h_1, \dots, h_k)$ such that for some $1 \leq i \leq k$, h_i has at least two occurrences of a factor of length $\frac{1}{8}\,\mathsf{min}(\vec{h})$,

and we want to verify that $\mathbb{P}_n(\mathcal{G}_1)$, $\mathbb{P}_n(\mathcal{G}_2)$, $\mathbb{P}_n(\mathcal{G}_3)$ and $\mathbb{P}_n(\mathcal{G}_4)$ all tend to 0 when n tends to infinity.

By Proposition 3.17, the set $\mathcal{G}_1$ is negligible as soon as $\lim_n \mathbb{P}_n(\mathsf{size}\,\alpha^{\frac{\mathsf{min}}{4}} > \eta_n) = 0$. This is true under the hypothesis in Equation (3) since $\mathsf{size}\,\alpha^{\frac{\mathsf{min}}{4}} < \mathsf{size}^2\,\mathsf{max}^2\,\alpha^{\frac{\mathsf{min}}{8}}$, and hence $\mathbb{P}_n(\mathsf{size}\,\alpha^{\frac{\mathsf{min}}{4}} > \eta_n) \leq \mathbb{P}_n(\mathsf{size}^2\,\mathsf{max}^2\,\alpha^{\frac{\mathsf{min}}{8}} > \eta_n)$.

Let now $X_{k,\ell,M}$ be the set of tuples $\vec{h} \in X_{k,\ell}$ such that $\mathsf{max}(\vec{h}) = M$. Let $1 \leq i < j \leq k$ and $\vec{h} \in X_{k,\ell,M}$. By Corollary 3.6, the probability that h_j has a given factor v of length $\frac{\ell}{8}$ is at most equal to $CM\alpha^{\frac{\ell}{8}}$. Summing this probability over all words v which occur as a factor of h_i (at most $|h_i| \leq M$ such words), it follows that the probability that h_i and h_j have a common factor of length $t = \frac{\ell}{8}$ is at most equal to $CM^2\alpha^{\frac{\ell}{8}}$. Summing now over the possible values of i and j, we find that $\mathbb{P}_n(\mathcal{G}_2 \cap X_{k,\ell,M}) \leq \min(1, k^2CM^2\alpha^{\frac{\ell}{8}})\,\mathbb{P}_n(X_{k,\ell,M})$ and therefore, as above

$$\mathbb{P}_n(\mathcal{G}_2) \;\leq\; \mathbb{P}_n(\mathsf{size}^2\,\mathsf{max}^2\,\alpha^{\frac{\mathsf{min}}{8}} > \eta_n) + C\,\eta_n.$$

It follows from Equation (3) that $\mathcal{G}_2$ is negligible.

By Lemma 3.7, the probability that h_i and h_i^{-1} have a common factor of length $\frac{\ell}{8}$ is at most $CM^2\alpha^{\frac{\ell}{8}}$. Summing over all choices of i, we find that

$$\mathbb{P}_n(\mathcal{G}_3) \;\leq\; \mathbb{P}_n(\mathsf{size}\,\mathsf{max}^2\,\alpha^{\frac{\mathsf{min}}{8}} > \eta_n) + C\,\eta_n.$$

Since $\mathsf{size}\,\mathsf{max}^2\,\alpha^{\frac{\mathsf{min}}{8}} < \mathsf{size}^2\,\mathsf{max}^2\,\alpha^{\frac{\mathsf{min}}{8}}$, we conclude that $\mathcal{G}_3$ is negligible.

Finally, we have $\mathbb{P}_n(\mathcal{G}_4) \leq \frac{C}{8}\,\mathsf{size}\,\mathsf{max}\,\mathsf{min}\,\alpha^{\frac{\mathsf{min}}{8}}$ by Lemma 3.9, and hence

$$\mathbb{P}_n(\mathcal{G}_4) \;\leq\; \mathbb{P}_n(\mathsf{size}\,\mathsf{max}\,\mathsf{min}\,\alpha^{\frac{\mathsf{min}}{8}} > \eta_n) + \frac{C}{8}\,\eta_n.$$

Since $\mathsf{size}\,\mathsf{max}\,\mathsf{min}\,\alpha^{\frac{\mathsf{min}}{8}} < \mathsf{size}^2\,\mathsf{max}^2\,\alpha^{\frac{\mathsf{min}}{8}}$, it follows as above that the set $\mathcal{G}_4$ is negligible. □

THEOREM 3.20 (Small cancellations property). *Let $(\mathbb{P}_n)_{n\geq 0}$ be a prefix-heavy sequence of measures on tuples of reduced words of parameters (C, α), such that $\liminf_n \mathbb{R}_n(\mathcal{C}_n) = p > 0$. For any $\lambda \in (0, \frac{1}{2})$, if there exists a sequence $(\eta_n)_{n\geq 0}$ of positive real numbers such that*

$$\lim_{n\to\infty} \mathbb{P}_n\left(\mathsf{size}^2\,\mathsf{max}^2\,\alpha^{\lambda\,\mathsf{min}} > \eta_n\right) = 0 \quad \text{and} \quad \lim_{n\to\infty} \eta_n = 0, \tag{4}$$

then the property $C'(\lambda)$ generically holds.

If the limits in Equation (4) converge polynomially (resp. super-polynomially, exponentially) fast, then Property $C'(\lambda)$ holds polynomially (resp. super-polynomially, exponentially) generically.

PROOF. A sufficient condition for a tuple of cyclically reduced words $\vec{h}$ to satisfy $C'(\lambda)$ is for every piece in $\vec{h}$ to have length less than $\lambda\,\mathsf{min}(\vec{h})$. Then the set $\mathcal{G}$ of tuples that fail to satisfy $C'(\lambda)$ is contained in the union $\mathcal{G}_1 \cup \mathcal{G}_2 \cup \mathcal{G}_3 \cup \mathcal{G}_4$ of the following sets:

- the set $\mathcal{G}_1$ of all tuples of cyclically reduced words $\vec{h} = (h_1, \ldots, h_k)$ such that for some $1 \leq i < j \leq k$, a word of length $\lambda \operatorname{\mathsf{min}}(\vec{h})$ occurs as a factor of h_i, and also of h_j or h_j^{-1},
- the set $\mathcal{G}_2$ of all tuples of cyclically reduced words $\vec{h} = (h_1, \ldots, h_k)$ such that for some $1 \leq i \leq k$, h_i has two non-overlapping occurrences of a factor of length $\lambda \operatorname{\mathsf{min}}(\vec{h})$,
- the set $\mathcal{G}_3$ of all tuples of cyclically reduced words $\vec{h} = (h_1, \ldots, h_k)$ such that for some $1 \leq i \leq k$, h_i has non-overlapping occurrences of a factor of length $\lambda \operatorname{\mathsf{min}}(\vec{h})$ and its inverse,
- the set $\mathcal{G}_4$ of all tuples of cyclically reduced words $\vec{h} = (h_1, \ldots, h_k)$ such that for some $1 \leq i \leq k$, h_i has overlapping occurrences of a factor of length $\lambda \operatorname{\mathsf{min}}(\vec{h})$,

and we want to verify that $\mathbb{P}_n(\mathcal{G}_1)$, $\mathbb{P}_n(\mathcal{G}_2)$, $\mathbb{P}_n(\mathcal{G}_3)$ and $\mathbb{P}_n(\mathcal{G}_4)$ all tend to 0 when n tends to infinity.

As in the proof of Theorem 3.19, we find that the probability that a tuple of reduced words $\vec{h}$ is such that a word of length $\lambda \operatorname{\mathsf{min}}(\vec{h})$ occurs as a factor of h_i, and also of h_j or h_j^{-1}, for some $i < j$ is at most $\mathbb{P}_n(\mathsf{size}^2 \mathsf{max}^2 \alpha^{\lambda \mathsf{min}} > \eta_n) + C\ \eta_n$. Reasoning as in the proof of Corollary 3.14, it follows that, for every $\delta > 1$,

$$\mathbb{P}_n(\mathcal{G}_1) \leq \frac{\delta}{p} \left(\mathbb{P}_n(\mathsf{size}^2 \mathsf{max}^2 \alpha^{\lambda \mathsf{min}} > \eta_n) + C\ \eta_n\right),$$

and it follows from Equation (4) that $\mathcal{G}_1$ is negligible.

Now using Corollary 3.14, we show that

$$\mathbb{P}_n(\mathcal{G}_2), \mathbb{P}_n(\mathcal{G}_3) \leq \frac{\delta}{p} \left(\mathbb{P}_n(\mathsf{size}(\mathsf{max}^2 + \mathsf{max}\,\mathsf{min})\alpha^{\lambda \mathsf{min}} > \eta_n) + C^2\eta_n\right),$$

$$\mathbb{P}_n(\mathcal{G}_4) \leq \frac{\delta}{p} \left(\mathbb{P}_n(\mathsf{size}(\mathsf{max}\,\mathsf{min} + \mathsf{min}^2)\alpha^{\lambda \mathsf{min}} > \eta_n) + 2C^2\eta_n\right).$$

Since $\mathsf{size}\,\mathsf{max}^2$, $\mathsf{size}\,\mathsf{max}\,\mathsf{min}$ and $\mathsf{size}\,\mathsf{min}^2$ are less than $\mathsf{size}^2\,\mathsf{max}^2$, the hypothesis in Equation (4) shows that $\mathcal{G}_2$, $\mathcal{G}_3$ and $\mathcal{G}_4$ are negligible, and this concludes the proof. □

3.6. Applications to the uniform distribution case. The few-generator model and the density model, based on the uniform distribution on reduced words of a given length and discussed in Section 2.2, are both instances of a prefix-heavy sequence of measures on tuples, for which the parameter α is $\alpha = \frac{1}{2r-1}$, see Examples 3.2 and 3.16. In this section, the measure $\mathbb{R}_n$ is the uniform distribution on $\mathcal{R}_n$.

The results of Section 3.5 above allow us to retrieve many of the results in Section 2.2 — typically the results on the small cancellation property $C'(\lambda)$ up to density $\frac{\lambda}{2}$, whether one considers tuples of cyclically reduced words of length n or of length at most n —, and to expand them. In particular, we show that the results on the central tree property and malnormality in the few-generator model can be extended to the density model, and that we have a phase transition theorem for the central tree property (at density $\frac{1}{4}$).

Small cancellation properties Let $0 < d < 1$. In the density model, at density d, we choose uniformly at random a $\nu(n)$-tuple of cyclically reduced words of length

n, with $\nu(n) = |\mathcal{C}_n|^d$. In particular, for every tuple $\vec{h}$ of that sort, we have $\mathsf{size}(\vec{h}) = \nu(n)$ and $\mathsf{max}(\vec{h}) = \mathsf{min}(\vec{h}) = n$.

Let $0 < \lambda < \frac{1}{2}$ and for each n, let

$$\eta_n = \left(\frac{2r}{2r-1}\right)^{2d} n^2 \, (2r-1)^{-(\lambda-2d)n} + \left(\frac{2r}{2r-1}\right)^{d} n^2 \, (2r-1)^{-(\lambda-d)n}.$$

Note that $|\mathcal{C}_n| < |\mathcal{R}_n| = \frac{2r}{2r-1}(2r-1)^n$. Therefore $\mathsf{size}^2 \, \mathsf{max}^2 \, \alpha^{\lambda \, \mathsf{min}} < \eta_n$ with probability 1. Now observe that η_n converges exponentially fast to 0 when $d < \frac{\lambda}{2}$. In view of Theorem 3.20, this provides a proof of part of Theorem 2.4 (2), namely, of the fact that, at density less than $\frac{\lambda}{2}$, Property $C'(\lambda)$ holds exponentially generically.

It is unclear whether the more difficult property, that hyperbolicity holds generically at density less than $\frac{1}{2}$, can be established with the same very general tools.

Observe that the set $\mathcal{R}_{\leq n}$ of reduced words of length at most n has cardinality $1 + \sum_{i=1}^{n} |\mathcal{R}_n| = \frac{r}{r-1}(2r-1)^n - \frac{1}{r-1}$. By the same reasoning as above, at density less than $\frac{\lambda}{2}$, a tuple of cyclically reduced words of length at most n exponentially generically has Property $C'(\lambda)$.

Properties of subgroups We now return to tuples of reduced words like in the few-generator model, but with a density type assumption on the size of the tuples. For $0 < d < 1$, we consider $|\mathcal{R}_{\leq n}|^d$-tuples of reduced words of length at most n, and the asymptotic properties of the subgroups generated by these tuples. For such tuples $\vec{h}$, we have $\mathsf{size}(\vec{h}) \leq \left(\frac{r}{r-1}\right)^d (2r-1)^{dn}$ and $\mathsf{max}(\vec{h}) = n$.

In addition, for every $0 < \mu < 1$, Proposition 2.1 shows that $\mathsf{min}(\vec{h}) > \mu n$, exponentially generically.

We first establish the central tree property.

PROPOSITION 3.21. *Let $0 < d < \frac{1}{4}$. At density d, a tuple of reduced words of length at most n chosen uniformly at random, exponentially generically has the central tree property, and in particular it is a basis of the subgroup it generates.*

If $d > \frac{1}{4}$, then at density d the central tree property exponentially generically does not hold.

PROOF. For a fixed $\mu < 1$, the following inequality holds exponentially generically:

$$\mathsf{size}^2 \, \alpha^{\frac{\mathsf{min}}{2}} \leq \left(\frac{r}{r-1}\right)^{2d} (2r-1)^{-(\frac{\mu}{2}-2d)n}.$$

At every density $d < \frac{1}{4}$, one can choose $\mu < 1$ such that $\frac{\mu}{2} - 2d > 0$ (say, $\mu = \frac{1+4d}{2}$). For such a value of μ, $\eta_n = \left(\frac{r}{r-1}\right)^{2d} (2r-1)^{-(\frac{\mu}{2}-2d)n}$ converges exponentially fast to 0 and, in view of Theorem 3.18, this proves the first part of the proposition.

If $d > \frac{1}{4}$, let d' be such that $\frac{1}{4} < d' < \min(\frac{1}{2}, d)$. By the classical Birthday Paradox[3], exponentially generically two words of the tuple share a prefix of length $2d'n$. This prove the second part of the proposition. $\square$

Along the same lines, we also prove the following result.

[3]If E is a set of size M and x is a uniform random tuple of E^m, the probability that the coordinates of x are pairwise distinct is $(1 - \frac{1}{M})(1 - \frac{2}{M}) \cdots (1 - \frac{m-1}{M})$, which is at most $\exp(-\frac{m(m-1)}{2M})$ by direct calculations.

PROPOSITION 3.22. *Let $0 < d < \frac{1}{16}$. At density d, a tuple of reduced words of length at most n chosen uniformly at random, exponentially generically generates a malnormal subgroup.*

PROOF. For a fixed $\mu < 1$, we have

$$\mathsf{size}^2 \, \mathsf{max}^2 \, \alpha^{\frac{\mathsf{min}}{8}} \leq \left(\frac{r}{r-1}\right)^{2d} n^2 (2r-1)^{-(\frac{\mu}{8}-2d)n},$$

exponentially generically.

If $d < \frac{1}{16}$, one can choose $\mu < 1$ such that $\frac{\mu}{8} - 2d > 0$ (say, $\mu = \frac{1+16d}{2}$), and we conclude as above, letting

$$\eta_n = \left(\frac{r}{r-1}\right)^{2d} n^2 (2r-1)^{-(\frac{\mu}{8}-2d)n}$$

and using Theorem 3.19. □

REMARK 3.23. Propositions 3.21 and 3.22 above generalize Corollary 2.2 (1) and (2), from the few generator case to an exponential number of generators — up to density $\frac{1}{4}$ and $\frac{1}{16}$, respectively (see Proposition 1.5).

Proposition 3.21 can actually be radically refined if the tuples have less than exponential size and if we drop the requirement of exponential genericity.

PROPOSITION 3.24. *Let f be an unbounded non-decreasing integer function. Let $k > 1$ be a fixed integer. Then a k-tuple $\vec{h}$ of reduced words of length at most n chosen uniformly at random, generically has the central tree property, with $\mathsf{lcp}(\vec{h}) \leq f(n)$.*

Let $c, c' > 0$ such that $c' \log(2r-1) > 2c$. Then an n^c-tuple $\vec{h}$ of reduced words of length at most n chosen uniformly at random, generically has the central tree property, with $\mathsf{lcp}(\vec{h}) \leq c' \log n$.

PROOF. If k is a fixed integer, then as in the proof of Proposition 3.21, we find that, for each $\mu < 1$, $\mathsf{size}^2 \, \alpha^{f(\mathsf{min})}$ is generically less than or equal to $\eta_n = k^2(2r-1)^{-f(\mu n)}$, which tends to 0. This concludes the proof on the size of the central tree of random k-tuples by Proposition 3.17.

If we now consider n^c-tuples, we find that, for each $\mu < 1$, $\mathsf{size}^2 \, \alpha^{c' \log(\mu n))}$ is generically less than or equal to $\eta_n = n^{2c}(2r-1)^{-c' \log n} = n^{-(c' \log(2r-1)-2c)}$, which tends to 0. By Proposition 3.17 again, this concludes the proof. □

4. Markovian automata

We now switch from the very general settings of the previous section to a specific and computable way to define prefix-heavy sequences of measures on reduced words.

We introduce Markovian automata (Section 4.1) which determine prefix-heavy sequences of measures under a simple and natural non-triviality assumption. These automata are a form of hidden Markov chain, and when they have a classical ergodicity property, then cyclically reduced words have asymptotically positive density. We are then able to generalize the results of Section 3.6 about central tree property and malnormality.

In the last part of the section, we give a generalization of Theorem 2.4 (2) and (3) on small cancellation and the degeneracy of a finite presentation.

4.1. Definition and examples. A *Markovian automaton*[4] $\mathcal{A}$ consists of

- a deterministic transition system $(Q, \cdot)$ on alphabet X, where Q is a finite non-empty set called the *state set*, and for each $q \in Q$, $x \in X$, $q \cdot x \in Q$ or $q \cdot x$ is undefined;
- an initial probability vector $\gamma_0 \in [0,1]^Q$, that is, a positive vector such that $\sum_{q \in Q} \gamma_0(q) = 1$;
- for each $p \in Q$, a probability vector $(\gamma(p, x))_{x \in X} \in [0,1]^X$, such that $\gamma(p, x) = 0$ if and only if $p \cdot x$ is undefined.

If $u = x_0 \cdots x_n \in X^*$ $(n \geq 0)$, we write $\gamma(q, u) = \gamma(q, x_0)\gamma(q \cdot x_0, x_1) \cdots \gamma(q \cdot (x_0 \cdots x_{n-1}), x_n)$. We let $\gamma(q, u) = 1$ if u is the empty word. We also write $\gamma_0(u) = \sum_{q \in Q} \gamma_0(q)\gamma(q, u)$.

Markovian automata are very similar to hidden Markov chain models, except that symbols are output on transitions instead of on states. We will discuss this further in Section 4.2 below. Markovian automata can be considered as more intuitive since sets of words (languages) are naturally described by automata.

We observe that, for each $n \geq 0$, $\sum_{|u|=n} \gamma(u) = 1$. Thus γ determines a probability measure $\mathbb{R}_n$ on the set of elements of X^* of length n: if $|u| = n$, then $\mathbb{R}_n(u) = \gamma(u)$.

In the sequel, we consider only Markovian automata on alphabet $\tilde{A}$, where only reduced words have non-zero probability. More precisely, the *support* of a Markovian automaton $\mathcal{A}$ is the set of words that can be read in $\mathcal{A}$, starting from a state q such that $\gamma_0(q) \neq 0$, that is, the set of all words u such that $\gamma(u) \neq 0$: we assume that our Markovian automata are such that their support is contained in $\mathcal{R}$.

EXAMPLE 4.1. *Uniform distribution on reduced words of length n.* It is immediately verified that the following Markovian automaton yields the uniform distribution on reduced words of each possible length. The state set is $Q = \tilde{A}$. For each $a \in \tilde{A}$, there is an a-labeled transition from every state except a^{-1}, ending in state a. All these transitions have the same probability, namely $\frac{1}{2r-1}$, and the initial probability vector is uniform as well, with each coordinate equal to $\frac{1}{2r}$.

One can also tweak these probabilities, to favor certain letters over others, or to favor positive letters (the letters in A) over negative letters.

EXAMPLE 4.2. *Distributions on rational subsets of $F(A)$.* The support of a Markovian automaton $\mathcal{A}$ is always rational and closed under taking prefixes, but it does not have to be equal to the set of all reduced words. We can consider a rational subset L of $F(A)$, or rather a deterministic transition system reading only reduced words, and impose probabilistic weights on its transitions to form a Markovian automaton. The resulting distribution gives non-zero weights only to prefixes of elements of L.

Figure 3 represents two such automata (transitions are labeled by a letter and a probability, and each state is decorated with the corresponding initial probability), which are related with the modular group, $PSL(2, \mathbb{Z}) = \langle a, b \mid a^2, b^3 \rangle$.

The support of the distribution defined by automaton $\mathcal{A}$ is the set of words over alphabet $\{a, b, b^{-1}\}$ without occurrences of the factors a^2, b^2, $(b^{-1})^2$, bb^{-1} and

[4]This notion is different from the two notions of probabilistic automata, introduced by Rabin [**26**] and Segala and Lynch [**28**], respectively.

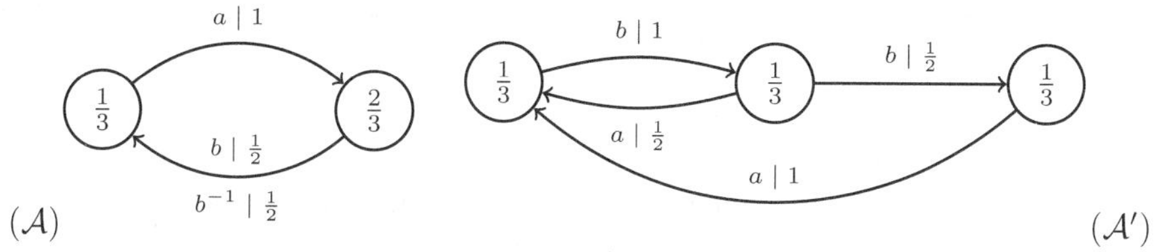

FIGURE 3. Markovian automata $\mathcal{A}$ and $\mathcal{A}'$.

$b^{-1}b$, and the support of the distribution defined by $\mathcal{A}'$ consists of the words on alphabet $\{a, b\}$, without occurrences of a^2 or b^3. Both are regular sets of unique representatives of the elements of $PSL(2, \mathbb{Z})$: the first is the set of geodesics of $PSL(2, \mathbb{Z})$, and also the set of Dehn-reduced words with respect to the given presentation of that group; the second is a set of quasi-geodesics of $PSL(2, \mathbb{Z})$. Notice that the distribution produced by $\mathcal{A}'$ is not uniform on words of length n of its support.

Example 4.1 shows that the sequence $(\mathbb{R}_n)_n$ of uniform measures on reduced words, discussed in Sections 2.2 and 3.6 can be specified by a Markovian automaton. We also know that this sequence is prefix-heavy (Example 3.2). This is a general fact, under mild assumptions on the Markovian automaton.

PROPOSITION 4.3. *Let $\mathcal{A}$ be a Markovian automaton and let $(\mathbb{R}_n)_n$ the sequence of probability measures it determines. If $\mathcal{A}$ does not have a cycle with probability 1, then $(\mathbb{R}_n)_n$ is a prefix-heavy sequence of measures, with computable parameters (C, α).*

PROOF. Let ℓ be the maximum length of an elementary cycle (one that does not visit twice the same state) and let δ be the maximum value of $\gamma(q, \kappa)$ where κ is an elementary cycle at state q. Under our hypothesis, $\delta < 1$.

Every cycle κ can be represented as a composition of at least $|\kappa|/\ell$ elementary cycles (here, the composition takes the form of a sequence of insertions of a cycle in another). Consequently $\gamma(q, \kappa) \leq \delta^{\frac{|\kappa|}{\ell}}$. Finally, every path can be seen as a product of cycles and at most $|Q|$ individual edges. So, if u is a word and $q \in Q$, then $\gamma(q, u) \leq \delta^{\frac{|u|-|Q|}{\ell}}$, that is $\gamma(q, u) \leq C\alpha^{|u|}$ where $C = \delta^{\frac{-|Q|}{\ell}}$ and $\alpha = \delta^{\frac{1}{\ell}}$.

Let u, v be reduced words such that uv is reduced and let $n \geq |uv|$. We have

$$\begin{aligned}
\mathbb{R}_n(\mathcal{P}(uv)) = \gamma_0(uv) &= \sum_{p \in Q} \gamma_0(p)\gamma(p, u)\gamma(p \cdot u, v) \\
&\leq \left(\sum_{p \in Q} \gamma_0(p)\gamma(p, u) \right) C\alpha^{|v|} \\
&= \gamma_0(u)\ C\alpha^{|v|} = \mathbb{R}_n(\mathcal{P}(u))\ C\alpha^{|v|},
\end{aligned}$$

and hence $\mathbb{R}_n(\mathcal{P}(uv) \mid \mathcal{P}(u)) \leq C\alpha^{|v|}$, which concludes the proof. □

REMARK 4.4. The parameters C and α described in the proof of Proposition 4.3 may be far from optimal. If $\beta < 1$ is a uniform bound on the probabilities of the transitions of $\mathcal{A}$, then $\gamma_0(v), \gamma(q, v) \leq \beta^{|v|}$ for each word v, and the computation in the proof above shows that $\mathbb{R}_n(\mathcal{P}(uv) \mid \mathcal{P}(u)) \leq \beta^{|v|}$. We will see in Section 4.2 that we can be more precise under additional hypotheses.

Now let $\mathcal{A}$ be a Markovian automaton without a probability 1 cycle, such that the sequence of probability measures it induces is prefix-heavy with parameters (C, α). If $0 < d < 1$, we say that a tuple $\vec{h}$ of reduced words of length at most (resp. exactly) n is chosen at random according to $\mathcal{A}$, *at α-density d* if $\vec{h}$ consists of α^{-dn} words. Observe that this generalizes the concept discussed in Section 2.2.2 and 3.6.

With the same proofs as in Section 3.6, we have the following generalization of Propositions 3.21 and 3.22 related to central tree property and malnormality.

COROLLARY 4.5. *Let $\mathcal{A}$ be a Markovian automaton without a probability 1 cycle, such that the induced sequence of probability measures is prefix-heavy with parameters (C, α). Then a tuple of reduced words of length at most n chosen at random according to $\mathcal{A}$, at α-density $d < \frac{1}{4}$, exponentially generically has the central tree property.*

At α-density $d < \frac{1}{16}$, it exponentially generically generates a malnormal subgroup.

4.2. Irreducible Markovian automata and coincidence probability. An (n, n)-matrix M is said to be *irreducible* if it has non-negative coefficients and, for every $i, j \leq n$, there exists $s \geq 1$ such that $M^s(i, j) > 0$. Equivalently, this means that M is not similar to a block upper-triangular matrix. We record the following general property of irreducible matrices.

LEMMA 4.6. *Let M be an irreducible matrix. Then its spectral radius ρ is a (positive) eigenvalue with a positive eingenvector. In particular, there exist positive vectors $\vec{v}_{\min}$ and $\vec{v}_{\max}$ such that, componentwise,*

$$\rho^n \vec{v}_{\min} \;\leq\; M^n \vec{1} \;\leq\; \rho^n \vec{v}_{\max} \quad \textit{for all } n > 0$$

where $\vec{1}$ is the vector whose coordinates are all equal to 1. Moreover, there exist $c_{\min}, c_{\max} > 0$ such that

$$c_{\min} \rho^n \;\leq\; \vec{1}^t M^n \vec{1} \;\leq\; c_{\max} \rho^n \quad \textit{for all } n > 0.$$

PROOF. We refer the reader to [**8**, chap. 13, vol. 2] for a comprehensive presentation of the properties of irreducible matrices and in particular for the Perron-Frobenius theorem, which establishes that the spectral radius of M is an eigenvalue with a positive eigenvector: let $\vec{v}_0$ be such an eigenvector, and let $\vec{v}_{\min}$ (resp. $\vec{v}_{\max}$) be appropriate multiples of $\vec{v}_0$ with all coefficients less than 1 (resp. greater than 1). Then we have, componentwise, $\rho^n \vec{v}_{\min} = M^n \vec{v}_{\min} \leq \vec{M}^n \vec{1} \leq M^n \vec{v}_{\max} = \rho^n \vec{v}_{\max}$.

Let $c_{\min}$ (resp. $c_{\max}$) be the sum of the coefficients of $\vec{v}_{\min}$ (resp. $\vec{v}_{\max}$). Then, summing over all components of $M^n \vec{v}_{\min}$ and $M^n \vec{v}_{\max}$, we get $c_{\min} \rho^n \leq \vec{1}^t M^n \vec{1} \leq c_{\max} \rho^n$. □

Going back to automata, we note that a *Markov chain* can be naturally associated with a Markovian automaton: if $\mathcal{A}$ is a Markovian automaton on alphabet $\tilde{A}$, with state set Q, we define the Markov chain $M(\mathcal{A})$ on Q as follows: its transition matrix is given by $M(p, q) = \sum_{a \in \tilde{A} \text{ s.t. } p \cdot a = q} \gamma(p, a)$ for all $p, q \in Q$, and its initial vector is γ_0.

We say that the Markov chain $M(\mathcal{A})$ (or, by extension, the Markovian automaton $\mathcal{A}$), is *irreducible* if this transition matrix is irreducible, which is equivalent to the strong connectedness of $\mathcal{A}$. We note that, in that case, if $\mathcal{A}$ does not consist of a simple cycle, then $\mathcal{A}$ does not have a cycle of probability 1. In view of Proposition 4.3, this implies that the sequence of probability measures determined by $\mathcal{A}$ is prefix-heavy. We will see below (Proposition 4.9) that we can give a precise evaluation of the parameters of this sequence.

To this end, we introduce the notion of local Markovian automata, where labels can be read on states instead of edges.

More precisely a Markovian automaton is *local* if all the incoming transitions into a given state are labeled by the same letter: for all states p, q and letters a, b, if $p \cdot a = q \cdot b$ then $a = b$. If $\mathcal{A}$ is a Markovian automaton, let $\mathcal{A}'$ denote the local Markovian automaton obtained as follows.

- its set of states is $Q' = \{(q, a) \in Q \times \tilde{A} \mid \exists p \in Q,\ p \cdot a = q\}$;
- its transition function $\star$ is given by $(p, a) \star b = (q, b)$ if $p \cdot b = q$;
- its initial probability vector γ_0' is given by

$$\gamma_0'\big((p,a)\big) = \begin{cases} \gamma_0(p) & \text{if } a \text{ is the least label of the transitions into } p \\ 0 & \text{otherwise} \end{cases}$$

  (we fix an arbitrary order on $\tilde{A}$)
- its transition probability vectors are given by $\gamma'\big((p,a), b\big) = \gamma(p, b)$.

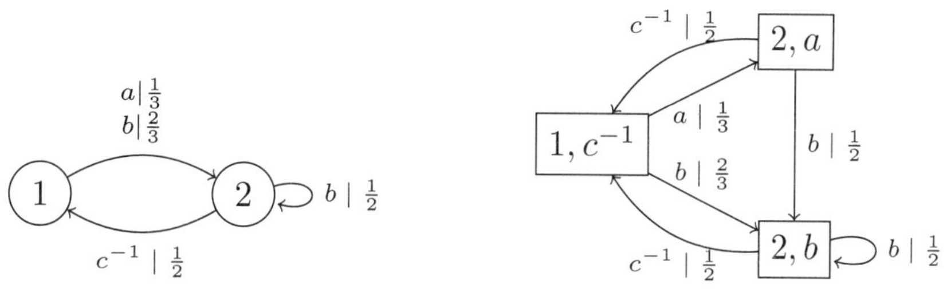

FIGURE 4. A Markovian automaton and its associated local automaton.

PROPOSITION 4.7. *Let $\mathcal{A}$ be a Markovian automaton. Then the associated local Markovian automaton $\mathcal{A}'$ assigns the same probability as $\mathcal{A}$ to every reduced word. Moreover, if $\mathcal{A}$ is irreducible, then so is $\mathcal{A}'$.*

PROOF. The first part of the statement follows directly from the definition, by a simple induction on the length of the words: indeed, we retrieve a path in $\mathcal{A}$ by forgetting the second coordinate on the states of $\mathcal{A}'$; and every path of $\mathcal{A}$ starting at some state q, can be lifted uniquely to a path in $\mathcal{A}'$ starting at any vertex of the form (q, a) of $\mathcal{A}'$.

Assume that $\mathcal{A}$ is irreducible and let (p, a) and (q, b) be states of $\mathcal{A}'$. By definition of $\mathcal{A}'$, there exists a state q' of $\mathcal{A}$ such that $q' \cdot b = q$. Moreover, since

$\mathcal{A}$ is irreducible, there exists a path from p to q' in $\mathcal{A}$, say $p \xrightarrow{a_1} q_1 \xrightarrow{a_2} \dots \xrightarrow{a_t} q'$. Then

$$(p, a) \xrightarrow{a_1} (q_1, a_1) \xrightarrow{a_2} \dots \xrightarrow{a_t} (q', a_t) \xrightarrow{b} (q', b)$$

is a path in $\mathcal{A}'$ from (p, a) to (q, b), so $\mathcal{A}'$ is irreducible as well. □

If $\mathcal{A}$ is a Markovian automaton, we denote by $\mathbb{M}_{\mathcal{A}}$ (or just $\mathbb{M}$ when there is no ambiguity) the stochastic matrix associated with its local automaton $\mathcal{A}'$:

$$\mathbb{M}\big((p,a),(q,b)\big) = \begin{cases} \gamma'\big((p,a),b\big) = \gamma(p,b) & \text{if } p \cdot b = q \\ 0 & \text{otherwise.} \end{cases}$$

We also denote by $\mathbb{M}_{[2]}$ and $\mathbb{M}_{[3]}$ the matrices defined by

$$\mathbb{M}_{[2]}\big((p,a),(q,b)\big) = \Big(\mathbb{M}\big((p,a),(q,b)\big)\Big)^2 \text{ and}$$

$$\mathbb{M}_{[3]}\big((p,a),(q,b)\big) = \Big(\mathbb{M}\big((p,a),(q,b)\big)\Big)^3,$$

and by $\alpha_{[2]}$ and $\alpha_{[3]}$ the largest eigenvalue of $\mathbb{M}_{[2]}$ and $\mathbb{M}_{[3]}$, respectively. The value $\alpha_{[2]}$ is called the *coincidence probability* of $\mathcal{A}$, and it will play an important role in the sequel.

Observe that if $\mathcal{A}$ is local, then $\mathcal{A}'$ is equal to $\mathcal{A}$, up to the name of the states. We are interested in local automata for the following properties.

LEMMA 4.8. *Let $\mathcal{A}$ be a local Markovian automaton. Then the following holds*

- *for all states p, q there is at most one transition from p to q;*
- *two paths starting from the same state are labeled by the same word if and only if they go through the same states in the same order;*
- *for every $\ell \geq 0$, we have $\mathbb{M}^\ell(p,q) = \sum_{u\in\mathcal{R}_\ell, p\cdot u = q} \gamma(p,u)$, $\mathbb{M}^\ell_{[2]}(p,q) = \sum_{u\in\mathcal{R}_\ell, p\cdot u = q} \gamma(p,u)^2$ and $\mathbb{M}^\ell_{[3]}(p,q) = \sum_{u\in\mathcal{R}_\ell, p\cdot u = q} \gamma(p,u)^3$.*

We can now give an upper bound for the parameters of the sequence of probability measures determined by an irreducible Markovian automaton.

PROPOSITION 4.9. *Let $\mathcal{A}$ be an irreducible Markovian automaton with coincidence probability $\alpha_{[2]}$, and let $(\mathbb{R}_n)_n$ be the sequence of probability measures it determines. If $\mathcal{A}$ does not consist of a single cycle, then there exists a constant $C > 0$ such that $(\mathbb{R}_n)_n$ is prefix-heavy with parameters $(C, \alpha_{[2]}^{1/2})$.*

PROOF. Let v be a reduced word of length ℓ and let $q \in Q$ be a state of $\mathcal{A}$. By Lemma 4.8, we have

$$\gamma(q,v) = \sqrt{\gamma(q,v)^2} \leq \sqrt{\mathbb{M}^\ell_{[2]}(q, q\cdot v)} \leq \sqrt{\vec{1}^t \mathbb{M}^\ell_{[2]} \vec{1}}.$$

Lemma 4.6 then shows that there exists $C > 0$ such that $\gamma(q,v) \leq C\alpha_{[2]}^{\frac{\ell}{2}}$. We can now conclude as in the proof of Proposition 4.3. □

This yields the following refinement of Corollary 4.5.

COROLLARY 4.10. *Let $\mathcal{A}$ be a Markovian automaton without a probability 1 cycle and with coincidence probability $\alpha_{[2]}$. Then a tuple of reduced words of length at most n chosen at random according to $\mathcal{A}$, at $\alpha_{[2]}$-density $d < \frac{1}{8}$ (resp. $d < \frac{1}{32}$), exponentially generically has the central tree property (resp. generates a malnormal subgroup).*

4.3. Ergodic Markovian automata. If the Markovian automaton $\mathcal{A}$ is irreducible and if, in addition, for all large enough n, $M(\mathcal{A})^n(q,q) > 0$ for each $q \in Q$, we say that $\mathcal{A}$ (resp. $M(\mathcal{A})$) is *ergodic*. This is equivalent to stating that $\mathcal{A}$ has a collection of loops of relatively prime lengths, or also that all large enough integral powers of $M(\mathcal{A})$ have only positive coefficients. If $\mathcal{A}$ is ergodic, we can apply a classical theorem on Markov chains, which states that there exists a *stationary vector* $\tilde{\gamma}$ such that the distribution defined by $\mathcal{A}$ converges to that stationary vector exponentially fast (see [**18**, Thm 4.9]). In the vocabulary of Markovian automata, this yields the following theorem.

If $u \in \tilde{A}^*$ has length n, let $Q_n^p(u) = p \cdot u$ be the state of $\mathcal{A}$ reached after reading the word u starting at state p. We treat Q_n^p as a random variable.

THEOREM 4.11. *Let $\mathcal{A}$ be an ergodic Markovian automaton on alphabet $\tilde{A}$, with state set Q ($|Q| \geq 2$). For each $q \in Q$, the limit $\lim_{n\to\infty} \mathbb{R}_n[Q_n^p = q]$ exists, and if we denote it by $\tilde{\gamma}(q)$, then $\tilde{\gamma}$ is a probability vector (called the* stationary vector*). In addition, there exist $K > 0$ and $0 < c < 1$, such that $|\mathbb{R}_n[Q_n^p = q] - \tilde{\gamma}(q)| < Kc^n$ for all n large enough.*

REMARK 4.12. The constant c in Theorem 4.11 is the maximal modulus of the non-1 eigenvalues of $M(\mathcal{A})$.

EXAMPLE 4.13. The Markovian automaton discussed in Example 4.1, relative to the uniform distribution on reduced words of length n, is ergodic. Its stationary vector $\tilde{\gamma}$ is equal to γ_0 ($\tilde{\gamma}(q) = \frac{1}{2r}$ for every state q), and the constant c is $\frac{1}{2r-1}$.

On the other hand, the Markovian automaton $\mathcal{A}$ in Example 4.2 is irreducible but not ergodic (loops have even lengths), and it does not have a stationary vector.

We use Theorem 4.11 to show that, under a very mild additional hypothesis, an ergodic Markovian automaton yields a prefix-heavy sequence of measures $(\mathbb{R}_n)_n$ such that $\liminf \mathbb{R}_n(\mathcal{C}) > 0$.

PROPOSITION 4.14. *Let $\mathcal{A}$ be an ergodic Markovian automaton, with initial vector γ_0 and stationary vector $\tilde{\gamma}$ and let $(\mathbb{R}_n)_n$ be the sequence of measures it induces on reduced words. If $\sum_{a\in\tilde{A}} \gamma_0(a)\tilde{\gamma}(a^{-1}) \neq 1$, then $\liminf \mathbb{R}_n(\mathcal{C}) > 0$.*

Observe that the sum $\sum_{a\in\tilde{A}} \gamma_0(a)\tilde{\gamma}(a^{-1})$ is less than 1, since we are dealing with probability vectors, unless there exists a (necessarily single) letter a such that $\gamma_0(a) = \tilde{\gamma}(a^{-1}) = 1$.

PROOF. The set $\mathcal{C}$ of cyclically reduced words is the complement in $\mathcal{R}$ of the disjoint union of the sets $a\tilde{A}^*a^{-1}$ ($a \in \tilde{A}$). Now we have

$$\begin{aligned}
\mathbb{R}_n(a\tilde{A}^*a^{-1}) &= \sum_{p\in Q} \gamma_0(p)\gamma(p,a) \left(\sum_{|u|=n-2} \gamma(p\cdot a, u)\gamma(p\cdot(au), a^{-1}) \right) \\
&= \sum_{p\in Q} \gamma_0(p)\gamma(p,a) \left(\sum_{q\in Q} \mathbb{R}_n(Q_{n-2}^{p\cdot a} = q)\gamma(q,a^{-1}) \right) \\
&= \sum_{p\in Q} \gamma_0(p)\gamma(p,a) \left(\sum_{q\in Q} (\tilde{\gamma}(q) + \varepsilon(q,n))\gamma(q,a^{-1}) \right),
\end{aligned}$$

where $|\varepsilon(q,n)| \leq Kc^{n-2}$, with K and c given by Theorem 4.11. Then we have

$$\mathbb{R}_n(a\tilde{A}^*a^{-1}) = \gamma_0(a)\tilde{\gamma}(a^{-1}) + \gamma_0(a)\left(\sum_{q\in Q}\varepsilon(q,n)\gamma(q,a^{-1})\right)$$

and $\lim \mathbb{R}_n(a\tilde{A}^*a^{-1}) = \gamma_0(a)\tilde{\gamma}(a^{-1})$. It follows that

$$\lim \mathbb{R}_n(\mathcal{C}) = 1 - \sum_{a\in\tilde{A}}\gamma_0(a)\tilde{\gamma}(a^{-1}),$$

thus concluding the proof. □

Proceeding as in Section 3.6, we can use Proposition 4.14, Corollary 3.14 and the results of Section 3.5, to generalize part of Theorem 2.4 (2), and show that, up to $\alpha_{[2]}$-density $\frac{\lambda}{4}$, a tuple of cyclically reduced words of length at most n chosen at random according to $\mathcal{A}$, exponentially generically satisfies the small cancellation property $C'(\lambda)$. We will now see (Theorem 4.15) that we can improve this bound, and go up to $\alpha_{[2]}$-density $\frac{\lambda}{2}$.

4.4. Phase transitions for the Markovian model. We can now state a phase transition theorem, which generalizes parts of Theorem 2.4. Let us say that an ergodic Markovian automaton is *non-degenerate* if its initial distribution γ_0 and its stationary vector $\tilde{\gamma}$ satisfy $\sum_{a\in\tilde{A}}\gamma_0(a)\tilde{\gamma}(a^{-1}) \neq 1$.

THEOREM 4.15. *Let $\mathcal{A}$ be a non-degenerate ergodic Markovian automaton with coincidence probability $\alpha_{[2]}$. Let $0 < d < 1$ and let G be the group presented by a tuple $\vec{h}$ of cyclically reduced words of length n, chosen independently and at random according to $\mathcal{A}$, at $\alpha_{[2]}$-density d. Then we have the following phase transitions:*

- *if $0 < \lambda < \frac{1}{2}$ and $0 < d < \frac{\lambda}{2}$, then exponentially generically $\vec{h}$ satisfies the small cancellation property $C'(\lambda)$; if $\lambda = \frac{1}{6}$, then G is generically infinite and hyperbolic;*
- *if $d > \frac{\lambda}{2}$ then exponentially generically $\vec{h}$ does not satisfy the small cancellation property $C'(\lambda)$;*
- *if $d > \frac{1}{2}$ then exponentially generically G is degenerated in a sense that is made precise in Proposition 4.23, and which implies that G is a free group or the free product of a free group with $\mathbb{Z}/2\mathbb{Z}$.*

The rest of the paper is devoted to the proof of Theorem 4.15. The first statement is established in Proposition 4.16, while the second and third statements are proved respectively in Propositions 4.22 and 4.23.

4.5. Long common factors at low density. In this section we estimate the probability that random words share a long common factor. More precisely, we show the following statement, the first part of Theorem 4.15.

PROPOSITION 4.16. *Let $\mathcal{A}$ be a non-degenerate ergodic Markovian automaton with coincidence probability $\alpha_{[2]}$. Let $\lambda \in (0, \frac{1}{2})$ and let $d \in (0, \frac{\lambda}{2})$. A tuple of cyclically reduced words of length n taken independently and randomly according to $\mathcal{A}$, at $\alpha_{[2]}$-density d, exponentially generically satisfies the small cancellation property $C'(\lambda)$.*

The structure of the proof of Proposition 4.16 resembles that of the proof of Theorem 3.20, and requires the consideration of several cases. This is the object of the rest of Section 4.5.

To this end, we introduce additional notation: let $\vec{\gamma}_q(n)$ be the vector of coordinates $\gamma(q,u)$ when u ranges over $\mathcal{R}_n$ in lexicographic order, and let $\|\vec{\gamma}_q(n)\|_k = (\sum_{u\in\mathcal{R}_n}\gamma(q,u)^k)^{1/k}$ be the ℓ_k-norm of this vector. We start with an elementary result.

LEMMA 4.17. *Let $\mathcal{A}$ be a Markovian automaton, let $0 < i,\ell < n$ be integers, and let $u \in \mathcal{R}_\ell$. The probability $\mathfrak{p}$ that u occurs as a cyclic factor at position i in a reduced word of length n is bounded above by*

$$\begin{cases} \sum_{q\in Q}\gamma(q,u) & \text{if } i \le n-\ell+1 \\ \sum_{q,q'\in Q}\gamma(q,u_1)\gamma(q',u_2) & \text{if } i > n-\ell+1 \text{ and } u = u_1u_2 \text{ with } |u_1| = n-i+1 \end{cases}$$

PROOF. If $i \le n-\ell+1$, then $\mathfrak{p} = \mathbb{R}_n(\tilde{A}^{i-1}u\tilde{A}^{n-\ell-i+1})$ is equal to

$$\begin{aligned}\sum_{p\in Q}\gamma_0(p)\sum_{w\in\mathcal{R}_{i-1}}\gamma(p,w)\gamma(p\cdot w,u) &= \sum_{p\in Q}\gamma_0(p)\sum_{q\in Q}\sum_{\substack{w\in\mathcal{R}_{i-1}\\ p\cdot w=q}}\gamma(p,w)\gamma(q,u)\\ &= \sum_{p\in Q}\gamma_0(p)\sum_{q\in Q}\mathbb{R}_{i-1}[Q^p_{i-1}=q]\gamma(q,u)\\ &\le \sum_{p,q\in Q}\gamma_0(p)\gamma(q,u) = \sum_{q\in Q}\gamma(q,u).\end{aligned}$$

If $i > n-\ell+1$ and $u = u_1u_2$ with $|u_1| = n-i+1$, then

$$\begin{aligned}\mathfrak{p} = \mathbb{R}_n(u_2\tilde{A}^{n-\ell}u_1) &= \sum_{q'\in Q}\gamma_0(q')\gamma(q',u_2)\sum_{w\in\mathcal{R}_{n-\ell}}\gamma(q'\cdot u_2,w)\gamma(q'\cdot u_2w,u_1)\\ &= \sum_{q'\in Q}\gamma_0(q')\gamma(q',u_2)\sum_{q\in Q}\sum_{\substack{w\in\mathcal{R}_{n-\ell}\\ q'\cdot u_2w=q}}\gamma(q'\cdot u_2,w)\gamma(q,u_1)\\ &= \sum_{q'\in Q}\gamma_0(q')\gamma(q',u_2)\sum_{q\in Q}\mathbb{R}_{n-\ell}[Q^{q'\cdot u_2}_{n-\ell}=q]\gamma(q,u_1)\\ &\le \sum_{q,q'\in Q}\gamma(q,u_1)\gamma(q',u_2),\end{aligned}$$

which concludes the proof. □

PROPOSITION 4.18. *Let $\mathcal{A}$ be an irreducible Markovian automaton with coincidence probability $\alpha_{[2]}$. Let n, ℓ, i and j be positive integers such that $\ell \le n$ and $i,j \le n$. Denote by $L(n,\ell,i,j)$ the probability that two reduced words of length n share a common cyclic factor of length ℓ at positions respectively i and j. Then there exists a positive constant K such that*

$$L(n,\ell,i,j) \le K\alpha_{[2]}^\ell.$$

PROOF. Without loss of generality (see Proposition 4.7), we may assume that $\mathcal{A}$ is local. The proof is based on a case study.
Case 1: $i,j \le n-\ell+1$. Using Lemma 4.17, we have

$$L(n,\ell,i,j) \le \sum_{p,q\in Q}\sum_{u\in\mathcal{R}_\ell}\gamma(p,u)\gamma(q,u).$$

By a repeated application of the Cauchy-Schwarz inequality, we get

$$L(n,\ell,i,j) \;\leq\; \sum_{p,q\in Q} \|\vec{\gamma}_p(\ell)\|_2 \|\vec{\gamma}_q(\ell)\|_2 \;\leq\; \sum_{q\in Q} \|\vec{\gamma}_q(\ell)\|_2^2. \tag{5}$$

Now, in view of Lemma 4.8 and since $\mathcal{A}$ is local, we have

$$\sum_{q\in Q} \|\vec{\gamma}_q(\ell)\|_2^2 = \sum_{q\in Q}\sum_{u\in\mathcal{R}_\ell} \gamma(q,u)^2 = \sum_{p\in Q}\sum_{q\in Q}\sum_{\substack{u\in\mathcal{R}_\ell\\ p\cdot u=q}} \gamma(p,u)^2 = \vec{1}^t\mathbb{M}_{[2]}^\ell\vec{1}. \tag{6}$$

Since $\mathbb{M}$ is irreducible, Lemma 4.6 shows that there exists a positive constant $K>0$ such that, for ℓ large enough, we have

$$L(n,\ell,i,j) \;\leq\; \sum_{q\in Q} \|\vec{\gamma}_q(\ell)\|_2^2 \;=\; \vec{1}^t\mathbb{M}_{[2]}^\ell\vec{1} \;\leq\; K\alpha_{[2]}^\ell,$$

which concludes the proof of the statement in that case.
Case 2: $i>n-\ell+1$ *and* $j\leq n-\ell+1$. (The case where $i\leq n-\ell+1$ and $j>n-\ell+1$ is symmetrical.) Let $k=n-i+1$ (so $1\leq k<\ell$). By Lemma 4.17, we have

$$\begin{aligned}
L(n,\ell,i,j) &\leq \sum_{\substack{u_1\in\mathcal{R}_k\\ u_2\in\mathcal{R}_{\ell-k}}} \sum_{p,p',q\in Q} \gamma(p,u_1)\gamma(p',u_2)\gamma(q,u_1u_2)\\
&\leq \sum_{\substack{u_1\in\mathcal{R}_k\\ u_2\in\mathcal{R}_{\ell-k}}} \sum_{p,p',q,q'\in Q} \gamma(p,u_1)\gamma(p',u_2)\gamma(q,u_1)\gamma(q',u_2)\\
&\leq \left(\sum_{u_1\in\mathcal{R}_k}\sum_{p,q\in Q}\gamma(p,u_1)\gamma(q,u_1)\right)\left(\sum_{u_2\in\mathcal{R}_{\ell-k}}\sum_{p',q'\in Q}\gamma(p',u_2)\gamma(q',u_2)\right).
\end{aligned}$$

By Cauchy-Schwarz, it follows that

$$\begin{aligned}
L(n,\ell,i,j) &\leq \left(\sum_{p,q\in Q}\|\vec{\gamma}_p(k)\|_2\|\vec{\gamma}_q(k)\|_2\right)\left(\sum_{p',q'\in Q}\|\vec{\gamma}_{p'}(\ell-k)\|_2\|\vec{\gamma}_{q'}(\ell-k)\|_2\right)\\
&\leq \left(\sum_{q\in Q}\|\vec{\gamma}_q(k)\|_2^2\right)\left(\sum_{q\in Q}\|\vec{\gamma}_q(\ell-k)\|_2^2\right)\\
&\leq \left(\vec{1}^t\,\mathbb{M}_{[2]}^k\,\vec{1}\right)\left(\vec{1}^t\,\mathbb{M}_{[2]}^{\ell-k}\,\vec{1}\right) \text{ by Equation (6).}
\end{aligned}$$

By Lemma 4.6, there exists a constant K_1 such that these two factors are bounded above, respectively, by $K_1\alpha_{[2]}^k$ and $K_1\alpha_{[2]}^{\ell-k}$. Therefore

$$L(n,\ell,i,j)\leq K_1^2\alpha_{[2]}^\ell$$

as announced.
Case 3: $i,j>n-\ell+1$. Without loss of generality, we may assume that $i<j$, and we let $k=n-j+1$ and $k'=\ell-(n-i+1)$. Then a word u of length ℓ occurs as a cyclic factor in two reduced words w_1 and w_2 of length n, at positions i and j respectively, if $u=u_1u_2u_3$ with $|u_1|=k$, $|u_2|=j-i$ and $|u_3|=k'$, and if

$w_1 \in u_3\tilde{A}^{n-\ell}u_1u_2$ and $w_2 \in u_2u_3\tilde{A}^{n-\ell}u_1$. Then we have

$$\begin{aligned}
L(n,\ell,i,j) &\leq \sum_{\substack{u_1\in\mathcal{R}_k\\ u_2\in\mathcal{R}_{j-i}\\ u_3\in\mathcal{R}_{k'}}} \sum_{\substack{p,p'\in Q\\ q,q''\in Q}} \gamma(q,u_1u_2)\gamma(q'',u_3)\ \gamma(p,u_1)\gamma(p',u_2u_3)\\
&\leq \sum_{\substack{u_1\in\mathcal{R}_k\\ u_2\in\mathcal{R}_{j-i}\\ u_3\in\mathcal{R}_{k'}}} \sum_{\substack{p,p',p''\in Q\\ q,q',q''\in Q}} \gamma(q,u_1)\gamma(q',u_2)\gamma(q'',u_3)\ \gamma(p,u_1)\gamma(p',u_2)\gamma(p'',u_3)\\
&\leq \sum_{\substack{u_1\in\mathcal{R}_k\\ p',q\in Q}} \gamma(q,u_1)\gamma(p',u_1) \sum_{\substack{u_2\in\mathcal{R}_{j-i}\\ p,q''\in Q}} \gamma(q'',u_2)\gamma(p,u_2) \sum_{\substack{u_3\in\mathcal{R}_{k'}\\ p'',q'\in Q}} \gamma(q',u_3)\gamma(p'',u_3).
\end{aligned}$$

By the Cauchy-Schwarz inequality, $L(n,\ell,i,j)$ is at most equal to

$$\sum_{p,q\in Q} \|\vec{\gamma}_p(k)\|_2\|\vec{\gamma}_q(k)\|_2 \sum_{p,q\in Q} \|\vec{\gamma}_p(j-i)\|_2\|\vec{\gamma}_q(j-i)\|_2 \sum_{p,q\in Q} \|\vec{\gamma}_p(k')\|_2\|\vec{\gamma}_q(k')\|_2$$

and hence to

$$\sum_{q\in Q} \|\vec{\gamma}_q(k)\|_2^2 \sum_{q\in Q} \|\vec{\gamma}_q(j-i)\|_2^2 \sum_{q\in Q} \|\vec{\gamma}_q(k')\|_2^2.$$

Lemma 4.6 shows that these three factors are bounded above, respectively, by $K_1\alpha_{[2]}^k$, $K_1\alpha_{[2]}^{j-i}$ and $K_1\alpha_{[2]}^{k'}$ for some constant K_1. Therefore

$$L(n,\ell,i,j) \leq K_1^3\alpha_{[2]}^{k+j-i+k'} = K_1^3\alpha_{[2]}^{\ell},$$

as announced. □

PROPOSITION 4.19. *Let $\mathcal{A}$ be an irreducible Markovian automaton with coincidence probability $\alpha_{[2]}$. Denote by $L^{(2)}(n,\ell,i,j)$ the probability for two reduced words of length n to have an occurrence of a factor of length ℓ in the first word at position i, and an occurrence of its inverse in the second word, at position j, with $\ell \leq n$ and $i,j \leq n-\ell+1$. Then there exists a positive constant K such that*

$$L^{(2)}(n,\ell,i,j) \leq K\alpha_{[2]}^{\ell}.$$

PROOF. The proof follows the same steps as that of Proposition 4.18. In the first case $(i,j \leq n-\ell+1)$, Lemma 4.17 shows that

$$L^{(2)}(n,\ell,i,j) \leq \sum_{p,q\in Q}\sum_{u\in\mathcal{R}_\ell} \gamma(p,u)\gamma(q,u^{-1}).$$

Since the set of reduced words of length ℓ and the set of their inverses are equal, we get, by the Cauchy-Schwarz inequality,

$$L^{(2)}(n,\ell,i,j) \leq \sum_{p,q\in Q} \|\vec{\gamma}_p(\ell)\|_2\|\vec{\gamma}_q(\ell)\|_2,$$

and the proof proceeds as in the corresponding case of Lemma 4.18.

In the second case ($i > n-\ell+1$ and $j \leq n-\ell+1$), if $k = n-i+1$, then we have

$$L^{(2)}(n,\ell,i,j) \leq \sum_{\substack{u_1\in\mathcal{R}_k\\ u_2\in\mathcal{R}_{\ell-k}}} \sum_{p,p',q\in Q} \gamma(p,u_1)\gamma(p',u_2)\gamma(q,u_2^{-1}u_1^{-1})$$

$$\leq \sum_{\substack{u_1\in\mathcal{R}_k\\ u_2\in\mathcal{R}_{\ell-k}}} \sum_{p,p',q,q'\in Q} \gamma(p,u_1)\gamma(p',u_2)\gamma(q,u_2^{-1})\gamma(q',u_1^{-1})$$

$$\leq \Big(\sum_{u_1\in\mathcal{R}_k} \sum_{p,q'\in Q} \gamma(p,u_1)\gamma(q',u_1^{-1})\Big)\Big(\sum_{u_2\in\mathcal{R}_{\ell-k}} \sum_{p',q\in Q} \gamma(p',u_2)\gamma(q,u_2^{-1})\Big)$$

and as in the previous case, the proof proceeds as in Lemma 4.18.

The situation is a little more complex in the last case ($i,j > n-\ell+1$). Without loss of generality, we may assume that $i < j$. With the same notation as in the proof of Lemma 4.18, we distinguish two cases. If $|u_3| < |u_2|$ (that is, $\ell-k < k'$, or $\ell+i+j < 2n+2$), we let $u_2 = u'_2u''_2$ with $|u'_2| = |u_3|$. Then $w_1 \in u_3\tilde{A}^{n-\ell}u_1u'2u''_2$ and $w_2 \in u_2'^{-1}u_1^{-1}\tilde{A}^{n-\ell}u_3^{-1}u_2''^{-1}$ and, as in the previous proof, we find that $L^{(2)}(n,\ell,i,j)$ is at most equal to the sum of the

$$\gamma(p,u_1)\gamma(q,u_1^{-1})\gamma(p',u'_2)\gamma(q',u_2'^{-1})\gamma(p'',u''_2)\gamma(q'',u_2''^{-1})\gamma(p''',u_3)\gamma(q''',u_3^{-1})$$

with $u_1 \in \mathcal{R}_{j-i}, u'_2 \in \mathcal{R}_{\ell-k}, u''_2 \in \mathcal{R}_{k'-(\ell-k)}, u_3 \in \mathcal{R}_{\ell-k}$, and $p,p',p'',p''',q,q',q'',q'''$ are states in Q. The proof then proceeds as before, with multiple applications of the Cauchy-Schwarz inequality.

The case where $|u_3| \geq |u_2|$ (that is, $\ell+i+j \geq 2n+2$) is handled in the same fashion. □

COROLLARY 4.20. *Let $\mathcal{A}$ be a non-degenerated ergodic Markovian automaton with coincidence probability $\alpha_{[2]}$. Let n,ℓ,i,j be positive integers such that $\ell \leq n$ and $i,j \leq n$. There exists a constant $K > 0$ such that the probability $\mathfrak{p}$ that two cyclically reduced words of length n have occurrences of the same word of length ℓ (resp. of a word of length ℓ and its inverse) as cyclic factors at positions respectively i and j, satisfies $\mathfrak{p} \leq K\alpha_{[2]}^{\ell}$.*

PROOF. The hypothesis on $\mathcal{A}$ guarantees that $\liminf \mathbb{R}_n(\mathcal{C}) = p > 0$ by Proposition 4.14. Our statement then follows from Propositions 4.18 and 4.19, in view of Lemma 3.13. □

We now consider the case of multiple occurrences of a length ℓ cyclic factor (or of such a word and its inverse) within a single reduced word.

PROPOSITION 4.21. *Let $\mathcal{A}$ be a non-degenerate ergodic Markovian automaton with coincidence probability $\alpha_{[2]}$. There exists a constant $K > 0$ such that the probability that a cyclically reduced word of length n has two occurrences of a length ℓ word as cyclic factors, or occurrences of a length ℓ word and its inverse as cyclic vactors, is at most $K\ell^2n^2\alpha_{[2]}^{\ell/2}$.*

PROOF. By Proposition 4.9, the sequence $(\mathbb{R}_n)_n$ induced by $\mathcal{A}$ is prefix-heavy with parameters $(C, \alpha_{[2]}^{1/2})$ for some C. The result then follows from Corollary 3.14. □

We can now proceed with the **proof of Proposition 4.16.** Let $N = \alpha_{[2]}^{-dn}$. An N-tuple of cyclically reduced words which fails to satisfy $C'(\lambda)$, must satisfy one of the following conditions: either two words in the tuple have occurrences of the same cyclic factor of length $\ell = \lambda n$ or occurrences of such a word and its inverse; or a word in the tuple has two occurrences of the same cyclic factor of length ℓ or occurrences of such a word and its inverse.

By Corollary 4.20, the first event occurs with probability at most

$$K\binom{N}{2}n^2\alpha_{[2]}^{\ell} \leq Kn^2\alpha_{[2]}^{(\lambda-2d)n}$$

for some $K > 0$. By Proposition 4.21, the second event occurs with probability at most

$$KN\ell^2n^2\alpha_{[2]}^{\frac{\ell}{2}} \leq Kn^4\alpha_{[2]}^{(\frac{\lambda}{2}-d)n},$$

for some $K > 0$. Thus both events occur with probabilities that vanish exponentially fast, and this concludes the proof of Proposition 4.16.

4.6. Long common prefixes at high density. In this section, we establish the following propositions corresponding respectively to the second and third statement of Theorem 4.15.

PROPOSITION 4.22. *Let $\mathcal{A}$ be a non-degenerate ergodic Markovian automaton with coincidence probability $\alpha_{[2]}$. Let $\lambda \in (0, \frac{1}{2})$ and let $d \in (\frac{\lambda}{2}, 1)$. A tuple of cyclically reduced words of length n taken independently and randomly according to $\mathcal{A}$, at density d, generically does not satisfy the small cancellation property $C'(\lambda)$.*

PROPOSITION 4.23. *Let $\mathcal{A}$ be a non-degenerate ergodic Markovian automaton with coincidence probability $\alpha_{[2]}$. Let E be the set of letters of $\tilde{A}$ which label a transition in $\mathcal{A}$ and let $D = A \setminus (E \cup E^{-1})$. Let $d > \frac{1}{2}$ and $N \geq \alpha_{[2]}^{-dn}$, and let G be a group presented by an N-tuple of cyclically reduced words chosen independently at random according to $\mathcal{A}$.*

If $E \cap E^{-1} = \emptyset$, then $G = F(|D| + 1)$ exponentially generically.

If $E \cap E^{-1} \neq \emptyset$, then exponentially generically $G = F(D) * \mathbb{Z}/2\mathbb{Z}$ (if n is even) or $G = F(D)$ (if n is odd).

Both proofs rely heavily on the methodology introduced by Szpankowski [**32**] to study the typical heigth of a random trie. We first establish simple lower and upper bounds for words to share a common prefix (Lemmas 4.24 and 4.25).

LEMMA 4.24. *Let $\mathcal{A}$ be an irreducible Markovian automaton with coincidence probability $\alpha_{[2]}$. Let $P(n, \ell)$ (resp. $P'(n, \ell)$) be the probability that two reduced (resp. cyclically reduced) words of length n share a common prefix of length ℓ. There exists a constant $K > 0$ such that $P(n, \ell) \geq K\alpha_{[2]}^{\ell}$.*

If $\mathcal{A}$ is non-degenerate and ergodic and t is large enough for all the coefficients of $\mathbb{M}^t$ to be positive, then K can be chosen such that $P'(n, \ell) \geq K\alpha_{[2]}^{\ell}$ when $n \geq \ell+t+1$.

PROOF. Let p be a state such that $\gamma_0(p) > 0$. To establish the announced lower bounds, we only need to consider the words that can be read from state p. More precisely, when considering reduced words, we have

$$P(n, \ell) \;\geq\; \gamma_0(p)^2 \sum_{u \in \mathcal{R}_\ell} \gamma(p, u)^2.$$

We observe that $\sum_{u\in\mathcal{R}_\ell}\gamma(p,u)^2$ is the p-component of $\mathbb{M}_{[2]}^{\ell}\vec{1}$, and by Lemma 4.6, it is greater than or equal to $\beta\alpha_{[2]}^{\ell}$, where β is the minimal component of $\vec{v}_{\min}$ (in the notation of Lemma 4.6). This completes the proof of the statement concerning $P(n,\ell)$.

We now consider cyclically reduced words, under the hypothesis that $\mathcal{A}$ is non-degenerate and ergodic. Let t be such that all the coefficients of $\mathbb{M}^t$ are positive, let $\bar{p}_{\min}$ be the least coefficient of this matrix, and let $p_{\min}$ be the least positive coefficient of $\mathbb{M}$. Finally, let $\mathfrak{p}=\liminf\mathbb{R}_n(\mathcal{C})$, which is positive by Proposition 4.14. Let X (resp. X_p) be the set of pairs of cyclically reduced words of length n that have a common prefix of length ℓ (resp. which can be read from state p). We note that

$$P'(n,\ell)\ =\ \frac{\mathbb{R}_n(X)}{\mathbb{R}_n(\mathcal{C})^2}\ \geq\ \frac{1}{\mathfrak{p}^2}\mathbb{R}_n(X)\ \geq\ \frac{1}{\mathfrak{p}^2}\mathbb{R}_n(X_p),$$

so we only need to find a lower bound for $\mathbb{R}_n(X_p)$.

Suppose that $n\geq\ell+t+1$. Then X_p contains the set of pairs of reduced words of the form $(uu_1u_1'a, uu_2u_2'a)$ which can be read from p, where a is the first letter of u, and u_1' and u_2' are words of length t such that $p\cdot(uu_1u_1')=p\cdot(uu_2u_2')=p$. Since these words start and end with the same letters, they are guaranteed to be cyclically reduced. Thus we have

$$\mathbb{R}_n(X_p)\ \geq\ \gamma_0(p)^2\sum_{u\in\mathcal{R}_\ell}\gamma(p,u)^2\ p_{\min}^2\ \bar{p}_{\min}^2\ \geq\ \beta\ \gamma_0(p)^2\ p_{\min}^2\ \bar{p}_{\min}^2\ \alpha_{[2]}^{\ell},$$

and this concludes the proof. □

LEMMA 4.25. *Let $\mathcal{A}$ be an irreducible Markovian automaton with coincidence probability $\alpha_{[2]}$. There exists a constant $K>0$ such that the probability that three reduced words share the same prefix of length ℓ is at most $K\alpha_{[3]}^{\ell}$.*

If $\mathcal{A}$ is non-degenerate and ergodic, the same holds for triples of cyclically reduced words.

PROOF. The probability $\mathfrak{p}(u)$ that three reduced words have a common prefix u is

$$\mathfrak{p}(u)=\sum_{p_1,p_2,p_3\in Q}\gamma_0(p_1)\ \gamma_0(p_2)\ \gamma_0(p_3)\ \gamma(p_1,u)\ \gamma(p_2,u)\ \gamma(p_3,u).$$

The probability we are interested in is obtained by summing over all $u\in\mathcal{R}_\ell$. It is bounded above by

$$\sum_{p_1,p_2,p_3\in Q}\ \sum_{u\in\mathcal{R}_\ell}\gamma(p_1,u)\ \gamma(p_2,u)\ \gamma(p_3,u).$$

By the Hölder and Cauchy-Schwarz inequalities, we have

$$\begin{aligned}\sum_{u\in\mathcal{R}_\ell}&\gamma(p_1,u)\ \gamma(p_2,u)\ \gamma(p_3,u)\\ &\leq\left(\sum_{u\in\mathcal{R}_\ell}\gamma(p_1,u)^3\right)^{\frac{1}{3}}\left(\sum_{u\in\mathcal{R}_\ell}\gamma(p_2,u)^{\frac{3}{2}}\ \gamma(p_3,u)^{\frac{3}{2}}\right)^{\frac{2}{3}}\\ &\leq\left(\sum_{u\in\mathcal{R}_\ell}\gamma(p_1,u)^3\right)^{\frac{1}{3}}\left(\sum_{u\in\mathcal{R}_\ell}\gamma(p_2,u)^3\right)^{\frac{1}{3}}\left(\sum_{u\in\mathcal{R}_\ell}\gamma(p_3,u)^3\right)^{\frac{1}{3}}.\end{aligned}$$

Moreover, we have

$$\sum_{p\in Q}\sum_{u\in\mathcal{R}_\ell}\gamma(p_1,u)^3=\vec{1}^t\ \mathbb{M}_{[3]}^\ell\ \vec{1}.$$

We now get the announced result using Lemma 4.6, Lemma 4.8 and the spectral properties of $\mathbb{M}_{[3]}^\ell$. The generalisation to cyclically reduced words follows from Lemma 3.13. □

We now build on the previous lemmas to show that, exponentially generically, large tuples of cyclically reduced words contain pairs of words with a common prefix of a prescribed length.

PROPOSITION 4.26. *Let $\mathcal{A}$ be an irreducible Markovian automaton with coincidence probability $\alpha_{[2]}$. Let $(\ell_n)_n$ be an unbounded, monotonous sequence of positive integers such that $\ell_n\le n$ for each n, and let $d>\frac{1}{2}$. Then an $\alpha_{[2]}^{-d\ell_n}$-tuple of reduced words of length n drawn randomly according to $\mathcal{A}$ generically contains two words with the same prefix of length ℓ_n.*

If $\mathcal{A}$ is non-degenerate and ergodic, the same holds for $\alpha_{[2]}^{-d\ell_n}$-tuples of cyclically reduced words.

PROOF. We use the so-called *second moment method*, as developed in [**32**], and we introduce the following notation to this end. Since the results of [**32**] are established for right-infinite words, we need to considered such words first; the result on words of length n directly follows by truncation. A right-infinite reduced word is an element u of $\tilde{A}^{\mathbb{N}}$ such that for every $i\in\mathbb{N}$, $u_i\neq u_{i+1}^{-1}$. We define the probability distribution $\mathbb{R}_\infty$ on right-infinite words induced by the Markovian automaton $\mathcal{A}$ by first setting $\mathbb{R}_\infty(\mathcal{P}_\infty(u))=\gamma(u)$, where $\mathcal{P}_\infty(u)$ is the set of right-infinite reduced words w such that the finite reduced word u is a prefix of w. The probability is then extended to the σ-algebra generated by the $\mathcal{P}_\infty(u)$, when u ranges over all finite reduced words (see [**34**] for more details on this kind of constructions). Let $N=\alpha_{[2]}^{-d\ell_n}$ and consider an N-tuple $\vec{h}=(h_i)_{1\le i\le N}$ of right-infinite reduced words, independently and randomly generated according to $\mathcal{A}$.

For $1\le i<j\le N$, let $X_{i,j}$ be the random variable computing the length of the longest common prefix of h_i and h_j. We want to show that, exponentially generically,

$$\max_{1\le i<j\le N}X_{i,j}\ge\ell_n.$$

Let us relabel the random variables $X_{i,j}$ $(i\neq j)$ as $Y_1,\dots,Y_m$, with $m=\binom{N}{2}$ and, say, $Y_1=X_{1,2}$. We are therefore computing the maximum of m random variables, which are identically distributed but not independent. Fortunately, they behave almost as if they were independent, as we will see.

Let d' be such that $\frac{1}{2}<d'<d$ and for each $m\ge1$, let

$$r_m=\log_{\alpha_{[2]}^{-2d'}}(m)=\frac{\log\binom{N}{2}}{\log\alpha_{[2]}^{-2d'}}\sim\frac{\log\alpha_{[2]}^{-2d\ell_n}}{\log\alpha_{[2]}^{-2d'}}=\frac{d\ell_n}{d'}.$$

In particular, r_m is asymptotically greater than ℓ_n, and we only need to show that

$$\lim_{n\to\infty}\mathbb{R}_\infty\left(\max_{k\in[m]}Y_k\ge r_m\right)=1. \tag{7}$$

Let $\nu(r_m)$ denote the quantity

$$\nu(r_m) = \sum_{k=2}^{m} \frac{\mathbb{R}_\infty(Y_1 \geq r_m, Y_k \geq r_m)}{m\,\mathbb{R}_\infty(Y_1 \geq r_m)^2}.$$

We use Lemma 3 in **[32]**, which states that the desired equation (7) holds if

$$\lim_{n\to\infty} m\,\mathbb{R}_\infty(Y_1 > r_m) = +\infty \text{ and } \lim_{n\to\infty} \nu(r_m) = 1.$$

We now proceed with the proof of these two equalities. By Lemma 4.24, we have $\mathbb{R}_\infty(Y_1 \geq r_m) \geq K\,\alpha_{[2]}^{r_m}$. Then

$$\begin{aligned}\log\left(m\mathbb{R}_\infty(Y_1 \geq r_m)\right) &\geq \log m + \log K + r_m \log \alpha_{[2]}\\ &= r_m \log(\alpha_{[2]}^{-2d'}) + \log K + r_m \log \alpha_{[2]}\\ &= r_m \log(\alpha_{[2]}^{1-2d'}) + \log K,\end{aligned}$$

which tends to $+\infty$, since $1 - 2d' < 0$ and $\alpha_{[2]} < 1$. Therefore,

$$\lim_{n\to\infty} m\,\mathbb{R}_\infty(Y_1 \geq r_m) = +\infty.$$

Let us now consider $\nu(r_m)$. Note that, if the Y_i were independent random variables, we would have $\nu(r_m) = \frac{m-1}{m}$, which tends to 1 when n tends to ∞.

Observe that if $2 < i < j \leq N$, then $X_{1,2}$ and $X_{i,j}$ are independant and identically distributed, so

$$\mathbb{R}_\infty(X_{1,2} \geq r_m, X_{i,j} \geq r_m) = \mathbb{R}_\infty(X_{1,2} \geq r_m)\,\mathbb{R}_\infty(X_{i,j} \geq r_m) = \mathbb{R}_\infty(Y_1 \geq r_m)^2.$$

Also, since h_1 and h_2 are drawn independently, we have $\mathbb{R}_\infty(X_{1,2} \geq r_m, X_{1,k} \geq r_m) = \mathbb{R}_\infty(X_{1,2} \geq r_m, X_{2,k} \geq r_m)$ for each $k \geq 3$. Therefore

$$\nu(r_m) = 2\sum_{k=3}^{N} \frac{\mathbb{R}_\infty(X_{1,2} \geq r_m, X_{1,k} \geq r_m)}{m\,\mathbb{R}_\infty(Y_1 \geq r_m)^2} + \binom{N-2}{2}\frac{1}{m}.$$

Since $m = \binom{N}{2}$, we have $\lim_n \binom{N-2}{2}\frac{1}{m} = 1$. Moreover, the joint probability $\mathbb{R}_\infty(X_{1,2} \geq r_m, X_{1,k} \geq r_m)$ is exactly the probability that three random reduced words share a common prefix of length r_m: by Lemma 4.25, this is at most equal to $K\,\alpha_{[3]}^{r_m}$ for some constant $K > 0$. Together with Lemma 4.24, this yields

$$\sum_{k=3}^{N} \frac{\mathbb{R}_\infty(X_{1,2} \geq r_m, X_{1,k} \geq r_m)}{m\,\mathbb{R}_\infty(Y_1 \geq r_m)^2} \leq \frac{K'}{N}\left(\frac{\alpha_{[3]}}{\alpha_{[2]}^2}\right)^{r_m},$$

for some $K' > 0$. In **[16]** it is proved that $(\alpha_{[m]})^{1/m}$ is a decreasing sequence, so we have $\alpha_{[3]}^{1/3} \leq \alpha_{[2]}^{1/2}$ and hence

$$\left(\frac{\alpha_{[3]}}{\alpha_{[2]}^2}\right)^{r_m} \leq \left(\frac{\alpha_{[2]}^{3/2}}{\alpha_{[2]}^2}\right)^{r_m} \leq \alpha_{[2]}^{-\frac{r_m}{2}}.$$

Therefore

$$\log\left(\frac{1}{N}\left(\frac{\alpha_{[3]}}{\alpha_{[2]}^2}\right)^{r_m}\right) = -\log N - \frac{r_m}{2}\log\alpha_{[2]} \leq -\frac{1}{2}\log m + K'' - \frac{r_m}{2}\log\alpha_{[2]}$$

for some constant K''. By definition of r_m, we have $\log m = -2d' r_m \log \alpha_{[2]}$ and it follows that

$$\log\left(\frac{1}{N}\left(\frac{\alpha_{[3]}}{\alpha_{[2]}^2}\right)^{r_m}\right) \le \frac{r_m}{2}(2d'-1)\log\alpha_{[2]} + K''.$$

This quantity tends to $-\infty$ when n tends to ∞ since $2d'-1>0$ and $\alpha_{[2]}<1$. This proves finally that $\lim_{m\to\infty}\nu(r_m)=1$ and establishes Equation (7). That is, the desired statement is proved for tuples of infinite reduced words. As $\ell_n \le n$, considering right-infinite words and truncating then at their prefix of length n yields the same result. By construction, the probability distribution induced on this truncated words is exactly $\mathbb{R}_n$, concluding the proof.

The generalisation to cyclically reduced words follows from Lemma 3.13. □

We now use Proposition 4.26 to prove Proposition 4.22.

PROOF OF PROPOSITION 4.22 Let $0<\lambda<\frac{1}{2}$. Proposition 4.26, applied to $\ell_n=\lambda n$ shows that, if $\frac{1}{2}<d<1$, then a random $\alpha_{[2]}^{-d\lambda n}$-tuple $\vec{h}$ of cyclically reduced words of length n, generically has two components h_i and h_j with the same prefix of length λn, which is sufficient to show that $\vec{h}$ does not satisfy Property $C'(\lambda)$. □

We now translate the result of Proposition 4.26 into a result on the group presented by a random $\alpha_{[2]}^{-dn}$-tuple, when $d>\frac{1}{2}$. We will use repeatedly Chernoff bounds [**20**, Th. 4.2 p.70], which state that, in a binomial distribution with parameters (k,p) — that is: X_k is the sum of k independent draws of 0 or 1 and p is the probability of drawing 1 —,

$$\mathbb{P}\left(X_k \le \frac{kp}{2}\right) \le \exp\left(-\frac{kp}{8}\right).$$

In other words,

$$\mathbb{P}\left(X_k \ge \frac{kp}{2}\right) \ge 1-\exp\left(-\frac{kp}{8}\right). \tag{8}$$

If $\vec{h}$ is a vector of cyclically reduced words, G is the group presented by $G=\langle A \mid \vec{h}\rangle$ and u,v are reduced words, we write that $u=_G v$ if u and v have the same projection in G (that is: if uv^{-1} lies in the normal closure of $\vec{h}$).

PROPOSITION 4.27. *Let $\mathcal{A}$ be an ergodic Markovian automaton with coincidence probability $\alpha_{[2]}$ and let $a,b\in\tilde{A}$ be labels of transitions in $\mathcal{A}$. Let $d>\frac{1}{2}$ and $N\ge\alpha_{[2]}^{-dn}$, and let G be a group presented by an N-tuple of cyclically reduced words chosen at random according to $\mathcal{A}$. Then $a=_G b$ exponentially generically.*

PROOF. Let $t>0$ be such that all the coefficients of $\mathbb{M}^t$ are positive (such an integer exists since $\mathbb{M}$ is ergodic) and let $\tau>0$ be the minimum coefficient of $\mathbb{M}^t$.

We proceed in two steps. First we consider transitions starting in the same state of the Markovian automaton and second we generalize the study to transitions beginning in different states of the automaton.

First step of the proof. We show that if $x=x_1\cdots x_s$ and $y=y_1\cdots y_s$ are reduced words of equal length $s\ge 1$ which label paths in $\mathcal{A}$ out of the same state q, then exponentially generically, we have $x_k=_G y_k$ for each $1\le k\le s$.

Recall that, in our model of Markovian automata, drawing a word of length n amounts to drawing a state $r\in Q$ according to γ_0, and then drawing a word of length n according to $\gamma(r,-)$. Thus, when drawing a tuple $\vec{h} = (h_i)_i$, we also draw a tuple $\vec{q} = (q_i)_i$ of states such that, in particular, $\gamma_0(q_i) > 0$ and $\gamma(q_i, h_i) > 0$.

Let r be a state such that $\gamma_0(r) > 0$. Let $T_0 = \{h_i \in \vec{h}$ such that $q_i = r\}$ and $N_0 = |T_0|$. Observe that drawing randomly and independently N words of length n in our model and then keeping only those starting in state r to obtain T_0 is the same as first choosing N_0 according to a binomial law of parameters $(\gamma_0(r), N)$ and then drawing randomly and independently N_0 words beginning in state r. Moreover Chernoff bounds (Equation (8) above, applied with $p = \gamma_0(r)$ and $k = N$) show that $\mathbb{P}\left(N_0 \geq \frac{\gamma_0(r)N}{2}\right) \geq \mathfrak{p}_0$ with $\mathfrak{p}_0 = 1 - \exp\left(-\frac{\gamma_0(r)N}{8}\right)$.

For each $s \geq 1$, we say that a pair of indices (i,j) is an *s-collision in T_0* if h_i and h_j belong to T_0 and have the same prefix of length $n-t-s$. Let e be such that $0 < e < d - \frac{1}{2}$ and let $N' = \alpha_{[2]}^{-(d-e)n}$. Then a random N_0-tuple of cyclically reduced words starting in r is obtained by drawing $\frac{N_0}{N'}$ times a random N'-tuple starting in state r. Moreover choosing a random word in a Markovian automaton given that the associated path begins in state r is the same as taking for initial probability vector γ_0 the probability vector such that $\gamma_0(r) = 1$. Since the conclusion of Proposition 4.26 does not depend on the initial probability vector and $d - e > \frac{1}{2}$, Proposition 4.26 applied to $\ell_n = n - t - s$ shows that a random N'-tuple of cyclically reduced words that starts in r generically exhibits at least one s-collision in T_0.

We assume that n is large enough so that the probability of an s-collision in T_0 of a random N'-tuple is at least $\frac{1}{2}$. Then Chernoff bounds (Equation (8), applied with $p = \frac{1}{2}$ and $k = N_0$) show that the set T_1 of s-collisions in T_0 of a random $\alpha_{[2]}^{-dn}$-tuple of cyclically reduced words of length n satisfies $|T_1| \geq \frac{1}{4}N_0$ with probability greater than or equal to $\mathfrak{p}_1 = 1 - \exp(-\frac{N_0}{16})$.

For each s-collision $(i,j) \in T_1$, we let $u(i,j)$ be the common length $n-t-s$ prefix of h_i and h_j. Then by a finiteness argument, there exists a state $q_1 \in Q$ and a set $T_2 \subset T_1$ such that, for every $(i,j) \in T_2$, $u(i,j)$ labels a path from r to q_1 in $\mathcal{A}$, and $|T_2| \geq \frac{|T_1|}{|Q|}$. Hence $|T_2| \geq \frac{N_0}{4|Q|}$ with probability greater than or equal to $\mathfrak{p}_1$.

Now let v be a reduced word of length t, labeling a path in $\mathcal{A}$ from q_1 to q: such a word exists since all the coefficients of $\mathbb{M}^t$ are positive, and we have $\gamma(q_1, v) \geq \tau$. For each $(i,j) \in T_2$, the probability that h_i starts with $u(i,j)v$ is $\gamma(q_1, v) \geq \tau$, and the probability that uv is a prefix of both h_i and h_j is at least τ^2. We can apply Chernoff bounds (8) again, with $p = \tau^2$ and $k = |T_2|$: then the subset $T_3 \subseteq T_2$ of pairs (i,j) such that $u(i,j)v$ is a prefix of both h_i and h_j, has cardinality $|T_3| \geq \frac{1}{2}|T_2|\tau^2$ with probability at least $\mathfrak{p}_2 = 1 - \exp(-\frac{\tau^2|T_2|}{8})$.

Finally, we note that $|u(i,j)v| = n - s$, so for each $(i,j) \in T_3$, we have $h_i = u(i,j)vx$ with probability $\gamma(q,x)$. Therefore the probability that $(h_i, h_j) = (u(i,j)vx, u(i,j)vy)$ is $\gamma(q,x)\gamma(q,y)$, which is positive by hypothesis. Applying Chernoff bounds one more time (with $k = |T_3|$ and $p = \gamma(q,x)\gamma(q,y)$) shows that $\vec{h}$ contains a pair of words of the form (wx, wy) with probability at least $\mathfrak{p}_3$ with $\mathfrak{p}_3 = \left(1 - \exp\left(-\frac{|T_3|\gamma(q,x)\gamma(q,y)}{8}\right)\right)$.

In conclusion, exponentially generically $N_0 \geq \gamma_0(r)\frac{N}{2}$ which implies that $\mathfrak{p}_1$ is exponentially close to 1. Hence $T_2 \geq \frac{\gamma_0(r)N}{8|Q|}$ exponentially generically, which implies that $\mathfrak{p}_2$ is exponentially close to 1. So $|T_3| \geq \frac{\gamma_0(r)N\tau^2}{16|Q|}$ exponentially generically, which implies that $\mathfrak{p}_3$ is exponentially close to 1. In particular, exponentially generically, $\vec{h}$ has a pair of the form (wx, wy), and hence we have $x =_G y$.

Applying this to the words x_1 and y_1, we find that $x_1 =_G y_1$. Next, considering the words x_1x_2 and y_1y_2, we find that $x_1x_2 =_G y_1y_2$, and hence $x_2 =_G y_2$. Iterating this reasoning, we finally show that $x_k =_G y_k$ for each $1 \leq k \leq s$.

Second step of the proof We now consider two transitions in $\mathcal{A}$, one labeled a from state q to state q' and another labeled b from state r to state r' $(a, b \in \tilde{A})$.

Let $q_0 \in Q$ be a state in $\mathcal{A}$ such that $\gamma_0(q_0) > 0$. Since $\mathcal{A}$ is irreducible, there exists a word w_1 which labels a loop at q_0 and visits every transition of $\mathcal{A}$. Moreover, since $\mathcal{A}$ is ergodic, there exists a word w_2 labeling another loop at q_0, such that $|w_1|$ and $|w_2|$ are relatively prime.

Since reading w_1 from q_0 visit all the transitions, let u_1 (resp. v_1) be a prefix of w_1 such that the last transition read after reading u_1 (resp. v_1) is the a-transition out of state q (resp. the b-transition out of state r). Then the Chinese remainder theorem shows that there exist words $x \in \{w_1, w_2\}^*u_1$ and $y \in \{w_1, w_2\}^*v_1$ of equal length.

Since a and b are the last letters of x and y, respectively, the first step of the proof shows that $a =_G b$, which concludes the proof of the proposition. □

We can now complete the **proof of Proposition 4.23.** By Proposition 4.27, exponentially generically, all the letters in E are equal in G. If $a, a^{-1} \in E$ for some letter a, then all these letters are equal to their own inverse in G, so the subgroup H of G generated by E is a quotient of $\mathbb{Z}/2\mathbb{Z}$. Since all the relators in the presentation have length n, it follows that H is isomorphic to $\mathbb{Z}/2\mathbb{Z}$ if n is even, and is trivial if n is odd. The result follows once we observe that the letters in D do not occur in any relator. □

Acknowledgments. The authors are thankful to the anonymous referee for her/his remarkably thorough reading of the first version of this paper and for his/her insightful and constructive suggestions. These helped simplify the presentation of Sections 3.2 and 3.3, sharpen some results in Section 3.6 and fix a technical mistake in the proof of Proposition 4.23.

References

[1] G. N. Arzhantseva and A. Yu. Ol′shanskiĭ, *Generality of the class of groups in which subgroups with a lesser number of generators are free* (Russian, with Russian summary), Mat. Zametki **59** (1996), no. 4, 489–496, 638, DOI 10.1007/BF02308683; English transl., Math. Notes **59** (1996), no. 3-4, 350–355. MR1445193

[2] F. Bassino, C. Nicaud, and P. Weil, *Generic properties of random subgroups of a free group for general distributions*, 23rd Intern. Meeting on Probabilistic, Combinatorial, and Asymptotic Methods for the Analysis of Algorithms (AofA'12), Discrete Math. Theor. Comput. Sci. Proc., AQ, Assoc. Discrete Math. Theor. Comput. Sci., Nancy, 2012, pp. 155–166. MR2957328

[3] F. Bassino, A. Martino, C. Nicaud, E. Ventura, and P. Weil, *Statistical properties of subgroups of free groups*, Random Structures Algorithms **42** (2013), no. 3, 349–373, DOI 10.1002/rsa.20407. MR3039683

[4] F. Bassino, C. Nicaud, and P. Weil, *Random generation of finitely generated subgroups of a free group*, Internat. J. Algebra Comput. **18** (2008), no. 2, 375–405, DOI 10.1142/S0218196708004482. MR2403827

[5] F. Bassino, C. Nicaud, and P. Weil, *On the genericity of Whitehead minimality*, J. Group Theory **19** (2016), no. 1, 137–159, DOI 10.1515/jgth-2015-0030. MR3441131

[6] M. R. Bridson and D. T. Wise, *Malnormality is undecidable in hyperbolic groups*, Israel J. Math. **124** (2001), 313–316, DOI 10.1007/BF02772626. MR1856523

[7] C. Champetier, *Propriétés statistiques des groupes de présentation finie* (French, with English summary), Adv. Math. **116** (1995), no. 2, 197–262, DOI 10.1006/aima.1995.1067. MR1363765

[8] F. R. Gantmacher. *The theory of matrices*. Chelsea, 1959.

[9] R. Gitik, M. Mitra, E. Rips, and M. Sageev, *Widths of subgroups*, Trans. Amer. Math. Soc. **350** (1998), no. 1, 321–329, DOI 10.1090/S0002-9947-98-01792-9. MR1389776

[10] M. Gromov, *Hyperbolic groups*, Essays in group theory, Math. Sci. Res. Inst. Publ., vol. 8, Springer, New York, 1987, pp. 75–263, DOI 10.1007/978-1-4613-9586-7_3. MR919829

[11] M. Gromov, *Asymptotic invariants of infinite groups*, Geometric group theory, Vol. 2 (Sussex, 1991), London Math. Soc. Lecture Note Ser., vol. 182, Cambridge Univ. Press, Cambridge, 1993, pp. 1–295. MR1253544

[12] T. Jitsukawa, *Malnormal subgroups of free groups*, Computational and statistical group theory (Las Vegas, NV/Hoboken, NJ, 2001), Contemp. Math., vol. 298, Amer. Math. Soc., Providence, RI, 2002, pp. 83–95, DOI 10.1090/conm/298/05115. MR1929717

[13] I. Kapovich. Musings on generic-case complexity. `arXiv:1505.03218`, 2015.

[14] I. Kapovich, A. Myasnikov, P. Schupp, and V. Shpilrain, *Generic-case complexity, decision problems in group theory, and random walks*, J. Algebra **264** (2003), no. 2, 665–694, DOI 10.1016/S0021-8693(03)00167-4. MR1981427

[15] I. Kapovich and A. Myasnikov, *Stallings foldings and subgroups of free groups*, J. Algebra **248** (2002), no. 2, 608–668, DOI 10.1006/jabr.2001.9033. MR1882114

[16] S. Karlin and F. Ost, *Counts of long aligned word matches among random letter sequences*, Adv. in Appl. Probab. **19** (1987), no. 2, 293–351, DOI 10.2307/1427422. MR820431

[17] O. Kharlampovich and A. Myasnikov, *Hyperbolic groups and free constructions*, Trans. Amer. Math. Soc. **350** (1998), no. 2, 571–613, DOI 10.1090/S0002-9947-98-01773-5. MR1390041

[18] D. A. Levin, Y. Peres, and E. L. Wilmer, *Markov chains and mixing times*, American Mathematical Society, Providence, RI, 2009. With a chapter by James G. Propp and David B. Wilson. MR2466937

[19] R. C. Lyndon and P. E. Schupp, *Combinatorial group theory*, Springer-Verlag, Berlin-New York, 1977. Ergebnisse der Mathematik und ihrer Grenzgebiete, Band 89. MR0577064

[20] R. Motwani and P. Raghavan, *Randomized algorithms*, Cambridge University Press, Cambridge, 1995. MR1344451

[21] A. Miasnikov, E. Ventura, and P. Weil, *Algebraic extensions in free groups*, Geometric group theory, Trends Math., Birkhäuser, Basel, 2007, pp. 225–253, DOI 10.1007/978-3-7643-8412-8_12. MR2395796

[22] J. Nielsen. Die Isomorphismen der allgemeinen, unendlichen Gruppe mit zwei Erzeugenden. *Mathematische Annalen*, 78, 1918.

[23] Y. Ollivier, *Sharp phase transition theorems for hyperbolicity of random groups*, Geom. Funct. Anal. **14** (2004), no. 3, 595–679, DOI 10.1007/s00039-004-0470-y. MR2100673

[24] Y. Ollivier, *A January 2005 invitation to random groups*, Ensaios Matemáticos [Mathematical Surveys], vol. 10, Sociedade Brasileira de Matemática, Rio de Janeiro, 2005. MR2205306

[25] A. Yu. Ol′shanskiĭ, *Almost every group is hyperbolic*, Internat. J. Algebra Comput. **2** (1992), no. 1, 1–17, DOI 10.1142/S0218196792000025. MR1167524

[26] M. O. Rabin. Probabilistic automata. *Information and Computation*, 6(3):230–245, 1963.

[27] A. Roig, E. Ventura, and P. Weil, *On the complexity of the Whitehead minimization problem*, Internat. J. Algebra Comput. **17** (2007), no. 8, 1611–1634, DOI 10.1142/S0218196707004244. MR2378055

[28] R. Segala and N. Lynch, *Probabilistic simulations for probabilistic processes*, Nordic J. Comput. **2** (1995), no. 2, 250–273. MR1346062

[29] J.-P. Serre, *Trees*, Springer Monographs in Mathematics, Springer-Verlag, Berlin, 2003. Translated from the French original by John Stillwell; Corrected 2nd printing of the 1980 English translation. MR1954121

[30] P. V. Silva and P. Weil, *On an algorithm to decide whether a free group is a free factor of another*, Theor. Inform. Appl. **42** (2008), no. 2, 395–414, DOI 10.1051/ita:2007040. MR2401269

[31] J. R. Stallings, *Topology of finite graphs*, Invent. Math. **71** (1983), no. 3, 551–565, DOI 10.1007/BF02095993. MR695906

[32] W. Szpankowski, *On the height of digital trees and related problems*, Algorithmica **6** (1991), no. 2, 256–277, DOI 10.1007/BF01759045. MR1093014

[33] N. W. M. Touikan, *A fast algorithm for Stallings' folding process*, Internat. J. Algebra Comput. **16** (2006), no. 6, 1031–1045, DOI 10.1142/S0218196706003396. MR2286421

[34] B. Vallée, J. Clément, J. A. Fill, and P. Flajolet, *The number of symbol comparisons in QuickSort and QuickSelect*, Automata, languages and programming. Part I, Lecture Notes in Comput. Sci., vol. 5555, Springer, Berlin, 2009, pp. 750–763, DOI 10.1007/978-3-642-02927-1_62. MR2544890

[35] P. Weil, *Computing closures of finitely generated subgroups of the free group*, Algorithmic problems in groups and semigroups (Lincoln, NE, 1998), Trends Math., Birkhäuser Boston, Boston, MA, 2000, pp. 289–307. MR1750503

Université Paris 13, Sorbonne Paris Cité, LIPN, CNRS UMR 7030, F-93430 Villetaneuse, France

E-mail address: bassino@lipn.univ-paris13.fr

Université Paris-Est, LIGM (UMR 8049), UPEMLV, F-77454 Marne-la-Vallée, France

E-mail address: nicaud@univ-mlv.fr

Univ. Bordeaux, LaBRI, CNRS UMR 5800, F-33400 Talence, France

E-mail address: pascal.weil@labri.fr

Contemporary Mathematics
Volume **677**, 2016
http://dx.doi.org/10.1090/conm/677/13620

A new multi-server scheme for private information retrieval

Chi Sing Chum and Xiaowen Zhang

ABSTRACT. Traditional private information retrieval (PIR) schemes have been around for about two decades. They mainly consider the effectiveness of the communication complexity in the sense to minimize the number of bits transferred between the user who does the query and the server which answers the query. In this paper, we introduce a new scheme which takes both the time and communication complexities into consideration. The scheme has a simple implementation and is especially suitable if there is only a small number of replicated servers available.

1. Introduction

Private information retrieval (PIR) deals with the privacy of a user when he queries a public database. It was first introduced by Chor et al. [**1**] in 1995. It is formalized as follows: given a database x which consists of n bits, $x = x_1 \dots x_n$, a user wants to inquire the ith bit without letting the database know any information about i. A trivial solution is to let the user download the entire database. In this case, the communication complexity, which is the number of bits transferred between the user and the database, is n. Chor et al. [**1**] proved that this trivial solution turned out to be optimal for a single database in the information theoretic setting. In the literature, PIR schemes were classified into information theoretic and computational. A PIR in an information theoretic setting provides perfect privacy against those databases with unlimited computational power, while computational PIR provides privacy against those databases with polynomial time computational power. This paper only considers PIR schemes under an information theoretic setting. For more information about computational PIR, please refer to [**3**].

The rest of the paper is organized as follow. Section 2 will introduce the PIR scheme that originally came from Chor et al. [**1**]. In section 3 we discuss the motivations and the proposal of the new scheme. We conclude the paper in section 4.

2010 *Mathematics Subject Classification.* Primary 68P20.
Key words and phrases. Private information retrieval, non-colluding servers, communication complexity.

©2016 American Mathematical Society

2. Private information retrieval schemes

Multi-server scheme: When the user downloads all the n bits, the database will not get any information about i even if the database has unlimited computational power. Chor et al. [1] showed that if we had more than one non-colluding servers with each having a complete database, we could reduce the communication complexity and preserve the perfect privacy as well.

Let $k = 2^d$ be the number of databases hosted on k different non-colluding servers, and for simplicity let $l = \sqrt[d]{n}$. Instead of considering the database as one dimensional, we think of it as d dimensional. Retrieving x_i, where $1 \le i \le n$, becomes the problem of retrieving $x_{i_1,\dots,i_d}$ where $1 \le i_1,\dots,i_d \le l$. The user generates d random binary strings of length l, i.e., $S_1^0, S_2^0, \dots, S_d^0 \in \{0,1\}^l$. Based on these d strings and $i_1,\dots,i_d$, the user constructs d binary strings S_1^1 by flipping the i_1th bit of S_1^0 (denoted by $S_1^1 = S_1^0 \oplus i_1$), S_2^1 by flipping the i_2th bit of S_2^0 (i.e. $S_2^1 = S_2^0 \oplus i_2$), ..., S_d^1 by flipping the i_dth bit of S_d^0 (i.e. $S_d^1 = S_d^0 \oplus i_d$). This way, the user gets d sets $\{S_1^0, S_1^1\}, \dots, \{S_d^0, S_d^1\}$, and each consists of 2 binary strings of length l. We choose 1 string from each set; then we have 2^d combinations. For each combination $\alpha = \alpha_1 \dots \alpha_d \in \{0,1\}^d$, the user will send $S_1^{\alpha_1}, S_2^{\alpha_2}, \dots, S_d^{\alpha_d}$ to the database $DB_{\alpha_1\dots\alpha_d}$.

The user sends out d strings of length l to each database. There are 2^d databases, so the total number of bits sent to the databases is $(dl)2^d$. Let $j \in S$ denote the jth position in S is 1. Each database DB_α will do the following calculation:

$$\bigoplus_{j_1 \in S_1^{\alpha_1},\dots,j_d \in S_d^{\alpha_d}} x_{j_1,\dots,j_d}, 1 \le j_1,\dots,j_d \le l. \tag{2.1}$$

That means the bit $x_{j_1,\dots,j_d}$ will be selected if $S_1^{\alpha_1}$ contains 1 in j_1th position, $\dots, S_d^{\alpha_d}$ contains 1 in the j_dth position. Here only 1 bit will be sent back to the user. So, 2^d bits will be sent back.

The total communication complexity equals to

$$2^d(dl+1) = 2^d(dn^{1/d}+1) = O(n^{1/d}). \tag{2.2}$$

For any bit $x_{j_1,\dots,j_d}$ which is different from $x_{i_1,\dots,i_d}$, even (possibly zero) of such bits will be sent back to the user and this will cancel out each other. Only $x_{i_1,\dots,i_d}$ will be sent back to the user one time. We will illustrate this in more detail in the following example.

Example. Consider that there are 8 databases (i.e., $d = 3$). Suppose the user wants to retrieve x_{i_1,i_2,i_3}. He first generates S_1^0, S_2^0, S_3^0, and then constructs

$$S_1^1 = S_1^0 \oplus i_1, \quad S_2^1 = S_2^0 \oplus i_2, \quad S_3^1 = S_3^0 \oplus i_3. \tag{2.3}$$

He then sends out

$$S_1^0, S_2^0, S_3^0 \text{ to } DB_{000}, \quad S_1^0, S_2^0, S_3^1 \text{ to } DB_{001}, \quad \dots, \quad S_1^1, S_2^1, S_3^1 \text{ to } DB_{111}. \tag{2.4}$$

3 binary strings of length l will be sent to each database. Altogether $8 \times 3 \times l$ bits will be sent. Each database sends back 1 bit, so 8 bits will be received by the user. The total number of bits exchanged is $8 \times 3 \times l + 8 = 8(3l+1) = 8(3n^{1/3}+1)$. Therefore the communication complexity is $O(n^{1/3})$.

Consider any bit x_{j_1,j_2,j_3}, and let $S(j)$ be the jth bit of the string S

(i) If $j_1 \neq i_1, j_2 \neq i_2$ and $j_3 \neq i_3$, then $S_1^0(j_1) = S_1^1(j_1), S_2^0(j_2) = S_2^1(j_2), S_3^0(j_3) = S_3^1(j_3)$. Either 0 or 8 of x_{j_1,j_2,j_3} will be sent back.

(ii) If two of js are different from the corresponding is, say $j_1 \neq i_1, j_2 \neq i_2$, then $S_1^0(j_1) = S_1^1(j_1), S_2^0(j_2) = S_2^1(j_2), S_3^0(j_3) \neq S_3^1(j_3)$. Either 0 or 4 of x_{j_1,j_2,j_3} will be sent back.

(iii) If one of js is different from the corresponding is, say $j_1 \neq i_1$, then $S_1^0(j_1) = S_1^1(j_1), S_2^0(j_2) \neq S_2^1(j_2), S_3^0(j_3) \neq S_3^1(j_3)$. Either 0 or 2 of x_{j_1,j_2,j_3} will be sent back.

(iv) If $j_1 = i_1, j_2 = i_2, j_3 = i_3$, then $S_1^0(j_1) \neq S_1^1(j_1), S_2^0(j_2) \neq S_2^1(j_2), S_3^0(j_3) \neq S_3^1(j_3)$. In this case only 1 of x_{j_1,j_2,j_3} will be sent back.

That means when the user XORs all the 8 bits from the 8 databases, he can get the desired bit x_{i_1,i_2,i_3}. □

The covering code scheme: Chor et al. [**1**] has a detailed discussion about the covering code scheme. Briefly, a DB_α can simulate any other $DB_{\alpha'}$ if the hamming distance between α and α' is 1. It begins with 2^d databases and the communication complexity $O(n^{1/d})$. The number of bits now sent back from DB_α will be changed from 1 to $O(n^{1/d})$. Since both the user and the databases send $O(n^{1/d})$ bits, the communication complexity remains $O(n^{1/d})$ but the number of databases can be reduced.

Suppose we have 2 databases DB_{000} and DB_{111}. DB_{000} works as before so it sends back 1 bit. In addition, it takes over the function of DB_{100}. DB_{000} receives S_1^0, S_2^0, S_3^0, while DB_{100} receives S_1^1, S_2^0, S_3^0. DB_{000} does not know what i_1 is, but it does know that there is only 1 bit difference between S_1^0 and S_1^1 at the i_1th bit, i.e., $S_1^0(i_1) \neq S_1^1(i_1)$. DB_{000} can simulate DB_{100} by flipping one bit at a time from S_0^0 until all l bits were flipped. It will send back l bits. The user knows what i_1 is, so he can pick up the right bit and ignore the others. DB_{000} can simulate DB_{010} and DB_{001} in the same manner. Same idea applies to DB_{111}. It can simulate DB_{011}, DB_{101}, and DB_{110}. All together $6l + 2$ bits will be sent back. Since the user sends out $6l$ bits, the total bits exchanged is $12l + 2 = 12n^{1/3} + 2$. Thus the communication complexity is $O(n^{1/3})$.

This technique shows how to reduce the number of databases while keeping the same communication complexity. In this case, we can reduce from 8 to 2 databases while keeping the communication complexity to $O(n^{1/3})$. In terms of communication complexity, this scheme achieves the optimal performance for 2 servers.

3. A new PIR scheme

Since the original paper by Chor et al. [**1**], there are a lot of improvements for the communication complexity. Please refer to [**2**] [**4**] for a detailed survey. However, it is our intention to look for a scheme which balances the communication complexity and computing efficiency, especially under a practical setting namely only a small number of servers, say 2 (another one as a backup) to 4, are available. Another reason is that we want to have a practical scheme which is easy to implement. Based on the above scheme, we use the same idea of adjusting the communication complexity of the user and servers to come up with the following new scheme. We divided a d-dimensional *DB* into many smaller d-dimensional *db*s. So the size of the random strings sent to the servers can be cut down as the databases *db*s are smaller. On the other hand, the number of bits sent back from the servers will be increased due to smaller databases *dbs*. If there is a scheme that the communication complexity from the user side is much greater than that from the servers side, we can use this approach to make a balance and hence to reduce the overall communication complexity.

Proposed new scheme: In the multi-server scheme, we consider the *DB* as d-dimensional, i.e., $n = \underbrace{n^{1/d} \times \ldots \times n^{1/d}}_{d\ times}$. And let $l = n^{1/d}$. We now divide each dimension into $n^{1/d(d+1)}$ equal parts. Each part has length $l^{'} = \frac{n^{1/d}}{n^{1/d(d+1)}} = n^{1/d+1}$. The original *DB* is now divided into smaller d-dimensional *dbs* of equal size which equals $\underbrace{n^{1/d+1} \times \ldots \times n^{1/d+1}}_{d\ times} = n^{d/d+1}$. The number of such *dbs* is $\frac{n}{n^{d/d+1}} = n^{1/d+1}$.

Suppose we want to retrieve $x_{i_1,\ldots,i_d}$, $1 \le i_1, \ldots, i_d \le l$, in the original *DB*. Now this bit will be in one of the *dbs* and let the position be $x_{i_1^{'},\ldots,i_d^{'}}$, $1 \le i_1^{'}, \ldots, i_d^{'} \le l^{'}$. We apply the original multi-server scheme to retrieve the bits at position $x_{i_1^{'},\ldots,i_d^{'}}$ from all the *dbs*. We just keep the desired bit and ignore those bits that we do not need. Since we apply the same random inquiry string to all *dbs*, the communication complexity will be $O(n^{1/d+1})$. There are $n^{1/d+1}$ *dbs*, so the communication complexity from the servers will also be equal to $O(n^{1/d+1})$. Hence the total communication complexity is $O(n^{1/d+1}) + O(n^{1/d+1}) = O(n^{1/d+1})$.

By breaking down the original d-dimensional *DB* into smaller d-dimensional *dbs* of equal size, the new scheme reduces the communication complexity from $O(n^{1/d})$ to $O(n^{1/d+1})$.

EXAMPLE 1. Let $d = 1$, that means there are 2 servers. In the new scheme, the original *DB* of length n (up-side) in Fig. 1 will be divided into $\sqrt{n}$ *dbs* each with size of $\sqrt{n}$, as shown in (low-side) Fig. 1. Suppose we want to retrieve ith bit $(1 \le i \le n)$ from the original *DB*. Now this bit lies within db_j $(1 \le j \le \sqrt{n})$ at position $i^{'}$th, $1 \le i^{'} \le \sqrt{n}$ as shown in Fig 1. Retrieving ith bit from the original *DB* becomes getting $i^{'}$th bit from all *dbs*. The user will ignore all the bits returned from the servers with the exception of the one from db_j. The communication complexity for both the user and the servers is equal to $O(n^{1/2})$. Therefore the overall communication complexity is also equal to $O(n^{1/2})$.

EXAMPLE 2. Let $d = 2$, i.e., there are 4 servers. In the new scheme, the original *DB*, viewed as a square of $\sqrt{n} \times \sqrt{n}$ (left-side) in Fig. 2, will now be divided into $n^{1/3}$ *dbs* (squares) of size $n^{2/3}$, as shown (right-side) in Fig. 2.

Suppose the bit that we want to retrieve (i, j) $(1 \le i, j \le \sqrt{n})$ (left-side) now lies within $db_{l,k}$ $(1 \le l, k \le n^{1/3})$ with position $(i^{'}, j^{'})$ $(1 \le i^{'}, j^{'} \le n^{1/3})$. In the new scheme, all the $(i^{'}, j^{'})$ bits of all *dbs* will be returned to the user, but only the $(i^{'}, j^{'})$ bit within $DB_{l,k}$ will be used and the others will be ignored. Since both the length of the *dbs* and the number of small *dbs* equal $n^{1/3}$, the overall communication complexity equals $O(n^{1/3})$.

REMARKS. We need to seek a point that balances the two complexities. That means both the user and server complexities are the same. After the balance point,

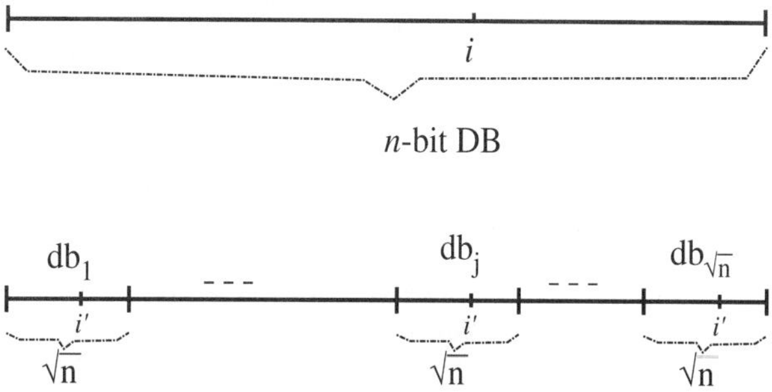

FIGURE 1. New 2 servers scheme for Example 1

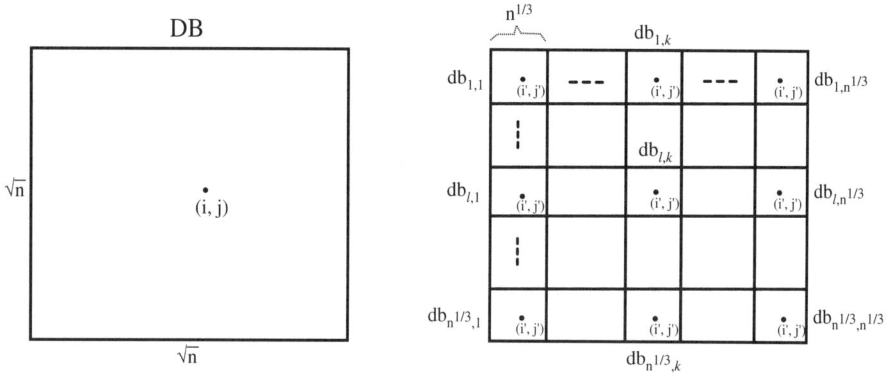

FIGURE 2. New 4 servers scheme for Example 2

if we keep on reducing the size of the random strings, then the server complexity will be greater than that of the user, and hence the overall communication complexity will be increased. Recall in the Example 1, if we divide the original DB into dbs of length $n^{1/3}$, instead of $n^{1/2}$, the number of dbs will be equal to $n^{2/3}$. Then the overall communication complexity will be $O(n^{1/3}) + O(n^{2/3}) = O(n^{2/3})$.

4. Conclusions

Based on Chor et al. [**1**], we come up with a new scheme which reduces the communication complexity from $O(n^{1/d})$ to $O(n^{1/d+1})$ when 2^d servers are available. This does not get too much noticeable advantages when d is large and in fact many existing schemes provide much better performances. However, if only a small number of servers are available, then this new scheme will provide a good balance between communication and computational efficiencies together with a simple implementation. For further research we would like to look for the possibility of applying this new technique to any existing scheme for improvements.

Acknowledgments

Authors would like to thank the referees for their valuable suggestions and comments. This work is supported, in part, by a PSC-CUNY Research Award.

References

[1] B. Chor, O. Goldreich, E. Kushilevitz, and M. Sudan, *Private Information Retrieval.* Proceedings of the 36th Annual IEEE Symposium on Foundations of Computer Science, (1995), 41–50.

[2] W. Gasarch, *A survey on private information retrieval*, Bull. Eur. Assoc. Theor. Comput. Sci. EATCS **82** (2004), 72–107. MR2132617

[3] E. Kushilevitz and R. Ostrovsky, *Replication Is Not Needed: Single Database, Computationally-Private Information Retrieval.* Proceedings of the 38th Annual IEEE Symposium on Foundations of Computer Science, (1997), 364–373.

[4] S. Yekhanin, *Private Information Retrieval.* Communications of the ACM, **53(4)**, (2010), 68–73.

Computer Science Dept., Graduate Center, CUNY, New York, New York
E-mail address: `CChum@gradcenter.cuny.edu`

Computer Science Dept., College of Staten Island, CUNY, Staten Island, New York. Computer Science Dept., Graduate Center, CUNY, New York, New York
E-mail address: `Xiaowen.Zhang@csi.cuny.edu`

Contemporary Mathematics
Volume **677**, 2016
http://dx.doi.org/10.1090/conm/677/13621

On secret sharing protocols

Chi Sing Chum, Benjamin Fine, Anja I. S. Moldenhauer,
Gerhard Rosenberger, and Xiaowen Zhang

ABSTRACT. Sharing a secret among n participants in such a way that any t of them but not less can access the secret is an extremely important cryptographic protocol. A beautiful solution to the general secret sharing problem was given by Shamir [**Sh**] and this has become the gold standard for secret sharing protocols yet there are continual improvements. In this paper, partly expository and partly research we examine many different cryptographic protocols and make some basic comparisons. We then introduce three new methods that are part of the Ph.D. thesis of A. Moldenhauer [**M**]. The first one is a combinatorial method while the second uses combinatorial group theory in particular Nielsen transformations. The last one applies similar techniques using solutions of the Hurwitz equation.

CONTENTS

1. Introduction

A **cryptographic protocol** consists of the collection of rules, formulas and methods to handle a cryptographic task. Extremely important along these lines are **secret sharing protocols**. These consist of methods to distribute a secret among a group of users by giving a share of the secret to each. The secret can be recovered only if a sufficient number of users (but perhaps not all) combine their pieces.

2010 *Mathematics Subject Classification.* Primary 20F67; Secondary 20F65, 20E06, 20E07.
Key words and phrases. Secret sharing protocol, Shamir scheme, verifiable secret sharing scheme, Nielsen transformations, polynomial interpolation, closest vector theorem.

©2016 American Mathematical Society

Formally we have the following. We have a secret $\mathcal{P}$ and a group of n participants. This group is called the **access control group**. A **dealer** allocates shares to each participant under given conditions. If a sufficient number of participants combine their shares then the secret can be recovered. If $t \leq n$ then an (t, n)-**threshold scheme** is one with n total participants and in which any t participants can combine their shares and recover the secret but not fewer than t. The number t is called the **threshold**. It is a **secure secret sharing scheme** if given less than the threshold there is no chance to recover the secret. If a measure is placed on the set of secrets, and on the set of shares, security can be made precise by saying that given less than the threshold all secrets are equally likely but given the threshold there is a unique secret. Secret sharing is an old idea but was formalized mathematically in independent papers in 1979 by Adi Shamir [**Sh**] and George Blakley [**B1**].

In the 1979 paper, Shamir [**Sh**] proposed a beautiful (t, n)-threshold scheme, based on polynomial interpolation, that has many desirable properties. We describe this in the next section. It is now the standard method for solving the (t, n)-secret sharing problem, although there are modifications for different situations (see [**CFZ**]). Blakely in his original paper proposed a geometric solution that is less space efficient, for computer storage, than Shamir's. In Blakely's scheme the distributed shares are larger than the secret, whereas in Shamir's scheme they are the same size. In this paper, which is part expository and part research, we look at many aspects of the secret sharing problem. We introduce two new secret sharing protocols, one based on combinatorial group theory and in particular on Nielsen transformations and one based on number theory. We consider the expository sections to be an excellent introduction to the whole area.

There are many different motivations for the secret sharing problem. One of the most important is the problem of maintaining sensitive information. There are two crucial issues here: availability and secrecy. If only one person keeps the entire secret, then there is a risk that the person might lose the secret or the person might not be available when the secret is needed. Hence it is often useful to utilize several people in order to access a secret. On the other hand, the more people who can access the secret, the higher the chance the secret will be leaked. By sharing a secret in a threshold scheme the availability and reliability issues can be addressed. The paper by Chum, Fine and Zhang [**CFZ**] contains a wealth of information on secret sharing in general and managing an access control group.

2. The Shamir Secret Sharing Scheme

Given a secret K, a (t, n)-**secret sharing threshold scheme** is a cryptographic primitive in which a secret is split into pieces (shares) and distributed among a collection of n participants $\{p_1, p_2, \ldots, p_n\}$ so that any group of t or more participants, with $(t \leq n)$, can recover the secret. Meanwhile, any group of $t-1$ or fewer participants cannot recover the secret. By sharing a secret in this way the availability and reliability issues can be solved.

Shamir solved the secret sharing problem in a very simple but beautiful manner using polynomial interpolation. The general idea in a Shamir (t, n)-threshold scheme is the following. Let F be any field and $(x_1, y_1), \ldots, (x_n, y_n)$ be n points in F^2 with pairwise distinct x_i. A polynomial $P(x)$ over F **interpolates** these points if $P(x_i) = y_i$ for $i = 1, \ldots, n$. The polynomial $P(x)$ is called an **interpolating**

polynomial for the given points. The crucial theoretical result is that for any n points (x_i, y_i) with distinct x_i there always exists a unique interpolating polynomial of degree $\leq n-1$.

THEOREM 2.1. *(Polynomial Interpolation Theorem) Let F be any field and $x_1, \ldots, x_n$ be n pairwise distinct elements of F and $y_1, \ldots, y_n$ any elements of F. Then there exists a* **unique** *polynomial of degree $\leq n-1$ that interpolates the n points $(x_i, y_i), i = 1, \ldots, n$.*

Using this theorem the Shamir (t, n)-threshold scheme works in the following manner. A field F with more than n elements is chosen. In general if F is a finite field we assume that the order of F is much much larger than n, the number of participants. The secret K is taken as an element of the field F and a polynomial $P(x)$ of degree $t-1$ is chosen with the secret K as its constant term. Then pairwise distinct elements of F, $x_1, \ldots, x_n$, are chosen with no $x_i = 0$. The points $(x_i, P(x_i))$ are distributed to each of the n participants. By the polynomial interpolation theorem, given above, any t participants can determine the interpolating polynomial $P(x)$ and hence recover the secret K. Given an infinite field and less than t people there are infinitely many polynomials of degree $t-1$ that can interpolate the given points and hence finding the correct polynomial has probability zero. In a finite field Shamir proved that under random choices for the x_i each secret in F is equally likely so guessing the secret is a random choice from F.

In the next section an alternative version to this Shamir scheme will be outlined that uses inner product spaces and the closest vector theorem rather than interpolation.

We now present a more explicit version of the Shamir scheme using the finite field $F = GF(q)$ where $q = p^k$ with $k \geq 1$ and p is a large prime. By using a finite field Shamir was able to place a finite measure on the set of plaintexts and ciphertexts and showed that with this scheme if there are less than t people all secrets are equally likely.

The Shamir (t, n)-Threshold Secret Sharing Scheme

Share distribution: Let K be the secret. The dealer generates a polynomial $P(x)$ of degree at most $t-1$ over $F = GF(q)$, where q is much larger than n as follows:

$$P(x) = a_0 + a_1 x + \cdots + a_{t-1} x^{t-1} \tag{1}$$

where $a_0 = K$ is the secret, $a_1, \ldots, a_{t-1} \in F$ and are generated randomly.

The dealer arbitrarily chooses pairwise distinct $x_i \in F \setminus \{0\}$, $i = 1, 2, \ldots, n$. Usually, $x_i = i$ will be chosen for simplicity. $x_1, x_2, \ldots, x_n$ are stored in a public area. The dealer calculates $y_i = P(x_i), i = 1, 2, \ldots, n$, and distributes to the n participants via a secure channel so that each participant p_i gets one share y_i. For the rest of the paper, we will not repeat the criteria of the generation of the coefficient a_i of the polynomial $P(x)$ and the calculation of the shares $P(x_i)$.

Secret Recovery (i): When any t participants join together, we have the following system of t equations. For simplicity, we assume the participants $p_1, p_2, \ldots, p_t$ join together.

$$y_1 = P(x_1) = a_0 + a_1 x_1 + \cdots + a_{t-1} x_1^{t-1},$$
$$y_2 = P(x_2) = a_0 + a_1 x_2 + \cdots + a_{t-1} x_2^{t-1},$$

$$\dots,$$

$$y_t = P(x_t) = a_0 + a_1 x_t + \cdots + a_{t-1} x_t^{t-1}.$$

In matrix representation, it will be:

$$\begin{pmatrix} 1 & x_1 & \cdots & x_1^{t-1} \\ 1 & x_2 & \cdots & x_2^{t-1} \\ \vdots & \cdots & \cdots & \vdots \\ 1 & x_t & \cdots & x_t^{t-1} \end{pmatrix} \begin{pmatrix} a_0 \\ a_1 \\ \vdots \\ a_{t-1} \end{pmatrix} = \begin{pmatrix} y_1 \\ y_2 \\ \vdots \\ y_t \end{pmatrix}$$

Let M be the above $t \times t$ Vandermonde matrix. Its determinant is

$$det(M) = \prod_{1 \le j < k \le t} (x_k - x_j).$$

Since we choose pairwise different points for the participants, that is, pairwise distinct x_i's, $det(M) \neq 0$, and this guarantees a unique solution. We can solve the system of equations by Gaussian elimination or Cramer's rule. Hence the secret can be recovered.

Secret Recovery (ii): Another method is to use Lagrange interpolation. We can construct the polynomial of degree at most $t-1$ by any t different points $(x_1, y_1), \dots, (x_t, y_t)$ as

$$P(x) = \sum_{i=1}^{t} y_i l_i(x),$$

where

$$l_i(x) = \prod_{j=1, j \neq i}^{t} \frac{x - x_j}{x_i - x_j} = \frac{(x - x_1) \cdots (x - x_{i-1})(x - x_{i+1}) \cdots (x - x_t)}{(x_i - x_1) \cdots (x_i - x_{i-1})(x_i - x_{i+1}) \cdots (x_i - x_t)}.$$

So, the secret a_0 will be

$$a_0 = P(0) = \sum_{i=1}^{t} y_i \prod_{j=1, j \neq i}^{t} \frac{-x_j}{x_i - x_j}.$$

Shamir in his original paper was able to prove that this secret sharing scheme is **perfect** in the sense that for $t-1$ participants any secret $K \in F$ is equally likely. If F is an infinite field, then the probability of correctly guessing the secret is zero.

3. Alternatives for Secret Sharing Protocols

In this section we describe several alternatives and enhancements to the Shamir Secret Sharing Scheme. The first alternative is a geometric method based on the closest vector theorem and introduced in [**CFRZ**]. It is similar to the original Blakely scheme but has the advantage that the secret is the same size as the shares. This was not the case in the Blakely protocol. This scheme, like the Shamir scheme, is also perfect in the sense that any t people out of the size n access control group can recover the secret but given less than t, then the participants have infinitely many subspaces of dimension t at choice in which the secret could lie in. The probability to choose the right subspace is negligible. Therefore, each possible secret is equally likely. The use of the closest vector theorem is prominent in lattice based cryptography. We refer to [**B**] for a discussion of this.

The second alternative we discuss is due to Panagopoulos and is based on combinatorial group theory. In particular he uses the solution to the word problem in finitely presented groups.

The third alternative is a new protocol due to A. Moldenhauer and developed in her Ph.D. thesis. The distribution of shares employs a modification of the Panagopoulos technique.

The final alternative is due to C. Chum and X. Zhang and is based on latin squares.

3.1. A Method Based on the Closest Vector Theorem. The geometric alternative scheme depends on the **closest vector theorem**. Recall the following facts. Let W be a real inner product space and V a subspace of finite dimension t. Suppose that $w \in W$ and $e_1, e_2, \ldots, e_t$ is an orthonormal basis of V. Then the **closest vector theorem** is the following (see [**A**]).

THEOREM 3.1. **Closest Vector Theorem**. *Let W be a real inner product space and V a subspace of finite dimension t. Suppose that $w \in W$, with w not in V, and $e_1, e_2, \ldots, e_t$ is an orthonormal basis of V. Then the unique vector $w^* \in V$ closest to w is given by*

$$w^* = < w, e_1 > e_1 + < w, e_2 > e_2 + \cdots + < w, e_t > e_t$$

where $<\ ,\ >$ is the inner product on W.

Notice that given any basis for the subspace V, the Gram-Schmidt orthonormalization procedure (see [**A**]) can be used to find an orthonormal basis for V. Hence given $w \in W$ we can algorithmically always find w^*, the unique vector in V closest to w. If a basis for V is not known and we only have knowledge or information on proper subspace spans in V of dimension less than t we cannot do this procedure. That is if we do not have complete knowledge of a basis for V we cannot apply the closest vector theorem. Further, since given a subspace of dimension less than t there are infinitely many subspaces of dimension t properly containing it, there is a negligible probability of obtaining the subspace V with only partial knowledge.

THE SECRET SHARING SCHEME

We start with an inner product space W of dimension m and an access control group of size n. We assume that the dimension m of W is much greater than n, that is $m >> n$. Within W there is a hidden subspace V of dimension $t < n$. The secret to be shared is given as an element in this hidden subspace, that is the secret $v \in V$ a vector in V.

The dealer distributes to each of the n members of the access control group, $i = 1, \ldots, n$, two vectors, v_i, w, where $v_i \in V$, and w is a vector in the big space W. The common vector w has the property that $w \notin V$ and v is the vector in V closest to w. In general the vector w can be given publically. The set $\{v_1, v_2, \ldots, v_n\}$ has the property that any subset of size t is independent. Hence any subset of size t determines a basis for V.

Suppose t valid users get together. They can determine a basis for V and hence using the Gram-Schmidt procedure (see [**A**]) determine an orthonormal basis. Since w is given, they can determine v by the closest vector theorem and recover the secret.

Given a subset of size less than t the given vectors generate a subspace of V of dimension less than t and hence in W there are infinitely many extensions to subspaces of dimension t. This implies that determining V with less than t elements of a basis has negligible probability.

As suggested by Shamir the secret should be altered periodically. In this method it is extremely easy to change the secret v without altering much of the scheme; simply send each user a new w.

This is a general method like the Shamir protocol. In [**FMR**], Fine, Moldenhauer and Rosenberger, compared several different secret sharing plans including the classic Shamir plan and the CFRZ plan.

3.2. A Method Based on Combinatorial Group Theory. The protocol in this section and the later new protocol using Nielsen transformations rely on group theory. For this material we refer to the book [**Ro**] for general group theoretic information and to the book [**MKS**] for material on combinatorial group theory.

Recall that a finitely presented group is a group with a presentation (see [**MKS**]) with finitely many generators and only finitely many relators. The word problem for group presentations is: Given a group presentation $G = \langle X; R \rangle$ and a word W in the generators of G to determine algorithmically if W represents the identity or not. It is known that the word problem is insolvable in general, that is, there exist group presentations where the word problem cannot be solved.

D. Panagopoulos [**P**] devised a secret sharing scheme based on the word problem in finitely presented groups. It is an (t, n)-threshold scheme and its main advantage over many other secret sharing schemes is that it does not require the secret message to be determined before each individual person receives his share of the secret. For this scheme it is assumed that the secret is in the form of a binary sequence. The scheme is as follows.

STEP 1. *A group G with finite presentation $G = \langle x_1, x_2, \ldots, x_k; r_1 = \cdots = r_m = 1 \rangle$ is chosen. It is assumed that the word problem is solvable for this presentation and that $m = \binom{n}{t-1}$.*

STEP 2. *Let $A_1, \ldots, A_m$ be an enumeration of the subsets of $\{1, \ldots, n\}$ with $t-1$ elements. Define n subsets of $\{r_1, \ldots, r_m\}$, $R_1, \ldots, R_n$ with $r_j \in R_i$ if and only if $i \notin A_j$, $j = 1, \ldots, m$, $i = 1, \ldots, n$. Then for every $j = 1, \ldots, m$, r_j is not contained in exactly $t-1$ of the subsets $R_1, \ldots, R_n$. It follows that r_j is contained in any union of t of them whereas if we take any $t-1$ of the $R_1, \ldots, R_n$ there exists a j such that r_j is not contained in their union.*

STEP 3. *Distribute to each of the n persons one of the $R_1, \ldots, R_n$. The set $\{x_1, \ldots, x_k\}$ is known to all of them.*

STEP 4. *If the binary sequence to be distributed is $a_1, \ldots, a_k$ construct and distribute a sequence of elements $w_1, \ldots, w_k$ of G such that $w_i = 1$ in G if and only if $a_i = 1, i = 1, \ldots, k$. The word w_i must involve most of the relations $r_1, \ldots, r_m$ if $w_i = 1$. Furthermore, all of the relations must be used at some point in the construction of some element.*

Any t of the n persons can obtain the sequence $a_1, \ldots, a_k$ by taking the union of the subsets of the relations of G that they possess and thus obtaining the presentation $G = \langle x_1, x_2, \ldots, x_k; r_1 = \cdots = r_m = 1 \rangle$ and solving the word problem $w_i = 1$ in G for $i = 1, \ldots, k$.

A collection of fewer than t persons cannot decode correctly the message since the union of fewer than t of the sets $R_1, \ldots, R_n$ contains some but not all of the relations $r_1, \ldots, r_m$. Thus such a collection could obtain a group presentation $G_1 = \langle x_1, x_2, \ldots, x_k; r_1 = \cdots = r_p = 1\rangle$ with $p < m$ and $G \neq G_1$, where $w_i = 1$ in G is not equivalent to $w_i = 1$ in G_1 in general.

Notice that the secret sequence to be shared is not needed until the final step. It is possible for someone to distribute the sets $R_1, \ldots, R_n$ and decide at a later time what the sequence would be. In that way the scheme can also be used so that t of the n persons can verify the authenticity of the message. In particular the binary sequence in step 4 could contain a predetermined subsequence (signature) along with the normal message. Then t persons may check whether this predetermined sequence is contained in the encoded message thus validating it.

In the paper by D. Panagopoulos, he also describes some methods for attacking this scheme and also makes some suggestions for possible group presentation types to use. We refer to [**P**] for these.

3.3. A Combinatorial Secret Sharing Scheme. In her Ph.D. thesis A. Moldenhauer [**M**] developed a combinatorial (t, n)-secret sharing scheme, where the secret is the sum of multiplicative inverses of elements in the natural numbers.

The method used by the dealer, to develop and distribute the shares, is patterned on the method of Panagopoulos as described in the last section.

Given n, the number of participants and t the threshold, the dealer does the following:

(1) The dealer first calculates the number $m = \binom{n}{t-1}$.

(2) He chooses m elements $a_1, a_2, \ldots, a_m \in \mathbb{N}$. From these elements he constructs the sets $R_1, R_2, \ldots, R_n$ analogously as in the Panagopoulos method. The secret S is the sum

$$S := \sum_{i=1}^{m} \frac{1}{a_i} \in \mathbb{Q}.$$

(3) Each participant P_i gets one share R_i, $1 \leq i \leq n$.

If t of the n participants come together they can reconstruct the secret. They first combine their t private sets R_i and get by construction the set $\tilde{R} = \{a_1, a_2, \ldots, a_m\}$. The secret is the sum of the inverse elements in the set $\tilde{R}$, that is

$$S = \sum_{i=1}^{m} \frac{1}{a_i}.$$

If the dealer needs a special secret $\tilde{S} \in \mathbb{Q}$ he gives every participant one more element $x \in \mathbb{Q}$ in each R_i, with

$$x := \frac{\tilde{S}}{S}.$$

The participants get $\tilde{S}$ by multiplying the reconstructed secret S with x.

The secret could also just be the sum of the elements, i.e., $S := \sum_{i=1}^{m} a_i$.

The security of this method is a consequence of the following. Each element a_j is exactly contained in $n-(t-1)$ subsets. Hence for each $j = 1, 2, \ldots, m$ the element a_j is not contained in $t-1$ subsets from $\{R_1, R_2, \ldots, R_n\}$. As a consequence, a_j is in each union of t subsets. On the other hand, if just $t-1$ arbitrary sets from

$\{R_1, R_2, \ldots, R_n\}$ are combined, there exist a j so that the element a_j is not included in the union of this sets.

If just one element a_j is absent, the participants do not get the correct sum S, and hence cannot compute the correct secret.

EXAMPLE 3.2. To illustrate the method we perform the steps for a $(3, 4)$-secret sharing protocol.

The dealer follows the steps:

(1) He first calculates $m = \binom{n}{t-1} = \binom{4}{2} = 6$.
(2) The dealer chooses the numbers $a_1 := 2, a_2 := 1, a_3 := 2, a_4 := 8$, $a_5 := 4$ and $a_6 := 2$. The secret is

$$S := \sum_{i=1}^{m} \frac{1}{a_i} = \frac{23}{8}.$$

(a) The six subsets with size 2 of the set $\{1, 2, 3, 4\}$ are

$$A_1 = \{1,2\}, \qquad A_2 = \{1,3\}, \qquad A_3 = \{1,4\},$$
$$A_4 = \{2,3\}, \qquad A_5 = \{2,4\}, \qquad A_6 = \{3,4\}.$$

With help of the A_i the dealer gets the sets R_1, R_2, R_3 and R_4, which contain elements from $\{a_1, \ldots, a_6\}$. He puts the element a_j for which i is not contained in the set A_j for $i = 1, \ldots, 4$ and $j = 1, \ldots, 6$, into the set R_i:

$$1 \notin A_4, A_5, A_6 \Longrightarrow R_1 = \{a_4, a_5, a_6\},$$
$$2 \notin A_2, A_3, A_6 \Longrightarrow R_2 = \{a_2, a_3, a_6\},$$
$$3 \notin A_1, A_3, A_5 \Longrightarrow R_3 = \{a_1, a_3, a_5\},$$
$$4 \notin A_1, A_2, A_4 \Longrightarrow R_4 = \{a_1, a_2, a_4\}.$$

(3) The dealer distributes the set R_i to the participant P_i, for $i = 1, \ldots, 4$.

If three of the four participants come together, they can calculate the secret S. For example the participants P_1, P_2 and P_3 have the set

$$\begin{aligned}\tilde{R} :=& R_1 \cup R_2 \cup R_3 \\ =& \{a_4, a_5, a_6\} \cup \{a_2, a_3, a_6\} \cup \{a_1, a_3, a_5\} \\ =& \{a_1, a_2, a_3, a_4, a_5, a_6\},\end{aligned}$$

and hence get the secret

$$S = \sum_{i=1}^{6} \frac{1}{a_i} = \frac{23}{8} \qquad \text{with } a_i \in \tilde{R}.$$

REMARK 3.3. It is important in terms of practicability, that the dealer calculates and distributes the shares for the participants long before the secret is needed by the participants. Hence the dealer has enough time to execute the share distribution and his computational cost should be of no consequence for the protocol. If t participants reconstruct the secret, they add up only m elements, which is feasible in linear time.

3.4. An Alternative Using Latin Squares. A second alternative is algebraic and relies on the use of Latin squares. A *Latin square* of order n is a 2-dimensional array that consists of n rows and n columns such that for any row and any column only one out of the n given symbols is filled in exactly once. For simplicity, we usually use $0, \ldots, n-1$ to represent the symbols so that each entry in a Latin square can be represented as a triple (i, j, k), where $0 \leq i, j, k \leq n-1$, and i, j, k are the row, the column and the symbol, respectively. In this subsection we introduce two kinds of Latin square based secret sharing schemes.

3.4.1. *Using an element of a union of critical sets as a share.* A *critical set* of a Latin square is a partial Latin square that leads to an unique full Latin square, which represents the secret. A critical set will become a partial Latin square if any element is removed. Cooper, Donovan and Seberry ([**CDS**]) proposed to use an element of a set S as a share for a participant in a secret sharing scheme. S is a union set of several critical sets, i.e., consisted of elements of several critical sets for a Latin square. Any subset of participants is an authorized subset if their shares form one of the critical sets in S.
Example: A $(2,3)$-threshold scheme is shown in Table 1.

TABLE 1. A $(2,3)$-threshold secret sharing scheme.

0		
	2	

C_1

	2	
		1

C_2

0		
		1

C_3

0	1	2
1	2	0
2	0	1

L

We can see that all the partial Latin squares C_1, C_2, C_3 are critical sets. They can be extended uniquely to the full Latin square L. Let S be the union of the three critical sets C_1, C_2, C_3. Then

$$S = \{(0,0,0), (1,1,2), (2,2,1)\}.$$

We distribute a triple to a participant as a share. Any two participants can recover the full Latin square. So we have a $(2,3)$-threshold scheme.

However there are serious limitations of using elements of critical sets as shares for a secret sharing scheme. For example, it is difficult to recover the full Latin square from a critical set, it is very hard to find or verify critical sets for a given Latin square. For detailed discussions on the limitations, please see ([**CZ**]).

3.4.2. *Applying hash function.* This was developed by Chum and Zhang ([**CZ1**]). Suppose we use the hash of a Latin square to represent a secret and its order, n, is made public. From the literature, we know that if the order n is increased by 1, the number of Latin squares will grow exponentially. For a large n, say $n \geq 10$, there are sufficient numbers of Latin squares of that order.

The idea of using Latin squares for a secret sharing scheme is to store the Latin square in a hash h. For simplicity, we assume that the size of the hash that represents the secret is the same as that of the Latin square (means the order n of the Latin square).

1) Scheme setup

(1) The dealer generates a random share (a random string) for each participant. Share s_i will be given to participant P_i, $i = 1, \ldots, n$.

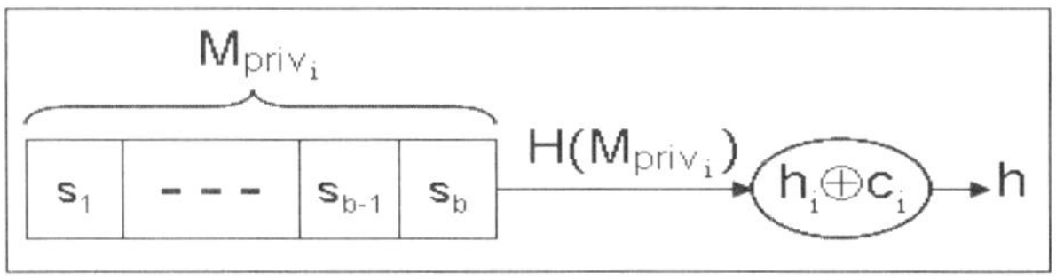

FIGURE 1. Secret recovery by combination of private and public information.

(2) The dealer decides the authorized subsets $A_1, \ldots, A_w$. Each participant holds a share and concatenation of the shares from any authorized subset leads to a private message M_{priv}. For example, if an authorized subset consists of P_1, P_3 and P_4, then the corresponding private message is $M_{priv} = s_1||s_3||s_4$, here $||$ symbolizes the concatenation between the shares.

(3) The dealer calculates the hashes for all w private messages

$$H(M_{priv_i}) = h_i,\ i = 1, \ldots, w.$$

Let h be the secret with the same size as h_i. Then the dealer generates a control c_i as public information for each authorized subset as follows (here $\oplus$ is bit-wise exclusive OR):

$$c_i = h_i \oplus h,\ i = 1, \ldots, w.$$

2) Secret recovering

Suppose authorized subset A_i consists of participants $P_1, \ldots, P_b$. Pooling their shares together they can recover the secret as follows, see Figure 1.

(1) Get the public information c_i.
(2) $H(s_1||s_2||\ldots||s_b) = h_i$, and $h_i \oplus c_i = h$.

This hash function based Latin square secret sharing scheme has the following features:

perfect: a participant in a minimal authorized subset is missing, the randomness property of a hash function makes it impossible to recover his/her share directly;

ideal: the hash of the private message has the same size as the hash of the secret;

fast: the scheme setup and secret recovery can be done very quickly;

flexible: no limit to the number of participants;

control area: used to make sure only authorized subsets can recover the secret.

For an example of the hash funtion method see [**CZ1**].

REMARK 3.4. The secret is a hash h (a binary sequence). Hence this secret sharing scheme also works without a Latin square. The Dealer needs as a secret just a hash h, which is a binary sequence. There is no need that it must be a hash of a Latin square. For each participant the dealer generates a random share s_j such that the hash h_i of the corresponding private messages to the authorized subsets has the same size than the hash h of the secret. Now the secret can be reconstruced with the public controls c_i, that correspond to the hashes h_i, as explained above.

3.5. Asymmetric and Ranked Secret Sharing. In many instances more involved secret sharing tasks must be handled. As an example consider the following situation. We are in a company that has directors and vice-directors. The directors and vice-directors are in the access control group but they do not have equal weight. Suppose that a secret can be recovered only if one of the following conditions is satisfied:

(a) two directors of the company cooperate;
(b) four vice-directors of the company cooperate;
(c) one director and two vice-directors of the company cooperate.

Thus here the threshold for the access control group differs depending on the status of the members. This is called an **asymmetric secret sharing scheme** or **ranked secret sharing scheme**.

In general, every (t, n)-secret sharing scheme can be converted into an asymmetric secret sharing protocol. In a standard (t, n)-secret sharing scheme every share is equivalent. To obtain an asymmetric secret sharing protocol every participant gets a different number of shares depending on the importance of the participant.

For example we can change a $(4, 8)$-secret sharing scheme into an asymmetric secret sharing protocol. The secret can be reconstructed if two presidents (D_1 and D_2) or four vice-presidents (V_1, V_2, V_3 and V_4) or one president and two vice-presidents get together (e.g. D_1, V_2 and V_4). The dealer distributes the shares as follow:

$$D_1 = (v_5, v_6), D_2 = (v_7, v_8);$$

$$V_1 = (v_1), V_2 = (v_2), V_3 = (v_3), V_4 = (v_4).$$

Here the $v_i, i = 1, 2, 3, 4, 5, 6, 7, 8$ are the shares.

Two of the presidents get together, so they have four different shares and with the protocol they can reconstruct the secret.

Every vice-president has only one share, hence four of them have to cooperate to determine the secret. It is sufficient though if one president and two vice-presidents get together to calculate the secret.

Less participants using this structure cannot determine the secret.

4. Comparison of Secret Sharing Protocols

In [**FMR**] a comparison was done between the Shamir scheme, the Panagopoulos scheme and the Closest Vector scheme. Here we present a brief recap of these results and add the combinatorial secret sharing scheme to the comparison.

Shamir in his paper [**Sh**] presents some useful properties for a secure (t, n)-secret sharing scheme:

(1) The size of each piece does not exceed the size of the secret.
(2) When t is kept fixed, pieces can be dynamically added or deleted. That is, when executives join or leave the company without affecting the other pieces. (A piece is deleted only when a leaving executive makes it completely inaccessible, even to himself.)
(3) It is easy to change the pieces without changing the original data (i.e. the secret). A frequent change of this type can greatly enhance security since the pieces exposed by security breaches cannot be accumulated.
(4) By using tuples of polynomial values as pieces, we can get a hierarchical scheme in which the number of pieces needed to determine the secret

depends on their importance. For example, if we give the company's president three values of $P(x)$, each vice-president two values of $P(x)$, and each executive one value of $P(x)$, then a $(3, n)$-threshold scheme enables checks to be signed either by any three executives, or by any two executives one of whom is a vice-president, or by the president alone.

All of these properties are satisfied by the standard Shamir interpolation scheme. An additional important property would be:

(5) It is easy to generate a new secret without changing the shares from the participants.

This last property is not true for Shamir's secret sharing scheme since the supporting points fixed the polynomial and the constant term (the secret) cannot be changed without redistributing the shares. In [**FMR**] the scheme of Panagopoulos and the closest vector scheme were analyzed relative to these criteria.

For the **Panagopoulos scheme**:

(1) The secret is a binary sequence and the shares are sets of relations. The length from the relations is not defined. In every set R_j are $\binom{n-1}{t-1}$ relations of the group. Therefore, the size of each share can exceed the size of the secret.
(2) The dealer creates the shares according to instructions. Hence he cannot add or delete shares, because the way he creates them depends on the number m of relations and the number n of participants.
(3) The dealer can change the shares if he changes the group G. He had to pay attention that the sent word in the new group is equivalent to 1 if and only if it is also in the previous group. Thus the secret is not changed.
(4) Every (t, n)-secret sharing scheme can be converted into an asymmetric secret sharing protocol.
(5) The secret, which is a binary sequence, can be changed at every time by sending new words to the participants.

For the **closest vector scheme**:

(1) The secret is a vector in the subspace V of the inner product space W. A share is a basis vector for the subspace V. The size of each share does not exceed the size of the secret.
(2) If we fix the number t of shares (we need at least to reconstruct the secret) we can arbitrarily add or delete many shares. The dealer has to pay attention to the construction, that every possible combination of t shares form a basis for the subspace V.
(3) We can change the shares without changing the secret. We need only another subspace, V_0, which contains the secret v. For this new subspace V_0 with dimension t we can calculate new shares, which are a set of vectors where every arbitrary t of them form a basis for V_0. A new vector w can be constructed (see [**FMR**]).
(4) An asymmetric system is possible.
(5) We can change the secret easily. Every vector in the subspace can be used as a new secret v_{new} (excluding the shares from the participants) and hence we can calculate the associated vector w_{new} as described in [**FMR**].

Thus the closest vector scheme satisfies all five of the important necessary properties. In the Shamir scheme it is not as easy to change the secret.

We add the comparison for the **combinatorial secret sharing scheme**:

(1) The secret is the sum over m elements: $\sum_{j=1}^{m} \frac{1}{a_j} \in \mathbb{Q}^+$. The shares are subsets R_j of $\{a_1, a_2, \ldots, a_m\}$, $a_i \in \mathbb{N}$, with $|R_j| = \binom{n-1}{t-1}$. Therefore, the size of each piece exceed the size of the secret.
(2) We use the method of D. Panagopoulos, hence this property is not valid due to the same reasons as for his secret sharing scheme.
(3) The shares are subsets of the set $\{a_1, a_2, \ldots, a_m\}$. If we choose a new set $\{a'_1, a'_2, \ldots, a'_m\}$ with the property $\sum_{j=1}^{m} \frac{1}{a_i} = \sum_{j=1}^{m} \frac{1}{a'_i}$ and give every participant subsets of this new set as a share, then the shares can be changed without changing the secret.
(4) An asymmetric system is possible.
(5) The secret cannot be changed easily, because it is the sum over all elements in the set $\{a_1, a_2, \ldots, a_m\}$ and hence depends on this set.

In [**FMR**] the time complexity for the participants of the three schemes was also compared. We summarize this below and add the combinatorial secret sharing scheme:

(1) **Shamir's scheme**: The polynomial interpolation has an quadratic running time, i.e. if we have t supporting points we have a complexity of $\mathcal{O}(t^2)$.
(2) **Panagopoulos' scheme**: The word problem in e.g. a Coxeter group is solvable in quadratic time. Because Coxeter groups are automatic and automatic groups have a solvable word problem with a quadratic running time.
(3) **The Closest Vector Scheme**: To orthonormalize t linear independent vectors in a real inner product space with dimension m we have a total running time of $\mathcal{O}(t^2 m)$. In the closest vector scheme the variable m depends on the number t, because $m > t$ is postulated. The total running time for this scheme is longer than for Shamir's.
(4) **Combinatorial secret sharing scheme:** For the reconstruction of the shares the participants only add up m elements. Therefore, for the participants it is just $\mathcal{O}(m)$, where $m = \binom{n}{t-1}$ is already previously calculated by the dealer, and hence m is fixed for the participants.

REMARK 4.1. In the special case of a $(t, t+1)$-secret sharing scheme the running time depends also only on t like in Sharmir's scheme:

$$\begin{aligned} m = \binom{n}{t-1} &= \binom{t+1}{t-1} = \binom{t+1}{2} \\ &= \frac{(t+1)\cdot t}{2} = \frac{t^2+t}{2} \\ &< \frac{2t^2}{2} = t^2. \end{aligned}$$

Hence the running time is also in $\mathcal{O}(t^2)$, but as shown above the participants only sum up m elements, which is a very easy operation to reconstruct the secret.

5. Verifying Secret Sharing Protocols (VSS)

In a standard secret sharing protocol it is assumed that the dealer and the participants are honest. To ensure the proper behavior of the dealer and the participants we enhance the standard secret sharing scheme. A **verifying secret sharing protocol**, denoted VSS, is one such enhancement. The aim of a VSS protocol is to be certain that the dealer and the participants behave correctly. A verifiable secret sharing protocol ensures that even if the dealer is dishonest there is a well-defined secret that the participants can recover. Verifiable secret sharing is important in secure multiparty computation.

There are many different verifiable secret sharing protocols (see [**CFZ**]). We first consider the case of a $(2,2)$-VSS and then give a generalization. We must construct a protocol where we can be certain that the dealer and the participants behave correctly. This first VSS protocol uses finite group theory. For a formal definition of homomorphism and isomorphism see [**Ro**].

(1) Let G and H be two groups and $\phi : G \to H$ be a group homomorphism so that
$$\phi(g_1 g_2) = \phi(g_1)\phi(g_2)$$
which is also a hash function.

(2) Let $s \in G$ be the secret. The dealer chooses $g_1, g_2 \in G$ with $g_1 g_2 = s$ and publishes $\phi(g_1), \phi(g_2)$. Nobody can efficiently calculate s from the published data since ϕ is a hash function. However, everyone can verify that $\phi(s) = \phi(g_1)\phi(g_2)$.

(3) The dealer tells g_1 to the participant A_1 and g_2 to the participant A_2. Each A_i can calculate $\phi(g_i)$ from his secret g_i and can check if the correct partial secret was received.

(4) When the participants reconstruct the secret s each A_i can prove if the other participant has exposed his correct partial secret by calculating $\phi(g_i)$ and comparing with the published values.

We now consider a more general (t, n)-VSS protocol. This example requires some elementary number theory. The required material can be found in the book [**FR**]. Recall that if p is a prime then an integer g is a primitive element modulo p if the order of g is $p-1$ in the multiplicative group $\mathbb{Z}_p^*$. For an integer g we let $\overline{g}$ denote its residue class modulo p.

In the situation from the $(2,2)$-VSS protocol above we choose for the group G the additive group modulo $(p-1)$, that is $(\mathbb{Z}_{p-1}, +)$. For H we choose the multiplicative group $(\mathbb{Z}_p^*, \cdot)$ where p is a sufficiently large prime.

Let g be a primitive element modulo p. As a hash function we choose
$$\phi : \mathbb{Z}_{p-1} \to \mathbb{Z}_p^*$$
by
$$a \mapsto \overline{g}^a \text{ for } a \in \{0, 1, \ldots, p-1\}.$$

The dealer chooses randomly a polynomial
$$f(x) = a_0 + a_1 x + \cdots + a_{t-1} x^{t-1}$$

where $a_0, a_1, \ldots, a_{t-1} \in \mathbb{Z}_{p-1} \setminus \{0, 1, \frac{p-1}{2}, p-1\}$ and $a_0 = s$ is the secret.

The dealer then publishes $\overline{g}^{a_i}$ for $i = 0, 1, \ldots, t-1$. Recall that $\overline{g}^{a_0} = \overline{g}^s$.

The dealer chooses randomly pairwise distinct elements $x_1, \ldots, x_n \in \mathbb{Z}_{p-1}$. He calculates $s_i = f(x_i)$ for $i = 1, \ldots, n$. He then publishes the values x_i and $\overline{g}^{s_i}$ for $i = 1, \ldots, n$.

Then:

(a) Each participant A_i can prove if they received s_i correctly by calculating $\overline{g}^{s_i}$ and comparing with the published values.

(b) Each participant can prove if

$$\prod_{j=0}^{t-1} (\overline{g}^{a_j})^{x_i^j} = \overline{g}^{f(x_i)} = \overline{g}^{s_i}.$$

Practically the dealer cannot cheat. All distributions from the dealer can be proved by the participants.

Finally we present another protocol similar to the one above, due to ElGamal, where the participants cannot cheat but discrepancies from the dealer are hard to discover. This is known as the **ElGamal** (t, n)-threshold signature protocol.

(1) The participants reach an agreement on two large primes p, q with $q|(p-1)$ and a hash function $h : \mathbb{Z}_p \times \mathbb{Z}_p \to \mathbb{Z}_p$. We assume that $\mathcal{M} = \mathbb{Z}_p$ is the set of plain text units.
(2) There exists a unique cyclic subgroup G_q of $\mathbb{Z}_p^*$ with q elements (see the ElGamal signature method). The participants reach an agreement on a generator $g \in G_q$.
(3) The dealer chooses randomly $a_0, \ldots, a_{t-1} \in \mathbb{Z}_q$ and defines the polynomial
$$f(x) = a_0 + a_1 x + \cdots + a_{t-1} x^{t-1} \in \mathbb{Z}_q[x].$$
(4) Let $s = f(0) = a_0$ be the secret where $s \in \mathbb{Z}_p \setminus \{0, 1, \frac{p-1}{2}, p-1\}$ and let $y \equiv g^s \pmod{p}$ be made public.
(5) The dealer chooses randomly pairwise distinct elements $\mu_i, x_i \in \mathbb{Z}_q^*$ for $i = 1, \ldots, n$ and calculates the partial secrets $s_i = \mu_i + f(x_i)$ for the participants A_i. The values x_i are published.
(6) The dealer calculates for each participant A_i the values $y_i \equiv g^{s_i} \pmod{p}$ and $z_i \equiv g^{\mu_i} \pmod{p}$ and makes these public.
(7) To sign the protocol, each participant A_i chooses randomly $k_i \in \mathbb{Z}_q^*$ and calculates $r_i \equiv g^{k_i} \pmod{p}$. The value r_i will be sent to the other participants.
(8) Suppose that t participants have sent their values, say, $A_1, \ldots, A_t$ have sent their values $r_1, \ldots, r_t$. Then participant A_i with $1 \le i \le t$ calculates the value
$$R = r_1 \cdots r_t \equiv g^{k_1 + \cdots + k_t} \pmod{p}$$
and the value
$$E \equiv h(m, R) \pmod{p}, m \in \mathcal{M}.$$
Then A_i has his partial signature
$$c_i \equiv s_i \prod_{j \ne i} \frac{-x_j}{x_i - x_j} + k_i E \pmod{q}.$$

(9) The dealer verifies the partial signatures by checking if

$$y_i^{\prod_{i\neq j}\frac{-x_j}{x_i-x_j}} r_i^E \equiv g^{c_i} \pmod p.$$

(Recall that $r_i \equiv g^{k_i} \pmod p$ and $y_i \equiv g^{s_i} \pmod p$.)

In this case the dealer then calculates $\sigma \equiv (c_1 + \cdots + c_t) \pmod q$. The signature of $m \in \mathcal{M}$ is then

$$(\{A_1, \ldots, A_t\}, R, \sigma).$$

(10) The verification of the signature is to calculate

$$T \equiv \prod_{i=1}^{t} z_i^{\prod_{i\neq j}\frac{-x_j}{x_i-x_j}} \pmod p$$

and check if

$$g^\sigma \equiv yTR^E \pmod p.$$

Notice that the set $\{A_1, \ldots, A_t\}$ has to be given. However, given such a set the participants practically cannot cheat and the dealer can realize this. The signature is correct. Further, in this protocol, which verifies that the participants cannot cheat, it is difficult to discover any violations to the protocol by the dealer.

6. A New Secret Sharing Protocol Using Combinatorial Group Theory

In her thesis [**M**] A. Moldenhauer suggested several ways to use combinatorial group theory to develop further cryptographic protocols. We describe one such new method that relies on Nielsen transformations. Nielsen transformations are the basis of a linear technique to study free groups and general infinite groups. For a complete discussion of these we refer to [**MKS**] and to the paper [**FRS**]. Below we review some basic definitions concerning regular Nielsen transformations and Nielsen reduced sets. (See [**CgrRR**], [**LS**], [**FRS**] or [**MKS**].)

Let F be a free group on the free generating set $X := \{x_1, x_2, \ldots\}$ and let $U := \{u_1, u_2, \ldots\} \subset F$.

Definition 6.1. An **elementary Nielsen transformation** on $U = \{u_1, u_2, \ldots\}$ is one of the following transformations

(T1) replace some u_i by u_i^{-1};
(T2) replace some u_i by $u_i u_j$ where $j \neq i$;
(T3) delete some u_i where $u_i = 1$.

In all three cases the u_k for $i \neq k$ are not changed. A (finite) product of elementary Nielsen transformations is called a **Nielsen transformation**. A Nielsen transformation is called **regular** if it is a finite product of the transformations $(T1)$ and $(T2)$, otherwise it is called **singular**. The set U is called **Nielsen-equivalent** to the set V, if there is a regular Nielsen transformation from U to V.

Definition 6.2. Consider elements v_1, v_2, v_3 of the form $u_i^{\pm 1}$, call U **Nielsen reduced** if for all such triples the following conditions hold:

(N0) $v_1 \neq 1$;
(N1) $v_1 v_2 \neq 1$ implies $|v_1 v_2| \geq |v_1|, |v_2|$;
(N2) $v_1 v_2 \neq 1$ and $v_2 v_3 \neq 1$ implies $|v_1 v_2 v_3| > |v_1| - |v_2| + |v_3|$.

Here $|\cdot|$ denotes the free length in F.

PROPOSITION 6.3. *If $U = \{u_1, u_2, \ldots, u_n\}$ is finite, then U can be carried by a Nielsen transformation into some V such that V is Nielsen reduced.*

For the secret sharing scheme based on Nielsen transformations we will only use regular Nielsen transformations.

We write $(T1)_i$ if we replace u_i by u_i^{-1} and we write $(T2)_{ij}$ if we replace u_i by $u_i u_j$. If we want to apply t-times one after the other the same Nielsen transformation $(T2)$ we write $[(T2)_{ij}]^t$ and hence replace u_i by $u_i u_j^t$. In all cases the u_k for $i \neq k$ are not changed.

We now describe a (t, n)-secret sharing scheme using Nielsen transformations. We consider free groups as abstract groups but also as subgroups of the special linear group of all 2×2 matrices over $\mathbb{Q}$, that is,

$$\mathrm{SL}(2, \mathbb{Q}) = \left\{ \begin{pmatrix} a & b \\ c & d \end{pmatrix} ; a, b, c, d \in \mathbb{Q} \text{ and } ad - bc = 1 \right\}.$$

We use the special linear group over the rational numbers because these numbers can be stored and computed more efficiently on a computer than irrational numbers.

Let F be a free group in $\mathrm{SL}(2, \mathbb{Q})$ of rank $m := \binom{n}{t-1}$. The dealer wants to distribute the shares for the participants as in the Panagopoulos scheme. The shares will be subsets of a free generating set of the group F.

Steps for the Dealer: The numbers n and t are given, whereby n is the number of participants and t is the threshold. We have $m := \binom{n}{t-1}$.

(1) The dealer chooses an abstract free generating set X for the free group F of rank m, it is

$$F = \langle X; \ \rangle \qquad \text{with } X := \{x_1, x_2, \ldots, x_m\}.$$

He also needs an explicit free generating set M, so it is

$$F = \langle M; \ \rangle \qquad \text{with } M := \{M_1, M_2, \ldots, M_m\}$$

and $M_i \in \mathrm{SL}(2, \mathbb{Q})$.

(2) With the known matrices in the set M he computes the secret

$$S := \sum_{j=1}^{m} \frac{1}{|a_j|} \in \mathbb{Q}^+ \qquad \text{with } a_j := tr(M_j) \in \mathbb{Q},$$

$tr(M_j)$ is the trace for the matrix $M_j := \begin{pmatrix} a & b \\ c & d \end{pmatrix} \in \mathrm{SL}(2, \mathbb{Q})$, that is, $tr(M_j) := a + d$. If the dealer needs a special secret he can act as described in Section 3.3.

(3) The dealer constructs the shares for the participants in the following way:
  (a) He first applies regular Nielsen transformations simultaneously for both sets X and M to get Nielsen equivalent sets U and N to X and M, respectively (see Figure 2).

$$
\begin{array}{ccc}
X := \{x_1, x_2, \dots, x_m\} & & M := \{M_1, M_2, \dots, M_m\} \\
\text{regular Nielsen transformation} \downarrow & & \text{regular Nielsen transformation} \downarrow \\
U := \{u_1, u_2, \dots, u_m\} & & N := \{N_1, N_2, \dots, N_m\}
\end{array}
$$

FIGURE 2. Simultaneously regular Nielsen transformations

The elements u_i are words in X and the elements N_i are words in M. Hence we have $N_i \in \mathrm{SL}(2, \mathbb{Q})$.

(b) The dealer now uses the method of Panagopoulos to split U and N and to get the share (R_i, S_j) for the participant P_k with $R_i \subset U$ and $S_j \subset N$.

(4) The dealer distributes the shares.

If t of the n participants combine their parts then they obtain the sets U and N. The secret can be recovered as follows:

(1) The participants apply regular Nielsen transformations in a Nielsen reduction manner for U and step by step simultaneously for N. By Proposition 6.3 they get Nielsen reduced sets $X^{\pm} = \{x_1^{\epsilon_1}, x_2^{\epsilon_2}, \dots, x_m^{\epsilon_m}\}$ and $M^{\pm} = \{M_1^{\delta_1}, M_2^{\delta_2}, \dots, M_m^{\delta_m}\}$ with $\epsilon_i, \delta_i \in \{+1, -1\}$, see Figure 3.

$$
\begin{array}{ccc}
U := \{u_1, u_2, \dots, u_m\} & & N := \{N_1, N_2, \dots, N_m\} \\
\text{regular Nielsen transformation} \downarrow & & \text{regular Nielsen transformation} \downarrow \\
X^{\pm} = \{x_1^{\epsilon_1}, x_2^{\epsilon_2}, \dots, x_m^{\epsilon_m}\} & & M^{\pm} = \{M_1^{\delta_1}, M_2^{\delta_2}, \dots, M_m^{\delta_m}\}
\end{array}
$$

FIGURE 3. Simultaneously regular Nielsen transformations

(2) With the knowledge of the set $M^{\pm}$ it is easy to reconstruct the secret

$$S = \sum_{j=1}^{m} \frac{1}{|a_j|} \in \mathbb{Q}^{+} \qquad \text{with } tr(M_j) = a_j \in \mathbb{Q}.$$

Recall that $tr(M_i^{\delta_i}) = tr(M_i)$ for $i = 1, \dots, m$.

Less than t participants can neither get the whole set U, which is Nielsen-equivalent to $X^{\pm}$, nor the set N, which is Nielsen-equivalent to $M^{\pm}$.
For the calculation of the secret, the participants need the set $M^{\pm}$, because the secret depends on the traces of the matrices $M_i \in M^{\pm}$. The participants need both sets U and N. If they just have one set U or N they cannot get information about the set $M^{\pm}$.
If the set U is known, it is only known which Nielsen transformation should be done to get the Nielsen equivalent set $X^{\pm}$, but it is unknown on which matrices they should be simultaneously done.

If only the set N is known, then the matrices in $\mathrm{SL}(2,\mathbb{Q})$ are known, but nobody knows which Nielsen transformations should be done on N to get the set $M^{\pm}$. It is also unknown how many Nielsen transformations were used.

REMARK 6.4. Analogously to the combinatorial secret sharing scheme, this scheme fulfills the same properties ((3) and (4)) of Section 4, because both schemes are based on the share distribution method of D. Panagopoulos. There is an idea to fulfill the property (5) by publishing the set N. If the Dealer changes the set N he changes the secret. But there is not enough security analysis about this idea, yet.
In [**Ste**] an algorithm, using elementary Nielsen transformations, is presented which, given a finite set S of m words of some free group, returns a set S' of Nielsen reduced words such that $\langle S\rangle = \langle S'\rangle$; the algorithm runs in $\mathcal{O}(\ell^2 m^2)$, where ℓ is the maximum length of a word in S.
In this protocol, the dealer fixes the number m, hence the running time depends only on the maximum length ℓ of the words in the Nielsen-equivalent set U to the set X. Thus the participants have a running time of $\mathcal{O}(\ell^2)$ to get the set $X^{\pm}$.
If the participants perform the associated elementary Nielsen transformations on the set N of matrices at the same time, then they perform either matrix multiplication or they calculate an inverse matrix. In order to multiply two 2×2 matrices in $SL(2,\mathbb{Q})$ they need 8 multiplications and 4 additions of rational numbers, hence 12 operations. The inverse matrix of

$$A = \begin{pmatrix} a & b \\ c & d \end{pmatrix} \in SL(2,\mathbb{Q})$$

is

$$A^{-1} = \frac{1}{ad-bc}\begin{pmatrix} d & -b \\ -c & a \end{pmatrix}.$$

The participants need 4 operations to get $\frac{1}{ad-bc}$; for the entries in the matrix A^{-1} they do not need any operations, they just swap two entries and write a minus in front of the other two entries.
All together the participants have a running time of $\mathcal{O}(\ell^2)$, where ℓ is the maximum length of the elements in U.

In the book of J. Lehner [**Leh**] on page 247 a method is given to explicitly obtain a free generating set M for a free group F on the abstract generating set $X := \{x_1, x_2, \ldots, x_m\}$:

EXAMPLE 6.5. Let F be a free group with countably many free generators $x_1, x_2, \ldots$. Corresponding to x_j define the matrix

$$M_j = \begin{pmatrix} -r_j & -1+r_j^2 \\ 1 & -r_j \end{pmatrix}$$

with $r_j \in \mathbb{Q}$ such that the following inequalities hold:

$$r_{j+1} - r_j \geq 3 \qquad \text{and} \qquad r_1 \geq 2. \tag{2}$$

The group G generated by $\{M_1, M_2, \ldots\}$ is isomorphic to F (see [**Leh**]).

We now present an example for this secret sharing scheme.

EXAMPLE 6.6. We perform the steps for an $(2,3)$-secret sharing scheme with the help of the computer program Maple 16. It is $n = 3$, $t = 2$ and hence $m = \binom{3}{1} = 3$. First the Dealer generates the shares for the participants.

(1) The dealer chooses an abstract presentation for the free group F of rank 3

$$F = \langle X; \ \rangle \qquad \text{with } X := \{x_1, x_2, x_3\}.$$

He takes an explicit presentation

$$F = \langle M; \ \rangle \qquad \text{with } M := \{M_1, M_2, M_3\},$$

$M_i \in \mathrm{SL}(2, \mathbb{Q})$ as above. We first mention that the inequalities (2) hold for

$$r_1 = \frac{7}{2}, \quad r_2 = \frac{15}{2}, \quad r_3 = 11$$

and hence the set of the matrices

$$M_1 = \begin{pmatrix} -\frac{7}{2} & -1 + \left(\frac{7}{2}\right)^2 \\ 1 & -\frac{7}{2} \end{pmatrix} = \begin{pmatrix} -\frac{7}{2} & \frac{45}{4} \\ 1 & -\frac{7}{2} \end{pmatrix},$$

$$M_2 = \begin{pmatrix} -\frac{15}{2} & -1 + \left(\frac{15}{2}\right)^2 \\ 1 & -\frac{15}{2} \end{pmatrix} = \begin{pmatrix} -\frac{15}{2} & \frac{221}{4} \\ 1 & -\frac{15}{2} \end{pmatrix},$$

$$M_3 = \begin{pmatrix} -11 & -1 + 11^2 \\ 1 & -11 \end{pmatrix} = \begin{pmatrix} -11 & 120 \\ 1 & -11 \end{pmatrix}$$

is a free generating set for a free group of rank 3.

(2) We have

$$a_1 := tr(M_1) = -7, \quad a_2 := tr(M_2) = -15, \quad a_3 := tr(M_3) = -22,$$

and hence the secret is

$$S := \sum_{j=1}^{3} \frac{1}{|a_j|} = \frac{589}{2310}.$$

(3) Construction of the shares for the participants:
   (a) First the dealer applies regular Nielsen transformations (NTs) simultaneously for both sets X and M to get Nielsen equivalent sets U and N to X or M, respectively. These transformations are shown in the Tables 2 and 3.

TABLE 2. Nielsen transformations (NTs) of the dealer I

NTs	theoretical set X	explicit set M
	$\{x_1, x_2, x_3\}$	$\left\{\begin{pmatrix} -\frac{7}{2} & \frac{45}{4} \\ 1 & -\frac{7}{2} \end{pmatrix}, \begin{pmatrix} -\frac{15}{2} & \frac{221}{4} \\ 1 & -\frac{15}{2} \end{pmatrix}, \begin{pmatrix} -11 & 120 \\ 1 & -11 \end{pmatrix}\right\}$
$(T1)_2$	$\{x_1, x_2^{-1}, x_3\}$	$\left\{\begin{pmatrix} -\frac{7}{2} & \frac{45}{4} \\ 1 & -\frac{7}{2} \end{pmatrix}, \begin{pmatrix} -\frac{15}{2} & -\frac{221}{4} \\ -1 & -\frac{15}{2} \end{pmatrix}, \begin{pmatrix} -11 & 120 \\ 1 & -11 \end{pmatrix}\right\}$
$(T2)_{12}$	$\{x_1x_2^{-1}, x_2^{-1}, x_3\}$	$\left\{\begin{pmatrix} 15 & 109 \\ -4 & -29 \end{pmatrix}, \begin{pmatrix} -\frac{15}{2} & -\frac{221}{4} \\ -1 & -\frac{15}{2} \end{pmatrix}, \begin{pmatrix} -11 & 120 \\ 1 & -11 \end{pmatrix}\right\}$
$[(T2)_{32}]^3$	$\{x_1x_2^{-1}, x_2^{-1}, x_3x_2^{-3}\}$	$\left\{\begin{pmatrix} 15 & 109 \\ -4 & -29 \end{pmatrix}, \begin{pmatrix} -\frac{15}{2} & -\frac{221}{4} \\ -1 & -\frac{15}{2} \end{pmatrix}, \begin{pmatrix} -8565 & -63664 \\ 799 & 5939 \end{pmatrix}\right\}$
$(T2)_{23}$	$\{x_1x_2^{-1}, x_2^{-1}x_3x_2^{-3}, x_3x_2^{-3}\}$	$\left\{\begin{pmatrix} 15 & 109 \\ -4 & -29 \end{pmatrix}, \begin{pmatrix} \frac{80371}{4} & \frac{597401}{4} \\ \frac{5145}{2} & \frac{38243}{2} \end{pmatrix}, \begin{pmatrix} -8565 & -63664 \\ 799 & 5939 \end{pmatrix}\right\}$
$(T1)_1$	$\{x_2x_1^{-1}, x_2^{-1}x_3x_2^{-3}, x_3x_2^{-3}\}$	$\left\{\begin{pmatrix} -29 & -109 \\ 4 & 15 \end{pmatrix}, \begin{pmatrix} \frac{80371}{4} & \frac{597401}{4} \\ \frac{5145}{2} & \frac{38243}{2} \end{pmatrix}, \begin{pmatrix} -8565 & -63664 \\ 799 & 5939 \end{pmatrix}\right\}$
$(T2)_{12}$	$\{x_2x_1^{-1}x_2^{-1}x_3x_2^{-3},$ $x_2^{-1}x_3x_2^{-3}, x_3x_2^{-3}\}$	$\left\{\begin{pmatrix} -\frac{3452369}{4} & -\frac{25661603}{4} \\ \frac{237917}{2} & \frac{1768447}{2} \end{pmatrix},\right.$ $\left.\begin{pmatrix} \frac{80371}{4} & \frac{597401}{4} \\ \frac{5145}{2} & \frac{38243}{2} \end{pmatrix}, \begin{pmatrix} -8565 & -63664 \\ 799 & 5939 \end{pmatrix}\right\}$

TABLE 3. Nielsen transformations (NTs) of the dealer II

NTs	theoretical set	explicit set
$(T1)_3$	$\{x_2x_1^{-1}x_2^{-1}x_3x_2^{-3},$ $x_2^{-1}x_3x_2^{-3}, x_2^3x_3^{-1}\}$	$\left\{\begin{pmatrix} -\frac{3452369}{4} & -\frac{25661603}{4} \\ \frac{237917}{2} & \frac{1768447}{2} \end{pmatrix}, \begin{pmatrix} \frac{80371}{4} & \frac{597401}{4} \\ \frac{5145}{2} & \frac{38243}{2} \end{pmatrix}, \begin{pmatrix} 5939 & 63664 \\ -799 & -8565 \end{pmatrix}\right\}$
$(T2)_{32}$	$\{x_2x_1^{-1}x_2^{-1}x_3x_2^{-3},$ $x_2^{-1}x_3x_2^{-3}, x_2^3x_3^{-1}x_2^{-1}x_3x_2^{-3}\}$	$\left\{\begin{pmatrix} -\frac{3452369}{4} & -\frac{25661603}{4} \\ \frac{237917}{2} & \frac{1768447}{2} \end{pmatrix}, \begin{pmatrix} \frac{80371}{4} & \frac{597401}{4} \\ \frac{5145}{2} & \frac{38243}{2} \end{pmatrix}, \begin{pmatrix} \frac{1132425929}{4} & \frac{8417369243}{4} \\ -\frac{152350279}{4} & -\frac{1132425989}{4} \end{pmatrix}\right\}$

The Dealer gets the sets

$$U = \{u_1, u_2, u_3\} := \{x_2x_1^{-1}x_2^{-1}x_3x_2^{-3}, x_2^{-1}x_3x_2^{-3}, x_2^3x_3^{-1}x_2^{-1}x_3x_2^{-3}\}$$

and

$$N = \{N_1, N_2, N_3\}$$
$$:= \left\{\begin{pmatrix} -\frac{3452369}{4} & -\frac{25661603}{4} \\ \frac{237917}{2} & \frac{1768447}{2} \end{pmatrix}, \begin{pmatrix} \frac{80371}{4} & \frac{597401}{4} \\ \frac{5145}{2} & \frac{38243}{2} \end{pmatrix}, \begin{pmatrix} \frac{1132425929}{4} & \frac{8417369243}{4} \\ -\frac{152350279}{4} & -\frac{1132425989}{4} \end{pmatrix}\right\}.$$

(b) He gets the share (R_i, S_j) for the participant P_k with $R_i \subset U$ and $S_j \subset N$ as follow:

(i) It is $m = \binom{n}{t-1} = \binom{3}{1} = 3$.

(ii) The dealer chooses the elements $\tilde{a}_1, \tilde{a}_2, \tilde{a}_3$ and gets the three sets

$$A_1 = \{1\}, \qquad A_2 = \{2\}, \qquad A_3 = \{3\}.$$

With the help of the A_i the dealer gets the sets R'_1, R'_2, and R'_3 which contain elements from the set $\{\tilde{a}_1, \tilde{a}_2, \tilde{a}_3\}$. He puts the element $\tilde{a}_j$ by which i is not contained in the set A_j for $i = 1, 2, 3$ and $j = 1, 2, 3$, into the set R'_i.

$$1 \notin A_2, A_3 \Longrightarrow R'_1 = \{\tilde{a}_2, \tilde{a}_3\},$$
$$2 \notin A_1, A_3 \Longrightarrow R'_2 = \{\tilde{a}_1, \tilde{a}_3\},$$
$$3 \notin A_1, A_2 \Longrightarrow R'_3 = \{\tilde{a}_1, \tilde{a}_2\}.$$

Now we apply this to U and N to create the share-sets for the participants, respectively:

$$R_1 = \{u_2, u_3\}, \qquad S_1 = \{N_2, N_3\},$$
$$R_2 = \{u_1, u_3\}, \qquad S_2 = \{N_1, N_3\},$$
$$R_3 = \{u_1, u_2\}, \qquad S_3 = \{N_1, N_2\}.$$

(4) The Dealer gives every participant P_k a tuple (R_i, S_j). Participant P_1 gets (R_1, S_2), P_2 gets (R_2, S_3) and P_3 gets (R_3, S_1).

Assume the participants P_1 and P_2 come together to reconstruct the secret. They generate the sets $U = \{u_1, u_2, u_3\}$ and $N = \{N_1, N_2, N_3\}$. The secret can be recovered as follow.

The participants apply regular Nielsen transformations step by step simultaneously for both sets U and N to get $X^{\pm}$ and $M^{\pm}$. The steps are shown in the Tables 4 and 5.

TABLE 4. Nielsen transformations (NTs) from the participants I

NTs	theoretical set U	explicit set N
	$\{x_2x_1^{-1}x_2^{-1}x_3x_2^{-3},$ $x_2^{-1}x_3x_2^{-3}, x_2^3x_3^{-1}x_2^{-1}x_3x_2^{-3}\}$	$\left\{\begin{pmatrix} -\frac{3452369}{4} & -\frac{25661603}{4} \\ \frac{237917}{2} & \frac{1768447}{2} \end{pmatrix}, \begin{pmatrix} \frac{80371}{4} & \frac{597401}{4} \\ \frac{5145}{2} & \frac{38243}{2} \end{pmatrix}, \begin{pmatrix} \frac{1132425929}{4} & \frac{8417369243}{4} \\ -\frac{152350279}{4} & -\frac{1132425989}{4} \end{pmatrix}\right\}$
$(T1)_2$	$\{x_2x_1^{-1}x_2^{-1}x_3x_2^{-3},$ $x_2^3x_3^{-1}x_2, x_2^3x_3^{-1}x_2^{-1}x_3x_2^{-3}\}$	$\left\{\begin{pmatrix} -\frac{3452369}{4} & -\frac{25661603}{4} \\ \frac{237917}{2} & \frac{1768447}{2} \end{pmatrix}, \begin{pmatrix} \frac{38243}{2} & -\frac{597401}{4} \\ -\frac{5145}{2} & \frac{80371}{4} \end{pmatrix}, \begin{pmatrix} \frac{1132425929}{4} & \frac{8417369243}{4} \\ -\frac{152350279}{4} & -\frac{1132425989}{4} \end{pmatrix}\right\}$
$(T2)_{32}$	$\{x_2x_1^{-1}x_2^{-1}x_3x_2^{-3},$ $\{x_2^3x_3^{-1}x_2,\ x_2^3x_3^{-1}\}$	$\left\{\begin{pmatrix} -\frac{3452369}{4} & -\frac{25661603}{4} \\ \frac{237917}{2} & \frac{1768447}{2} \end{pmatrix}, \begin{pmatrix} \frac{38243}{2} & -\frac{597401}{4} \\ -\frac{5145}{2} & \frac{80371}{4} \end{pmatrix}, \begin{pmatrix} 5939 & 63664 \\ -799 & -8565 \end{pmatrix}\right\}$
$(T1)_2$	$\{x_2x_1^{-1}x_2^{-1}x_3x_2^{-3},$ $x_2^{-1}x_3x_2^{-3}, x_2^3x_3^{-1}\}$	$\left\{\begin{pmatrix} -\frac{3452369}{4} & -\frac{25661603}{4} \\ \frac{237917}{2} & \frac{1768447}{2} \end{pmatrix}, \begin{pmatrix} \frac{80371}{4} & \frac{597401}{4} \\ \frac{5145}{2} & \frac{38243}{2} \end{pmatrix}, \begin{pmatrix} 5939 & 63664 \\ -799 & -8565 \end{pmatrix}\right\}$
$(T2)_{23}$	$\{x_2x_1^{-1}x_2^{-1}x_3x_2^{-3},$ $x_2^{-1}, x_2^3x_3^{-1}\}$	$\left\{\begin{pmatrix} -\frac{3452369}{4} & -\frac{25661603}{4} \\ \frac{237917}{2} & \frac{1768447}{2} \end{pmatrix}, \begin{pmatrix} -\frac{15}{2} & -\frac{221}{4} \\ -1 & -\frac{15}{2} \end{pmatrix}, \begin{pmatrix} 5939 & 63664 \\ -799 & -8565 \end{pmatrix}\right\}$

TABLE 5. Nielsen transformations (NTs) from the participants II

NTs	theoretical set	explicit set
$(T2)_{13}$	$\{x_2x_1^{-1}x_2^{-1}, x_2^{-1}, x_2^3x_3^{-1}\}$	$\left\{\begin{pmatrix} \frac{653}{2} & \frac{9679}{4} \\ -45 & -\frac{667}{2} \end{pmatrix}, \begin{pmatrix} -\frac{15}{2} & -\frac{221}{4} \\ -1 & -\frac{15}{2} \end{pmatrix}, \begin{pmatrix} 5939 & 63664 \\ -799 & -8565 \end{pmatrix}\right\}$
$(T1)_2$	$\{x_2x_1^{-1}x_2^{-1}, x_2, x_2^3x_3^{-1}\}$	$\left\{\begin{pmatrix} \frac{653}{2} & \frac{9679}{4} \\ -45 & -\frac{667}{2} \end{pmatrix}, \begin{pmatrix} -\frac{15}{2} & \frac{221}{4} \\ 1 & -\frac{15}{2} \end{pmatrix}, \begin{pmatrix} 5939 & 63664 \\ -799 & -8565 \end{pmatrix}\right\}$
$(T2)_{12}$	$\{x_2x_1^{-1}, x_2, x_2^3x_3^{-1}\}$	$\left\{\begin{pmatrix} -29 & -109 \\ 4 & 15 \end{pmatrix}, \begin{pmatrix} -\frac{15}{2} & \frac{221}{4} \\ 1 & -\frac{15}{2} \end{pmatrix}, \begin{pmatrix} 5939 & 63664 \\ -799 & -8565 \end{pmatrix}\right\}$
$(T1)_1$	$\{x_1x_2^{-1}, x_2, x_2^3x_3^{-1}\}$	$\left\{\begin{pmatrix} 15 & 109 \\ -4 & -29 \end{pmatrix}, \begin{pmatrix} -\frac{15}{2} & \frac{221}{4} \\ 1 & -\frac{15}{2} \end{pmatrix}, \begin{pmatrix} 5939 & 63664 \\ -799 & -8565 \end{pmatrix}\right\}$
$(T2)_{12}$	$\{x_1, x_2, x_2^3x_3^{-1}\}$	$\left\{\begin{pmatrix} -\frac{7}{2} & \frac{45}{4} \\ 1 & -\frac{7}{2} \end{pmatrix}, \begin{pmatrix} -\frac{15}{2} & \frac{221}{4} \\ 1 & -\frac{15}{2} \end{pmatrix}, \begin{pmatrix} 5939 & 63664 \\ -799 & -8565 \end{pmatrix}\right\}$
$(T1)_3$	$\{x_1, x_2, x_3x_2^{-3}\}$	$\left\{\begin{pmatrix} -\frac{7}{2} & \frac{45}{4} \\ 1 & -\frac{7}{2} \end{pmatrix}, \begin{pmatrix} -\frac{15}{2} & \frac{221}{4} \\ 1 & -\frac{15}{2} \end{pmatrix}, \begin{pmatrix} -8565 & -63664 \\ 799 & 5939 \end{pmatrix}\right\}$
$[(T2)_{32}]^3$	$\{x_1, x_2, x_3\}$	$\left\{\begin{pmatrix} -\frac{7}{2} & \frac{45}{4} \\ 1 & -\frac{7}{2} \end{pmatrix}, \begin{pmatrix} -\frac{15}{2} & \frac{221}{4} \\ 1 & -\frac{15}{2} \end{pmatrix}, \begin{pmatrix} -11 & 120 \\ 1 & -11 \end{pmatrix}\right\}$

With the knowledge of the set $M^{\pm} = \left\{\begin{pmatrix} -\frac{7}{2} & \frac{45}{4} \\ 1 & -\frac{7}{2} \end{pmatrix}, \begin{pmatrix} -\frac{15}{2} & \frac{221}{4} \\ 1 & -\frac{15}{2} \end{pmatrix}, \begin{pmatrix} -11 & 120 \\ 1 & -11 \end{pmatrix}\right\}$ the participants can reconstruct the secret easily. It is

$$a_1 := tr(M_1) = -7, \quad a_2 := tr(M_2) = -15, \quad a_3 := tr(M_3) = -22,$$

and hence it is

$$S := \sum_{j=1}^{3} \frac{1}{|a_j|} = \frac{1}{7} + \frac{1}{15} + \frac{1}{22} = \frac{589}{2310}.$$

In general we can use any free matrix group F of rank $m := \binom{n}{t-1}$ for a (t, n)-secret sharing scheme as it is described in this section. The shares can be generated by the method of D. Panagopoulos and are tuples (R_i, S_j) with $R_i \subset U$ and $S_j \subset N$.

Some other ideas for the secret S are

$$S := \prod_{i=1}^{m} |tr(M_i)| \text{ or } S := \sum_{i=1}^{m} |tr(M_i)| \text{ or}$$

$$S := \prod_{i=1}^{m} (tr(M_i))^2 \text{ or } S := \sum_{i=1}^{m} (tr(M_i))^2 \text{ or}$$

$$S := \prod_{i=1}^{\frac{m}{2}} tr([M_{2i-1}, M_{2i}]) \text{ if } m \text{ is even or } S := \sum_{i=1}^{m} tr(M_i^2).$$

7. A Variation of the Secret Sharing Scheme based on Nielsen Transformations

We now present a variation of the secret sharing protocol given in the last section. Let F be a finitely generated free group with the abstract free generating set $X := \{x_1, x_2, \ldots, x_q\}$, $q \in \mathbb{N}$, that is,

$$F = \langle X; \quad \rangle.$$

In this variation we just work with respect to the given basis elements of a finitely generated free group. For a (t, n)-secret sharing scheme the dealer chooses a Nielsen reduced set $U \subset F$ with $U = \{u_1, u_2, \ldots, u_m\}$. The u_i are given as words in X. The secret is the sum

$$S := \sum_{i=1}^{m} \frac{1}{|u_i|},$$

with $|u_i|$ the free length of the word u_i.

The dealer does a regular Nielsen transformation on the set U to get the Nielsen-equivalent set V as shown in Figure 4.

$$U := \{u_1, u_2, \ldots, u_m\}$$

$\downarrow$ regular Nielsen transformation

$$V := \{v_1, v_2, \ldots, v_m\}$$

FIGURE 4. Regular Nielsen transformation

Each participant P_i , $1 \leq i \leq n$, gets one set $R_i \subset V$, which was generated with the method of D. Panagopoulos.

If t of the n participants come together to reconstruct the secret, they combine their shares and get the set $V = \{v_1, v_2, \ldots, v_m\}$. They have to find a Nielsen-reduced set $U' := \{u'_1, u'_2, \ldots, u'_m\}$ to V. They apply Nielsen transformations in a Nielsen reducing manner as described in [**CgrRR**] and [**LS**] and get from V a Nielsen-reduced set U'. The secret is the sum

$$S = \sum_{i=1}^{m} \frac{1}{|u_i'|},$$

because for each i we have $|u_i'| = |u_j|$ for some j (see the proof of Corollary 3.1 in [**MKS**]). From U' we get U by permutations and length preserving Nielsen transformations.

REMARK 7.1. Analogously to the combinatorial secret sharing scheme, this scheme fulfills the same properties ((3) and (4)) of Section 4, because both schemes are based on the share distribution method of D. Panagopoulos.
In [**Ste**] an algorithm, using elementary Nielsen transformations, is presented which, given a finite set S of m words of some free group, returns a set S' of Nielsen reduced words such that $\langle S \rangle = \langle S' \rangle$; the algorithm runs in $\mathcal{O}(\ell^2 m^2)$, where ℓ is the maximum length of a word in S.
In this protocol, the dealer fixes the number m, hence the running time depends only on the maximum length ℓ of the words in the Nielsen-equivalent set V to the set U. Thus the participants have a running time of $\mathcal{O}(\ell^2)$ to get the set U'. The secret is then the above sum, which is computable in linear time.
As an example we may take the respective Example 6.6 but now the participants do not need the single elementary Nielsen transformations, they just have to Nielsen reduce the set V (see [**Ste**]).

8. A Secret Sharing Protocol based on the Hurwitz Equation

A generalized Hurwitz equation is given by

$$a_1 x_1^2 + \cdots + a_m x_m^2 = d x_1 \cdots x_m - k$$

where $m \geq 3$, $k \in \mathbb{N} \cup \{0\}$, $a_1, \ldots, a_m, d \in \mathbb{N}$ with $a_i | d$ for $i = 1, \ldots, m$ and the a_i are pairwise coprime. Hurwitz [**H**] considered the special case $a_1 = \cdots = a_m = 1$. For $m = 3$ the equation is often called the Markov-Rosenberger equation and is quite well understood. Such an equation occurs in connection with many different mathematical theories and problems (see [**FKMR**] and the references there).

Related to the general Hurwitz-equation is the Baragar and Umeda diophantine equation

$$ax^2 + by^2 + cz^2 = dxyz + e$$

with $a, b, c, d, e \in \mathbb{N}$ such that $a|d$, $b|d$, $c|d$ and $\gcd(a, b, c) = 1$. Further we assume that $\gcd(a, b, c) = 1$ because if $\gcd(a, b, c) = t$ then $t|e$. In [**FKMR**] we considered various aspects of the integer solutions to these equations and certain variations of them. As a by-product, we observed that the equations could be used as the basis of a (t, n)-secret sharing protocol.

The secret sharing protocol works as follows:

Consider the basic Hurwitz equation

$$x_1^2 + x_2^2 + \cdots + x_m^2 = x_1 \cdots x_m - k.$$

We now consider this equation over a field K, for example $K = \mathbb{Q}$ or $K = F$ where F is a big finite field, with $k \neq 0$.

The secret in this protocol is the element

$$S = x_1^2 + x_2^2 + \cdots + x_m^2 - x_1 \cdots x_m$$

The shares for the participants are subsets from $\{x_1, \ldots, x_m\}$. To generate these shares the method from D. Panagopoulos is again used:

(1) $m = \binom{n}{t-1}$ is the number of elements the participants need to know to reconstruct the secret, that is, they have to know the set $\{x_1, \ldots, x_m\}$.

(2) Let $A_1, A_2, \ldots, A_m$ be an enumeration of the subsets of $\{1, 2, \ldots, n\}$ with $t-1$ elements. Define n subsets $\{R_1, R_2, \ldots, R_n\}$ of $\{x_1, x_2, \ldots, x_m\}$ with the property

$$x_j \in R_i \Longleftrightarrow i \notin A_j$$

for $j = 1, \ldots, m$ and $i = 1, \ldots, n$.

(3) Each of the n participants gets one of the sets $\{R_1, R_2, \ldots, R_n\}$.

Each element x_j is exactly contained in $n - (t-1)$ subsets. Hence for each $j = 1, \ldots, m$ the element x_j is not contained in $t-1$ subsets from $R_1, \ldots, R_n$. As a consequence, x_j is in each union of t subsets. On the other hand if just $t-1$ arbitrary sets from $\{R_1, \ldots, R_n\}$ are combined, there exist a j so that the element x_j is not included in the union of this sets. If just one element x_j is absent, the participants cannot get the element S and hence cannot compute the secret.

If t of n participants come together they get by construction the set

$$\{x_1, x_2, \ldots, x_m\}$$

and hence they can calculate the secret

$$S = x_1^2 + x_2^2 + \cdots + x_m^2 - x_1 \cdots x_m.$$

REMARK 8.1. Analogously to the combinatorial secret sharing scheme, this scheme fulfills the same properties ((3) and (4)) of Section 4, because both schemes are based on the share distribution method of D. Panagopoulos.
The time complexity for the participants is as in the combinatorial secret sharing scheme. Therefore the running time is linear in m. As an example we can take the respective variation of Example 3.2. With the numbers in Example 3.2 the secret here is

$$S = 2^2 + 1^2 + 2^2 + 8^2 + 4^2 + 2^2 - 2 \cdot 1 \cdot 2 \cdot 8 \cdot 4 \cdot 2 = -163.$$

References

[A] K. E. Atkinson, *An introduction to numerical analysis*, 2nd ed., John Wiley & Sons, Inc., New York, 1989. MR1007135

[B] G. Baumslag, B. Fine, M. Kreuzer, G. Rosenberger; **A Course in Mathematical Cryptography**; De Gruyter, 2015.

[B1] G. Blakley; Safeguarding cryptographic keys; Proceedings of the National Computer Conference, 48, 313-317.

[Bo] D. Bogdanov; Foundations and properties of Shamir's secret sharing scheme; University of Tartu, available online http://www.cs.ut.ee/peeter l/teaching /seminar07k/bogdanov.pdf, 2007.

[CgrRR] T. Camps, V. große Rebel, and G. Rosenberger, *Einführung in die kombinatorische und die geometrische Gruppentheorie* (German), Berliner Studienreihe zur Mathematik [Berlin Study Series on Mathematics], vol. 19, Heldermann Verlag, Lemgo, 2008. MR2378619

[CDS] J. Cooper, D. Donovan, and J. Seberry, *Secret sharing schemes arising from Latin squares*, Bull. Inst. Combin. Appl. **12** (1994), 33–43. MR1301402

[CZ] C. Chum, X. Zhang; Improving Latin square based secret sharing schemes; Contemporary Mathematics, Volume 582 - Computational and Combinatorial Group Theory and Cryptography (AMS), December 2012, pp. 51-64.

[CZ1] C. Chum, X. Zhang; Hash function based secret sharing scheme designs; Security and Communication Networks (Wiley), 6(5), 2013, pp. 584-592.

[CFZ] C. Chum, B. Fine, X. Zhang; Shamir's Threshold Scheme and Its Enhancements; to appear.

[CFRZ] C. S. Chum, B. Fine, G. Rosenberger, and X. Zhang, *A proposed alternative to the Shamir secret sharing scheme*, Computational and combinatorial group theory and cryptography, Contemp. Math., vol. 582, Amer. Math. Soc., Providence, RI, 2012, pp. 47–50, DOI 10.1090/conm/582/11559. MR2987383

[FR] B. Fine and G. Rosenberger, *Number theory*, Birkhäuser Boston, Inc., Boston, MA, 2007. An introduction via the distribution of primes. MR2261276

[FMR] B. Fine, A. I. S. Moldenhauer, and G. Rosenberger, *A secret sharing scheme based on the closest vector theorem and a modification to a private key cryptosystem*, Groups Complex. Cryptol. **5** (2013), no. 2, 223–238. MR3245108

[FKMR] B. Fine, G. Kern-Isberner, A. I. S. Moldenhauer, and G. Rosenberger, *On the generalized Hurwitz equation and the Baragar-Umeda equation*, Results Math. **69** (2016), no. 1-2, 69–92, DOI 10.1007/s00025-015-0500-0. MR3449356

[FRS] B. Fine, G. Rosenberger, and M. Stille, *Nielsen transformations and applications: a survey*, Groups—Korea '94 (Pusan), de Gruyter, Berlin, 1995, pp. 69–105. MR1476950

[H] A. Hurwitz; Über eine Aufgabe der unbestimmten Analysis; Archiev Math. Phys. 3, 185–196, 1907.

[Leh] J. Lehner, *Discontinuous groups and automorphic functions*, Mathematical Surveys, No. VIII, American Mathematical Society, Providence, R.I., 1964. MR0164033

[LSc] M. Lohrey and S. Schleimer; Effcient computation in groups via compression; Computer Science Theory and Applications Lecture Notes in Computer Science, (4649). 249-258, 2007.

[LS] R. C. Lyndon and P. E. Schupp, *Combinatorial group theory*, Classics in Mathematics, Springer-Verlag, Berlin, 2001. Reprint of the 1977 edition. MR1812024

[MKS] W. Magnus, A. Karrass and D. Solitar; **Combinatorial Group Theory**; Wiley Interscience, New York, 1968.

[M] A. I. S. Moldenhauer; Cryptographic protocols based on inner product spaces and group theory with a special focus on the use of Nielsen transformations; Ph.D. thesis, University of Hamburg, 2016.

[P] D. Panagopoulos; A secret sharing scheme using groups; CoRR, 2010.

[Ro] J. Rotman; **The Theory of Groups**; W. C. Brown, 1984.

[Sh] A. Shamir, *How to share a secret*, Comm. ACM **22** (1979), no. 11, 612–613, DOI 10.1145/359168.359176. MR549252

[Ste] I. A. Stewart; Obtaining Nielsen Reduced Sets in Free Groups; Technical Report Series No 293, University of Newcastle upon Tyne, 1989.

[St] D. R. Stinson, *An explication of secret sharing schemes*, Des. Codes Cryptogr. **2** (1992), no. 4, 357–390, DOI 10.1007/BF00125203. MR1194776

Computer Science Department, Graduate Center, CUNY
E-mail address: cchum@gradcenter.cuny.edu

Department of Mathematics. Fairfield University, Fairfield, Connecticut 06430
E-mail address: fine@fairfield.edu

Fachbereich Mathematik, University of Hamburg, Bundesstrasse 55, 20146 Hamburg, Germany
E-mail address: anja.moldenhauer@uni-hamburg.de

Fachbereich Mathematik, University of Hamburg, Bundesstrasse 55, 20146 Hamburg, Germany
E-mail address: Gerhard.Rosenberger@math.uni-hamburg.de

Computer Science Department, College of Staten Island, Staten Island, New York
E-mail address: xiaowen.zhang@csi.cuny.edu

Contemporary Mathematics
Volume **677**, 2016
http://dx.doi.org/10.1090/conm/677/13622

A verifiable secret sharing scheme using non-abelian groups

Maggie E. Habeeb

ABSTRACT. A verifiable secret sharing scheme is a secret sharing scheme in which the participants can check that their shares are correct, solving the problem of a dishonest dealer. In this paper, we propose a verification protocol to be utilized with the Habeeb-Kahrobaei-Shpilrain [**2**] secret sharing scheme, resulting in a verifiable secret sharing scheme using non-abelian groups.

1. Introduction

Secret sharing schemes provide a method of distributing a secret among a group of n participants in such a way that a certain number of participants, say t, must work together to obtain the secret; if any less than t participants work together they gain no information about the secret. A scheme satisfying these properties is called a (t, n)-threshold scheme. In 1979 Shamir [**3**] proposed a (t, n)-threshold scheme in which the secret is an element $x \in \mathbb{Z}_p$ and polynomial interpolation is used to recover the secret. Generally, in a secret sharing scheme there is a dealer that distributes shares of the secret to the n participants. In order to distribute the secret in Shamir's scheme, the dealer begins by choosing a polynomial f of degree $t-1$ such that $f(0) = x$. Then he sends the value $x_i = f(i)$ secretly to participant P_i, giving him a point $(x_i, f(x_i))$ on the polynomial. In order for the participants to recover the secret, they use polynomial interpolation to recover the polynomial, f, and hence the secret $f(0)$. In this setting, no $t-1$ participants can gain any information about the secret while any t of them can (see [**3**] or [**4**]).

In general, the dealer is assumed to be honest. What if the dealer is dishonest? In this case, one may utilize a verifiable secret sharing scheme in which participants can check correctness of their shares and overcome the problem of the dishonest dealer. In 1987 Feldman [**1**] proposed the first efficient verifiable secret sharing (VSS) scheme. The scheme is a non-interactive verifiable secret sharing scheme; that is, a scheme in which the dealer and players need not interact with each other in order to verify their shares are correct. The scheme is based on Shamir's scheme, also requiring the dealer to choose a polynomial f known only to him. In Feldman's scheme, the dealer sends with each share a commitment of the coefficients of the polynomial f. With this commitment, each player can check the correctness of his share.

2010 *Mathematics Subject Classification.* Primary 68P25, 94A60; Secondary 20F05, 20F10.

©2016 American Mathematical Society

More recently, a secret sharing scheme using non-abelian groups has been proposed by Habeeb-Kahrobaei-Shpilrain [**2**]. In this scheme group presentations and the *word problem* are exploited. In this paper, we extend this scheme with a verification protocol where the participants can verify that their shares are, in fact, correct.

2. Feldman's VSS scheme

Here, we briefly describe Feldman's VSS scheme. Let p and q be primes and m an integer such that $p = mq + 1$. Let $g \in \mathbb{Z}_p$ be of order q. The dealer chooses a polynomial f over $\mathbb{Z}_q$ of degree t with coefficients $a_0, a_1, \ldots, a_t$. As in Shamir's scheme, the secret is the constant term a_0 in the polynomial f.

(1) The dealer broadcasts $g \in \mathbb{Z}_p$ and the commitments, $g^{a_0}, g^{a_1}, \ldots, g^{a_t}$, to the polynomial f.
(2) The dealer then secretly transmits the value $x_i = f(i) \mod q$ to participant P_i.
(3) Each participant, P_i can verify his own share by checking that the following equation holds.

$$g^{x_i} = (g^{a_0})(g^{a_1})^i(g^{a_2})^{i^2} \cdots (g^{a_t})^{i^t}$$

(4) If the equation holds, P_i broadcasts a message stating that his share is valid. If all participants verify their shares are correct, the dealing phase is completed successfully.

Upon completion of the dealing phase, the participants can then pool together to recover the secret as in Shamir's scheme. We would like to note that if an adversary could solve the discrete log problem, then he could recover the secret as the values g and g^{a_0} are public.

3. Habeeb-Kahrobaei-Shpilrain Scheme

We begin this section with the group theory required to understand the HKS secret sharing scheme. Let $X = \{x_1, x_2, \ldots, x_m\}$ be a set of letters. The free group of rank m generated by X is the set of all reduced words in the alphabet $X^{\pm 1} = \{x_1^{\pm 1}, x_2^{\pm 1}, \ldots, x_m^{\pm 1}\}$, where a word is reduced if it contains no subwords of the form $x_i x_i^{-1}$ or $x_i^{-1} x_i$. Given a set of words R from F_m the group $G = \langle X | R \rangle$ is defined as the factor group F_m/N, where N is the normal subgroup generated by R. The set X is called the set of generators of G and the set R is called the set of relators of G. Given a group presentation $G = \langle X | R \rangle$, we define the word problem as follows.

Word Problem: Given a word $w \in G$, determine if $w =_G 1$.

The scheme proposed by HKS relies on group presentations and the word problem. In order for the scheme to be feasible, the word problem must be efficiently solvable.

In [**2**], the secret is an element $x \in \mathbb{Z}_p$ and the dealer chooses a polynomial f of degree $t - 1$ such that $f(0) = x$, just as in Shamir's scheme. Additionally, the dealer computes integers $y_i = f(i)$ that need to be distributed to participants P_i, $1 \le i \le n$ and a set of generators $\{x_1, \cdots, x_m\}$ is made public. The scheme is as follows.

(1) The dealer distributes over a secure channel to each participant P_j a set of relators R_j such that each group $G_j = \langle x_1, \cdots, x_m | R_j \rangle$ has efficiently solvable word problem.

(2) The dealer then distributes over open channels k-column vectors

$$b_j = \begin{pmatrix} b_{1j} \\ b_{2j} \\ \vdots \\ b_{kj} \end{pmatrix} \text{ for } 1 \leq j \leq n \text{ of words in } x_1, \cdots, x_m \text{ to each participant.}$$

The column vectors b_j are chosen so that after replacing them by bits ("1" if $b_{ij} = 1$ in the group G_j and "0" otherwise), the resulting bit column represents the integer y_j.

(3) Participant j then checks if each word $b_{ij} = 1$ in his/her group G_j, obtaining a binary representation of the number y_j.

(4) Each participant now has a point $f(i) = y_i$ of the polynomial. Using polynomial interpolation, any t participants can now recover the polynomial f, and hence the secret $f(0)$.

3.1. Platform Group. In [**2**], small cancellation groups were proposed as the platform group for the above protocol. We begin with some definitions.

Let $F(X)$ be the free group on generators $X = \{x_1, \ldots, x_n\}$. A word $w(x_1, \ldots, x_n) = x_{i_1}^{\epsilon_1} \cdots x_{i_n}^{\epsilon_n}$ where $\epsilon_i = \pm 1$ for $1 \leq i \leq n$ and x_{i_j} not necessarily distinct is *cyclically reduced* if $x_{i_j}^{\epsilon_j} \neq x_{i_{j+1}}^{-\epsilon_{j+1}}$ for $1 \leq j \leq n-1$ and $x_{i_1}^{\epsilon_1} \neq x_{i_n}^{-\epsilon_n}$.

A set R is called *symmetrized* if it contains only cyclically reduced words and is closed under taking cyclic permutations and inverses. Given a set R of relators, a non-empty word $w \in F(X)$ is called a *piece* if there are two distinct relators $r_1, r_2 \in R$ such that w is an initial segment of both r_1 and r_2; that is, $r_1 = wv_1$ and $r_2 = wv_2$ for some $v_1, v_2 \in F(X)$ and there is no cancellation between w and v_1 or w and v_2.

In the definition below, $|w|$ denotes the lexicographic length of a word w.

DEFINITION 3.1. Let R be a symmetrized set of relators, and let $0 < \lambda < 1$. A group $G = \langle X | R \rangle$ with the set X of generators and the set R of relators is said to satisfy the small cancellation condition $C'(\lambda)$ if for every $r \in R$ such that $r = wv$ and w is a piece, one has $|w| < \lambda |r|$. In this case, we say that G belongs to the class $C'(\lambda)$.

Groups satisfying the the small cancellation property were proposed because groups in the class $C'(\frac{1}{6})$ have the word problem solvable in linear time by Dehn's algorithm, which is necessary for the feasibility of the scheme.

4. Verification Protocol

As noted in the previous section, the scheme proposed by [**2**] has sets of relators as the shares for each participant. A cheating dealer may give the participants incorrect shares, causing the participants to reconstruct the secret incorrectly. Here, we suggest a method motivated by Feldman's scheme that allows the participants to check that their shares are valid.

The verification protocol is as follows.

(1) The dealer makes public a generator $g \in \mathbb{Z}_p^*$ and a set of generators $\{x_1, x_2, \ldots, x_m\}$. Additionally, the dealer randomly chooses $y_i \in \mathbb{Z}_p^*$, for $1 \leq i \leq n$.

(2) The dealer distributes over a secure channel to each participant P_j a set of relators R_j (participant P_j's secret share) such that each group $G_j = \langle x_1, \cdots, x_m | R_j \rangle$ has efficiently solvable word problem.
(3) The dealer then distributes over open channels k-column vectors $b_j = \begin{pmatrix} b_{1j} \\ b_{2j} \\ \vdots \\ b_{kj} \end{pmatrix}$ for $1 \leq j \leq n$ of words in $x_1, \cdots, x_m$ to each participant. The column vectors b_j are chosen so that the resulting bit column represents the integer y_j. The dealer also distributes commitments, g^{y_j}, for each participant P_j.
(4) Participant j then checks if each word $b_{ij} = 1$ in his/her group G_j, obtaining a bit representation of a number z_j. He then verifies that $g^{z_j} = g^{y_j}$.
(5) If the equation $g^{z_j} = g^{y_j}$ holds, P_j broadcasts a message stating that his share is correct. If all participants verify their shares are correct, the verification protocol is completed successfully.

Upon completion of the verification protocol, when one is ready to share a secret, the dealer may continue from step (2) in the HKS protocol. Hence, the participants can check their shares are correct and a secret can be distributed at a later time.

We would like to note, as in Feldman's protocol, if an adversary could solve the discrete log problem, then some information about each participant's share can be gained. In this case, the shares are relators of the individual's group. If one could solve the discrete log problem, then upon recovering y_j from g^{y_j} the adversary would have the corresponding bit column, and know which words would be trivial in the group. As proposed in [**2**], we suggest small cancellation groups as a platform for the protocol to ensure that each participant has a group presentation, $\langle G|R \rangle$, with an efficiently solvable word problem.

5. Conclusion

We have proposed a verification protocol utilizing the word problem to be used with the HKS secret sharing scheme. The verification protocol allows participants to check that their shares are correct, protecting against a dishonest dealer. Motivated by Feldman's scheme, the dealer simply sends a commitment to the relators sent to each participant.

Acknowledgement

The author is grateful to Professor Delaram Kahrobaei for her continuous support and guidance.

References

[1] P. Feldman. A practical scheme for non-interactive verifiable secret sharing. *Proceedings of the 28th Annual Symposium on Foundations of Computer Science*, SFCS '87, Washington, D.C.:427–438, 1987.
[2] M. Habeeb, D. Kahrobaei, and V. Shpilrain, *A secret sharing scheme based on group presentations and the word problem*, Computational and combinatorial group theory and cryptography, Contemp. Math., vol. 582, Amer. Math. Soc., Providence, RI, 2012, pp. 143–150, DOI 10.1090/conm/582/11557. MR2987392
[3] A. Shamir, *How to share a secret*, Comm. ACM **22** (1979), no. 11, 612–613, DOI 10.1145/359168.359176. MR549252

[4] D. R. Stinson, *Cryptography*, 3rd ed., Discrete Mathematics and its Applications (Boca Raton), Chapman & Hall/CRC, Boca Raton, FL, 2006. Theory and practice. MR2182472

California University of Pennsylvania
E-mail address: Habeeb@CalU.edu

Contemporary Mathematics
Volume **677**, 2016
http://dx.doi.org/10.1090/conm/677/13623

Non-associative public-key cryptography

Arkadius Kalka

ABSTRACT. We give an overview on non-associative public-key cryptography (PKC) which generalizes the concept of non-commutative PKC. In particular, we introduce a generalized Anshel-Anshel-Goldfeld (AAG) key establishment protocol (KEP) for magmas. Left selfdistributive systems appear in a natural special case of a generalized AAG-KEP for magmas, and we discuss concrete realizations using f-conjugacy in groups and shifted conjugacy in braid groups and the advantages of our schemes compared with the classical AAG-KEP.

1. Introduction

Currently public key cryptography relies mainly on a few number-theoretic problems, namely integer factorization [**RSA78**] and the computation of discrete logarithms in $\mathbb{Z}_p^\times$ and over elliptic curves. The systems based on these problems remain unbroken. Nevertheless, after the advent of quantum computers, systems like RSA [**RSA78**] and its variants (e.g. [**Ra79**]), Diffie-Hellman (DH) [**DH76**], ElGamal [**El85**] and ECC [**Mi85, Ko87**] will be broken easily [**Sh97, PZ03**].

Under the label Post Quantum Cryptography, there have been several efforts to develop new cryptographic primitives which may also serve for the post quantum computer era. Here we focus on key establishment protocols (KEPs) as cryptographic primitives, because they are the most important and the hardest to construct. Note that, using hash functions, it is easy to build public key encryption schemes from KEPs.

One approach became known as *non-commutative cryptography*. Recall that the involved algebraic structures in the number-theoretic systems are commutative groups and rings. In non-commutative cryptography these are replaced by non-commutative groups and rings, and we consider computational problems therein. One may say that, roughly, the discrete logarithm problem is replaced by the conjugacy problem and its variants. After some precursors, in particular [**WM85**], non-commutative cryptography was mainly established in a few seminal papers around the turn of the millenium [**AAG99, KL+00, CK+01**]. Of particular importance is the ingenious Anshel-Anshel-Goldfeld (AAG) Commutator KEP which only exists in the non-commutative setting, while the systems in [**KL+00, CK+01**]

2010 *Mathematics Subject Classification.* Primary 20N02, 20F36.
Key words and phrases. Non-commutative cryptography, key establishment protocol, magma (grupoid), left selfdistributive system, braid group.

©2016 American Mathematical Society

may be considered as straightforward non-commutative analogues of the classical DH-KEP.

Since they admit efficiently computable normal forms and a supposedly hard conjugacy problem, braid groups were explicitly suggested as platform groups for these systems. Nevertheless, explicit specifications of these systems in braid groups as well as most other non-commutative cryptosystems have been broken in the first decade. This led to a decline of interest in non-commutative cryptography inside the main cryptographic community. A revival of non-commutative cryptography may be achieved by means of research in one of the following two directions.

The first approach is to stick with the suggested protocols and search for better platform groups. One may even keep braid groups as platforms and search for families of hard instances of the conjugacy problem that can be efficiently generated. Note that the main reason why braid-based cryptosystems have been broken is the fact that "randomly" generated keys turned out to be a very bad choice. This situation is quite typical for public-key cryptography. Consider, for example, the familiar RSA scheme where the keys have to be chosen with care.

Another approach is to construct new or generalized non-commutative cryptosystems which are based on other or supposedly harder computational problems. In [**Ka07**, **KaT13a**, **KaT13b**] we pursued the latter approach. In particular, we broadened the scope of non-commutative cryptography as we go beyond non-commutative, associative binary operations - we utilize non-associative binary operations, i.e. *magmas* (often also called *grupoids*), thus hoping to establish the field of *non-associative public-key cryptography.* In particular, we generalize the AAG-KEP for monoids to a general AAG-KEP for magmas and give a lot of examples.

Summary. The main purpose of this paper is to popularize the notion of non-associative public-key cryptography by giving a survey on non-associative KEPs. We also provide a general framework for non-associative and non-commutative KEPs by utilizing the unifying approach that stems from the general AAG-KEP for magmas. We argue for the superiority of the non-associative schemes introduced in section 4 compared to classical non-commutative AAG commutator KEP. In our opinion the field of non-commutative cryptography lacked over the last years supply of new innovative cryptosystems. We hope that non-associative cryptography will contribute to revived interest in non-commutative cryptography.

Outlook. However, this is not the end, rather the beginning of the story of non-associative cryptography.

In [**KaT13a**], by introducing a small asymmetry in the non-associative AAG protocol for magmas, we constructed non-associative KEPs for all LD- and multi-LD-systems (in general: sets with distributive operations). We consider the systems and instances given in [**KaT13a**], and in particular the one in [**KaT13b**] as much more practical and interesting than the one given in this paper. In particular, since these systems work for all LD- and multi-LD-systems, they deploy two further advantages. First, we may consider encryption functions using iterated multiplication (in the magma) from the left. Therefore, in order to obtain the secret key an attacker has to solve an iterated f- or shifted conjugacy problem. Second, for a given (partial) multi-LD-system it turns out that even the used operations can be hidden, i.e., they are part of the secret key.

However, since these systems were described sufficiently detailed in [**KaT13a**, **KaT13b**], we do not discuss them here further. In this survey paper we only describe simple examples of non-associative KEPs that were so far not published in any journal, i.e. examples from my thesis [**Ka07**] enriched with some more examples.

Historical remarks. Non-associative structures, in particular quasigroups seem to have a long history in cryptography. For an overview on cryptographic applications of quasigroups and Latin squares, see [**Shc09, GS10, Shc12**]. In particular, we mention the work of Denes and Keedwell [**DK74, DK91, DK92, DK02**]. Nevertheless, except for authentication schemes and zero-knowledge protocols, most of these applications are in classical (i.e. symmetric key) cryptography. The earliest quasigroup-based public-key cryptosystem that we are aware of is due to Koscielny and Mullen [**KM99**].

Non-associative cryptography that goes beyond quasigroups, in particular, the generalized AAG-KEP for magmas were introduced by the author in his PhD thesis in 2007 [**Ka07**]. During a postdoctoral stay at the Bar-Ilan University, hosted by M. Teicher and B. Tsaban, we had the opportunity to refine and improve our non-associative systems. In particular, we developed the non-associative KEPs for all distributive systems [**KaT13a, KaT13b**].

Other non-associative cryptosystems that came up during the last few years include [**GMK08, MZ12**].

Outline. The paper is structured as follows. In section 2 we emphasize the important and integrating role of the AAG protocol in non-commutative and commutative cryptography. In particular, we introduce a generalized notion of AAG-KEP for monoids (section 2.1), and we show that not only the AAG commutator KEP for groups [**AAG99**] (section 2.2), but also the Ko-Lee et al. protocol, the group Diffie-Hellman protocol (section 2.3), and even the classical DH-KEP (section 2.5) are special instances of that generalized AAG scheme. Furthermore, we also subsume the Sakalauskas, Tvarijonas and Raulynaitis KEP(STR-KEP), a natural hybrid of the classical DH-KEP and the Ko-Lee-KEP, as a further instance (section 2.6).

The main part of this paper is contained in the sections 3 and 4. In particular, in section 3.1 we extend the generalized AAG-KEP from monoids to magmas. Here finitely generated submonoids are replaced by f.g. submagmas, and Alice and Bob know their secret key submagma elements as products of the generators, including planar rooted binary trees describing the bracket structure of such products. The first example of an instance of the generalized AAG-KEP for magmas is a non-associative KEP based on simultaneous double coset problem a (see sections 3.2).

The most interesting and natural instances of the generalized AAG-KEP for magmas come from left-selfdistributive (LD) systems and their generalizations (section 4). In section 4.1 we introduce LD- and multi-LD-systems with f-conjugacy in groups and shifted conjugacy in braid groups as key examples for LD-operations. The nonassociative AAG f-commutator KEP (section 4.2) and the AAG shifted commutator KEP (section 4.3) are discussed as major examples. We note that for some instances we may even drop the simultaneity of the underlying base problem, namely the shifted conjugacy problem, because here submagmas generated by one element still have a rich and complicated structure and a hard membership problem.

In section 5 we discuss generalizations, like AAG-schemes over non-associative magmas, open problems and further work.

Implementation. Most of our non-associative KEPs have been implemented using MAGMA [**BCP97**], in particular the most advanced systems described in [**KaT13b**] (see [**KaT16**]).

2. Anshel-Anshel-Goldfeld key establishment

2.1. AAG key establishment protocol for monoids. Here we use and describe a slightly generalized version of the *AAG key establishment protocol for monoids* [**AAG99**]. Though it is easy to introduce further generalizations, the following notion will suffice for our purposes.

For this *general AAG key establishment protocol for monoids* we need sets S_1, S_2, two feasible monoids $(M, \cdot_M), (N, \cdot_N)$, and functions

$$\beta_i : S_i \times M \longrightarrow N, \quad \gamma_i : S_i \times N \longrightarrow N, \quad \pi_i : S_i \longrightarrow M \quad (i = 1, 2)$$

which satisfy the following conditions:

(1) For $i = 1, 2$, $\beta_i(x, \cdot) : M \to N$ is for all $x \in S_i$ a monoid homomorphism, i.e.

$$\forall x \in S_i, y_1, y_2 \in M : \quad \beta_i(x, y_1 \cdot_M y_2) = \beta_i(x, y_1) \cdot_N \beta_i(x, y_2).$$

(2) For $i = 1, 2$, it is, in general, not feasible to determine a secret $x \in S_i$ from the knowledge of $y_1, y_2, \ldots, y_k \in M$ and $\beta_i(x, y_1), \ldots, \beta_i(x, y_k) \in N$.

(3) For all $x \in S_1$ and $y \in S_2$: $\quad \gamma_1(x, \beta_2(y, \pi_1(x))) = \gamma_2(y, \beta_1(x, \pi_2(y)))$.

Furthermore Alice and Bob select elements $s_1, \ldots, s_m, t_1, \ldots, t_n \in M$. These elements are public, and they define submonoids $S_A = \langle s_1, \ldots, s_m \rangle$ and $S_B = \langle t_1, \ldots, t_n \rangle$ of M. Now Alice and Bob have to perform the following protocol steps:

1. Alice generates an element $a \in S_1$ such that $\pi_1(a) \in S_A$, and Bob chooses a $b \in S_2$ s.t. $\pi_2(b) \in S_B$.
2. Alice computes the elements $\beta_1(a, t_1), \ldots, \beta_1(a, t_n)$ and publicly announces this list. This list is her public key. Analogously Bob computes the elements $\beta_2(b, s_1), \ldots, \beta_2(b, s_m)$ and publishes this list.
3. Knowing that $\pi_1(a) = r_1 \cdots r_k$ with $r_i \in \{s_1, \ldots, s_m\}$ for some $k \in \mathbb{N}$ and $i = 1, \ldots, k$, Alice computes from Bob's public key $\beta_2(b, \pi_1(a)) = \beta_2(b, r_1 \cdots r_k) \stackrel{(1)}{=} \beta_2(b, r_1) \cdots \beta_2(b, r_k)$.

   And Bob, knowing $\pi_2(b) = u_1 \cdots u_{k'}$ with $u_j \in \{t_1, \ldots, t_n\}$ for some $k' \in \mathbb{N}$ and $j = 1, \ldots, k'$, computes from Alice's public key $\beta_1(a, \pi_2(b)) = \beta_1(a, u_1 \cdots u_{k'}) \stackrel{(1)}{=} \beta_1(a, r_1) \cdots \beta_1(a, u_{k'})$.
4. Alice computes $K_A = \gamma_1(a, \beta_2(b, \pi_1(a)))$, and symmetrically Bob computes $K_B = \gamma_2(b, \beta_1(a, \pi_2(b)))$.

Because of (3), the equivalence $K_A = K_B$ holds in the monoid N. Now any key extractor ϕ defined on the monoid N provides a shared key $\phi(K_A)$. Here a *key extractor* is any effectively computable function from a monoid to any keyspace[1] (compare with [**AAG03**]). A key extractor may be given by a normal form algorithm in the monoid, but in general the key extractor map needs not be injective. Anyway, for brevity we will refer in the sequel to the monoid element $K := K_A \in N$ as the shared key.

Alice's secret key is the pair $(a, I) \in S_1 \times \{1, \ldots, m\}^k$ where I denotes the index vector $(I_1, \ldots, I_k)$ such that $r_i = s_{I_i}$ for $i = 1, \ldots, k$, i.e., I determines a

[1] A standard key space is the semigroup of bitstrings $\{0, 1\}^*$.

word over $\{s_1, \ldots, s_m\}$ representing $\pi_1(a) \in S_A$. Analogously Bob's secret key is a pair $(b, J) \in S_2 \times \{1, \ldots, n\}^{k'}$.

The AAG key agreement scheme is formulated in a too general manner to be applied. For practical purposes we have to specify the sets S_1, S_2, the monoids M, N and the functions β_i, γ_i, π_i for $i = 1, 2$.

Setting $S_1 = S_2 = M$, $\beta_1 = \beta_2$ and $\pi_1 = \pi_2 = \mathrm{id}_M$, we recover the original *AAG key establishment protocol for monoids* [**AAG99**] as a special case of this generalized notion.

2.2. AAG commutator KEP for groups. The *AAG commutator KEP for groups* [**AAG99**] is determined by the following specifications: Let $S_1 = S_2 = M = N = G$ be a group, and S_A and S_B are assumed to be subgroups of G^2. We have $\pi_1 = \pi_2 = \mathrm{id}_G$ and $\beta_1 = \beta_2 =: \beta$. The functions $\beta, \gamma_1, \gamma_2 : G^2 \to G$ are defined by

$$\beta(x, y) = x^{-1}yx, \quad \gamma_1(x, y) = x^{-1}y, \quad \gamma_2(x, y) = y^{-1}x.$$

Note that the shared key is the commutator

$$\begin{aligned} K_A &= \gamma_1(a, \beta_2(b, \pi_1(a))) = \gamma_1(a, b^{-1}ab) = a^{-1}(b^{-1}ab) = [a, b] \\ &= (a^{-1}ba)^{-1}b = \gamma_2(b, a^{-1}ba) = \gamma_2(b, \beta_1(a, \pi_2(b))) = K_B. \end{aligned}$$

If the group elements are given by representative words (over some alphabet of generators) as usual in combinatorial group theory, then multiplication is defined by simple concatenation of words. Therefore Alice and Bob have to publish the words representing the elements $\beta(a, t_i) = a^{-1}t_ia$ and $\beta(b, s_j) = b^{-1}s_jb$ in a disguised form. Therefore the question, whether one can efficiently disguise elements by using defining relations [**SZ06**], is very important for any platform group. One way is to use efficiently computable normal forms.

Such efficiently computable normal forms exist in many groups, e.g., in braid groups. Furthermore, the conjugator search, i.e. determining x from $\beta(x, y) = x^{-1}yx$, was assumed to be hard in braid groups. Therefore, Anshel, Anshel and Goldfeld suggested braid groups as platform groups for the AAG commutator KEP [**AAG99**].

2.3. Group Diffie-Hellman key establishment. In 2000 Ko, Lee, Cheo, Han, Kang and Park introduced a new key agreement scheme based on braid groups [**KL+00**]. Here we describe a generalized version of this KEP [**CK+01**] for a general platform group G. Since this KEP is a non-abelian generalization of the classical Diffie-Hellman (DH) key agreement in the abelian group $\mathbb{F}_p^\times$ [**DH76**], we call it the *group Diffie-Hellman (DH) key establishment protocol.* Let (A_1, B_1) and (A_2, B_2) be two pairs of public, commuting subgroups of a given group G, i.e., we have $[A_i, B_i] = 1$ for $i = 1, 2$. Furthermore, let x be a "generic" element in G. Alice and Bob have to perform the following protocol steps.

1. Alice generates her secret key $(a_1, a_2) \in A_1 \times A_2$. And Bob selects his private key $(b_1, b_2) \in B_1 \times B_2$.
2. Alice computes $y_A = a_1xa_2$ and sends it to Bob. And Bob computes $y_B = b_1xb_2$ and submits it to Alice.
3. Alice receives y_B and computes $K_A := a_1y_Ba_2$. And Bob receives y_A and computes the shared key

$$K_B := b_1y_Ab_2 = b_1(a_1xa_2)b_2 = a_1(b_1xb_2)a_2 = a_1y_Ba_2 = K.$$

[2]Now r_i and u_j are elements from $\{s_1^{\pm 1}, \ldots, s_m^{\pm 1}\}$ and $\{t_1^{\pm 1}, \ldots, t_n^{\pm 1}\}$, respectively.

For $a_1 = a_2^{-1}$ and $b_1 = b_2^{-1}$ we obtain the original *Ko-Lee et al. protocol* [**KL+00**]. In [**AAG03**] it is shown that the Ko-Lee et al. protocol may be seen also as an instance of the Anshel-Anshel-Goldfeld KEP for monoids. The following proposition is a straightforward generalization of that claim from [**AAG03**] using the same proof idea[3].

PROPOSITION 2.1.. *The group Diffie-Hellman key establishment protocol is an instance of the general Anshel-Anshel-Goldfeld KEP for monoids.*

PROOF. Here we set $S_1 = A_1 \times A_2$, $S_2 = B_1 \times B_2$, $M = G$ and $N = \{g_1 x g_2 \mid (g_1, g_2) \in G^2\}$. On N we define the following "forgetful" binary opperation:

$$1 \cdot_N u = u \cdot_N 1 = u \quad (\forall x \in N) \quad \text{and} \quad u \cdot_N v = u \quad (\forall u, v \in N, u \neq 1, v \neq 1).$$

This turns N into a monoid. We define the functions $\beta_1 : (A_1 \times A_2) \times G \to N$ and $\beta_2 : (B_1 \times B_2) \times G \to N$ by

$$\beta_1((u_1, u_2), v) = \beta_2((u_1, u_2), v) = u_1 x u_2.$$

Then condition (1) is satisfied obviously. Indeed, given the forgetful operation on N, any constant function $\beta(u) : G \to N$ provides a monoid homomorphism. Further, condition (2) holds, because it is assumed to be hard for the group G to determine $a = (a_1, a_2) \in A_1 \times A_2$ from $\beta((a_1, a_2), b) = a_1 x a_2$. The computational problem is a search version of the *Double Coset Problem* or *Decomposition Problem* (DCP) - see also section 2.4. We define the functions $\gamma_1 : (A_1 \times A_2) \times N \to N$ and $\gamma_2 : (B_1 \times B_2) \times N \to N$ by

$$\gamma_1((u_1, u_2), v) = \gamma_2((u_1, u_2), v) = u_1 v u_2.$$

Then (3) is satisfied, because we have for all $a = (a_1, a_2) \in A_1 \times A_2$, $b = (b_1, b_2) \in B_1 \times B_2$ (recall $[A_i, B_i] = 1$ for $i = 1, 2$):

$$\begin{aligned} \gamma_1(a, \beta_2(b, \pi_1(a))) &= \gamma_1((a_1, a_2), b_1 x b_2) = a_1(b_1 x b_2) a_2 \\ = b_1(a_1 x a_2) b_2 &= \gamma_2((b_1, b_2), a_1 x a_2) = \gamma_2(b, \beta_1(a, \pi_2(b))). \end{aligned}$$

This proves that the conditions (1)-(3) are fulfilled. It remains to show that the protocol steps 1.-3. of the DH-KEP are specializations of the protocol steps 1.-4. of the general AAG-KEP. Set $S_A = S_B = \langle x \rangle$ and define, for $i = 1, 2$, $\pi_i : S_i \longrightarrow M$ by $u \mapsto x$, i.e., π_1, π_2 are constant functions.

1. Alice generates an element $a = (a_1, a_2) \in S_1 = A_1 \times A_2$ such that $\pi_1(a) = x \in \langle x \rangle = S_A$, and Bob chooses a $b = (b_1, b_2) \in S_2 = B_1 \times B_2$ s.t. $\pi_2(b) = x \in \langle x \rangle = S_B$.
2. Alice computes the element $\beta_1(a, x) = a_1 x a_2 = y_A$ and publicly announces this element. This element is her public key. Analogously Bob computes the element $\beta_2(b, x) = b_1 x b_2 = y_B$ and publishes this element.
3. Knowing that $\pi_1(a) = x$, Alice computes from Bob's public key $\beta_2(b, \pi_1(a)) = \beta_2(b, x) = b_1 x b_2$. Indeed, this computation is trivial, because here $\beta_2(b, \pi_1(a))$ is Bob's public key.

   And Bob, knowing $\pi_2(b) = x$, computes from Alice's public key $\beta_1(a, \pi_2(b)) = \beta_1(a, x) = a_1 x a_2$. Also this computation is trivial. Therefore, here protocol step 3 becomes redundant.
4. Alice computes $K_A = \gamma_1(a, \beta_2(b, \pi_1(a))) = a_1 y_B a_2$, and symmetrically Bob computes $K_B = \gamma_2(b, \beta_1(a, \pi_2(b))) = b_1 y_A b_2$. □

---

[3]It is also a corrected reformulation of Proposition 5.1 in [**Ka07**].

We have proven that the group DH-KEP is a special case of the AAG-KEP for monoids. Nevertheless, not every special case is obvious. Indeed, the group DH-KEP does not use the homomorphy property (1) at all. Therefore step 3 in this specification of the general AAG-KEP (see proof above) became trivial. This observation motivates us to introduce the following somehow informal notion:

DEFINITION 2.2. We call a key establishment protocol *AAG-like* if it is an instance of the general AAG-KEP *and* it utilizes property (1) in a non-trivial way.

According to this notion, and contrary to the AAG commutator KEP, the group Diffie-Hellman KEP is *not* an AAG-like KEP, though it can be formally considered as an instance of the general AAG-KEP (see Proposition 2.1).

2.4. Base Problems. The following search problems are related with the group based protocols from the previous sections. Let G be a group.

CSP (*Conjugacy Search Problem*):
INPUT: $(s, s^x) \in G^2$. (s^x denotes $x^{-1}sx$.)
OBJECTIVE: Find $x' \in G$ such that $s^{x'} = s^x$.

l-simCSP (*l-Simultaneous Conjugacy Search Problem*):
INPUT: $\{(s_i, s_i^x) \in G^2 | i = 1, \ldots, l\}$.
OBJECTIVE: Find $x' \in G$ such that $s_i^{x'} = s_i^x \quad \forall i = 1, \ldots, l$.

subCSP (*Subgroup Conjugacy Search Problem*): Let H be a subgroup of G.
INPUT: $(s, s^x) \in G^2$ with $x \in H \leq G$.
OBJECTIVE: Find $x' \in H$ such that $s^{x'} = s^x$.

l-ssCSP (*l-Simultaneous Subgroup Conjugacy Search Problem*):
INPUT: $\{(s_i, s_i^x) \in G^2 | i = 1, \ldots, m\}$ with $x \in H \subset G$.
OBJECTIVE: Find $x' \in H$ such that $s_i^{x'} = s_i^x \quad \forall i = 1, \ldots, l$.

AAGP (*Anshel-Anshel-Goldfeld Problem*): Let $A = \langle a_1, \ldots, a_k \rangle$ and $B = \langle b_1, \ldots, b_m \rangle$ be two f.g. subgroups of G.
INPUT: $\{(a_i, a_i^y) \in G^2 | i = 1, \ldots, k\} \cup \{(b_j, b_j^x) \in G^2 | j = 1, \ldots, m\}$ with $x \in A$ and $y \in B$.
OBJECTIVE: Find $K := x^{-1}y^{-1}xy$.

KLP (*Ko-Lee Problem - a Diffie-Hellman version of the GCSP or CDP*): Let $A, B \leq G$ with $[A, B] = 1$.
INPUT: $(s, s^x, s^y) \in G^3$ with $x \in A, y \in B$.
OBJECTIVE: Find $K := x^{-1}y^{-1}sxy$.

DCP (*Double Coset or Decompositon Problem*): Let $H_1, H_2 \leq G$.
INPUT: $(s, x_1 s x_2) \in G^2$ for some $x_1 \in H_1$ and $x_2 \in H_2$.
OBJECTIVE: Find $(x_1', x_2') \in H_1 \times H_2$ such that $x_1' s x_2' = x_1 s x_2$.

CDP (*Conjugacy Decompositon Problem*):
INPUT: $(s, s^x) \in G^2$ with $x \in H \leq G$.
OBJECTIVE: Find $(x_1', x_2') \in H^2$ such that $x_1' s x_2' = s^x$.

DH-DCP (*Diffie-Hellman Decompositon Problem*): A_1, A_2, B_1, B_2 subgroups of G such that $[A_i, B_i] = 1$ for $i = 1, 2$.
INPUT: $(s, x_1 s x_2, y_1 s y_2) \in G^3$ with $x_1 \in A_1$, $x_2 \in A_2$, $y_1 \in B_1$, $y_2 \in B_2$.
OBJECTIVE: Find $K := x_1 y_1 s x_2 y_2$.

Indeed, the AAG commutator KEP, the Ko-Lee protocol and the group DH-KEP are based on the AAGP, KLP and DH-DCP, respectively.
Now, let P_1, P_2 be two computational problems. We say P_1 is harder than P_2 or P_1 implies P_2, written $P_1 \to P_2$, if a P_1-oracle provides a solution to problem P_2.

PROPOSITION 2.3.. *We have the following hierarchy of search problems:*

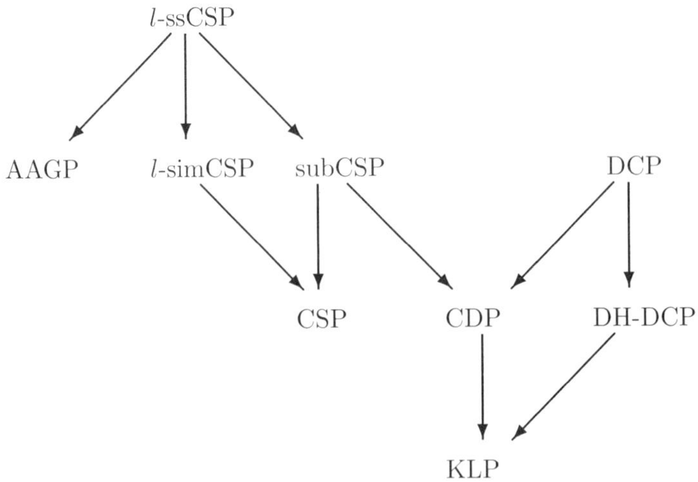

PROOF. Most of the sketched implications are obvious consequences of the definitions. We just prove CDP $\to$ KLP and l-ssCSP $\to$ AAGP:

(1) (see [**KL+00**]) The input is a triple $(s, s^x, s^y) \in G^3$ with $x \in A, y \in B$, and $A, B \subset G$ with $[A, B] = 1$. A CDP-oracle provides $(x_1, x_2) \in A^2$ with $x_1 s x_2 = s^x$. Now we can compute the shared key

$$x_1 s^y x_2 = x_1 y^{-1} s y x_2 = y^{-1}(x_1 s x_2) y = y^{-1}(x^{-1} s x) y = K.$$

(2) Here the input is $\{(a_i, a_i^y) \in G^2 | i = 1, \ldots, k\} \cup \{(b_j, b_j^x) \in G^2 | j = 1, \ldots, m\}$ with $x \in A = \langle a_1, \ldots, a_k \rangle$ and $y \in B = \langle b_1, \ldots, b_m \rangle$. A m-ssCSP-oracle provides a $x' \in A$ with $x'^{-1} b_j x' = b_j^x$ for all $j = 1, \ldots, m$. And a k-ssCSP-oracle provides a $y' \in B$ with $y'^{-1} a_i y' = a_i^y$ for all $i = 1, \ldots, k$. Now, since $x'^{-1} b_j x' = b_j^x \Leftrightarrow [x' x^{-1}, b_j] = 1 \, \forall j$, we have $x' = c_b x$ for some $c_b \in C_G(B)$. Here $C_G(B)$ denotes the intersection of all centralizers $C_G(b_j)$ of b_j $(j = 1, \ldots, m)$ in G. Analogously, we can write $y' = c_a y$ with $c_a \in C_G(A) = \bigcap_{i=1}^k C_G(a_i)$.

Now, $x' \in A$ implies $c_b \in A$. Therefore we have $[c_a, c_b] = 1$, and we can compute the shared key

$$\begin{aligned} K' := x'^{-1} y'^{-1} x' y' = (c_b x)^{-1} (c_a y)^{-1} c_b x c_a y &= x^{-1} c_b^{-1} y^{-1} c_a^{-1} c_b x c_a y \\ = x^{-1} y^{-1} c_b^{-1} c_a^{-1} c_b c_a x y &\stackrel{!}{=} x^{-1} y^{-1} x y = K. \quad \square \end{aligned}$$

We see, that solving the classical CSP is insufficient for breaking the AAG protocol or the Ko-Lee protocol. Furthermore, it is, in general, insufficient to solve the l-simCSP to obtain the shared key K of the AAG protocol [**SU06**]:

Let $x' = c_b x \in G$ and $y' = c_a y \in G$ with $c_a \in C_G(A), c_b \in C_G(B)$ be the output of a m-simCSP-oracle and a k-simCSP-oracle, respectively. Then we have $K' = K$ if and only if $[c_a, c_b] = 1$. A necessary condition for $[c_b, c_a] \neq 1$ is $c_b \notin A \wedge c_a \notin B$, which implies $x' \notin A \wedge y' \notin B$. Otherwise, if $x' \notin A$, but $y' \in B$ (or vice versa), the adversary gets $K' = K$.

Alternatively, the adversary could solve the simCSP and the

MSP (*Membership Search Problem*):

INPUT: $x, a_1, \ldots, a_k \in G$.

OBJECTIVE: Find an expression of x as a word in $a_1, \ldots, a_k$ (notation $x = x(a_1, \ldots, a_k)$), if it exists, i.e. if $x \in \langle a_1, \ldots, a_k \rangle$.

to break the AAG key agreement scheme [**SU06**]:

If a m-simCSP-oracle outputs a $x' = c_b x \in A$, then the MSP-oracle provides the word expression $x'(a_1, \ldots, a_k)$. Now the adversary can compute the shared key

$$x'^{-1} x'(a_1^y, \ldots, a_k^y) = x'^{-1} x'^y = (x^{-1} c_b) y^{-1} (c_b x) y = [x, y] = K.$$

But we have shown above, that it is not necessary to solve the MSP.

2.5. Diffie-Hellman key establishment protocol. Recall the classical *Diffie-Hellman key establishment protocol* [**DH76**]. Let G be a cyclic group and x an element of big order in G. Alice and Bob have to perform the following protocol steps.

1. Alice chooses a $k \in \mathbb{Z}$, computes $y_A = x^k$, and sends it to Bob. And Bob chooses a $l \in \mathbb{Z}$, computes $y_B = x^l$, and submits it to Alice.
2. Alice receives y_B and computes $K_A := y_B^k$. And Bob receives y_A and computes the shared key $K_B := y_A^l = (x^k)^l = (x^l)^k = y_B^k = K_A$.

PROPOSITION 2.4.. *The Diffie-Hellman key establishment protocol is an instance of the Anshel-Anshel-Goldfeld KEP for monoids. Furthermore, it is a AAG-like KEP.*

PROOF. Here we set $S_1 = S_2 = \mathbb{Z}$ and $M = N = S_A = S_B = \langle x \rangle$. For $i = 1, 2$, we define the functions $\beta_i, \gamma_i : \mathbb{Z} \times \langle x \rangle \to \langle x \rangle$ and $\pi_i : \mathbb{Z} \to \langle x \rangle$ by

$$\beta_i(k, y) = y^k, \quad \gamma_i(k, y) = y \quad \text{and} \quad \pi_i(k) = x^k.$$

Then, for $i = 1, 2$, condition (1) holds for all $y_1, y_2 \in M$, because $M = \langle x \rangle$ is cyclic, and therefore abelian:

$$\beta_i(k, y_1 \cdot y_2) = (y_1 y_2)^k = y_1^k y_2^k = \beta_i(k, y_1) \cdot \beta_i(k, y_2).$$

Note that exponentiation is only a homomorphism if the monoid M is abelian. Furthermore, condition (2) holds, because it is assumed to be hard to determine $k \in \mathbb{Z}$ from $\beta(k, x) = x^k$. The computational problem is well known as the *Discrete Logarithm Problem* (DLP).

And (3) is satisfied, because we have for all $k, l \in \mathbb{Z}$:

$$\gamma_1(k, \beta_2(l, \pi_1(k))) = \beta_2(l, x^k) = (x^k)^l = (x^l)^k = \beta_1(k, x^l) = \gamma_2(l, \beta_1(k, \pi_2(l))).$$

This proves that the conditions (1)-(3) are fulfilled. It remains to show that the protocol steps 1.-2. of the Diffie-Hellman KEP are specializations of the protocol steps 1.-4. of the general AAG-KEP.

1. Alice generates an element $k \in S_1 = \mathbb{Z}$ such that $\pi_1(k) = x^k \in \langle x \rangle = S_A$, and Bob chooses a $l \in S_2 = \mathbb{Z}$ s.t. $\pi_2(l) = x^l \in \langle x \rangle = S_B$.
2. Alice computes the element $\beta_1(k, x) = x^k = y_A$ and publicly announces this element. This element is her public key. Analogously Bob computes the element $\beta_2(l, x) = x^l = y_B$ and publishes this element.
3. Knowing that $\pi_1(k) = x^k$, Alice computes from Bob's public key $\beta_2(l, \pi_1(k)) = \beta_2(l, x^k) = (x^k)^l = (x^l)^k = y_B^k$. And Bob, knowing $\pi_2(l) = x$, computes from Alice's public key $\beta_1(k, \pi_2(l)) = \beta_1(k, x^l) = (x^l)^k = (x^k)^l = y_A^l$.
4. Alice computes $K_A = \gamma_1(k, \beta_2(l, \pi_1(k))) = \beta_2(l, \pi_1(k)) = y_B^k$, and symmetrically Bob computes $K_B = \gamma_2(l, \beta_1(k, \pi_2(l))) = \beta_1(k, \pi_2(l)) = y_A^l$. Since this is exactly the output of the computation in step 3, here step 4 is redundant or trivial.

Let us recall and emphasize that in step 3 the homomorphy property (1) is used in a nontrivial way. For example, Alice knowing $\pi_1(k) = x^k = \underbrace{x \cdots x}_{k \text{ times}} =: w_k(x)$ can compute

$$\begin{aligned} y_B^k &= (\beta_2(l, x))^k = (x^l)^k = w_k(x^l) = \underbrace{x^l \cdots x^l}_{k \text{ times}} \overset{(1)}{=} (\underbrace{x \cdots x}_{k \text{ times}})^l \\ &= (w_k(x))^l = (x^k)^l = \beta_2(l, w_k(x)) = \beta_2(l, \pi_1(k)). \end{aligned}$$

Therefore, we may view the classical DH-KEP as an AAG-like KEP. □

2.6. Sakalauskas, Tvarijonas and Raulynaitis Key Establishment Protocol (STR-KEP). The following KEP is a natural hybrid of the classical DH-KEP and the Ko-Lee-KEP. It was introduced in 2007 by Sakalauskas, Tvarijonas and Raulynaitis in [**STR07**].

Let G be a (noncommutative) group and A, B a pair of commuting subgroups in G. Furthermore, let x be a "generic" element in G. Alice and Bob have to perform the following protocol steps.

1. Alice generates her secret key $(k, a) \in \mathbb{Z} \times A$. And Bob selects his private key $(l, b) \in \mathbb{Z} \times B$.
2. Alice computes $y_A = a^{-1}x^k a$ and sends it to Bob. And Bob computes $y_B = b^{-1}x^l b$ and submits it to Alice.
3. Alice receives y_B and computes $K_A := a^{-1}y_B^k a$. And Bob receives y_A and computes the shared key

$$\begin{aligned} K_B &:= b^{-1}y_A^l b = b^{-1}(a^{-1}x^k a)^l b = b^{-1}(a^{-1}(x^k)^l a)b \\ &= a^{-1}(b^{-1}(x^l)^k b)a = a^{-1}(b^{-1}x^l b)^k a = a^{-1}y_B^k a = K_A. \end{aligned}$$

PROPOSITION 2.5.. *The Sakalauskas, Tvarijonas and Raulynaitis Key Establishment Protocol is an instance of the Anshel-Anshel-Goldfeld KEP for monoids. Furthermore, it is an AAG-like KEP.*

PROOF. Here we set $S_1 = \mathbb{Z} \times A$, $S_2 = \mathbb{Z} \times B$, $M = S_A = S_B = \langle x \rangle$ and $N = G$. For $i = 1, 2$, we define the functions $\beta_i : S_i \times \langle x \rangle \to G$ and $\pi_i : S_i \to \langle x \rangle$ by

$$\beta_i((k, z), y) = z^{-1}y^k z, \quad \gamma_i((k, z), y) = z^{-1}yz \quad \text{and} \quad \pi_i(k) = x^k.$$

Then, for $i = 1, 2$, condition (1) holds for all $y, y' \in M$:

$$\beta_i((k, z), y \cdot y') = z^{-1}(yy')^k z = z^{-1}y^k z \cdot z^{-1}y'^k z = \beta_i((k, z), y) \cdot \beta_i((k, z), y').$$

Furthermore, condition (2) holds, because it is assumed to be hard to determine $k \in \mathbb{Z}$ and $z \in G$ from $\beta((k,z),x) = z^{-1}x^k z$. The computational problem is a "mixed problem" requiring to solve simultaneously the DLP and the CSP (see [**STR07**]).

And (3) is satisfied, because we have for all $k, l \in \mathbb{Z}$, $a \in A$, $b \in B$:

$$\begin{aligned} \gamma_1((k,a), \beta_2((l,b), \pi_1(k,a))) &= a^{-1}\beta_2((l,b), x^k)a = a^{-1}(b^{-1}(x^k)^l b)a \\ = b^{-1}a^{-1}(x^l)^k ab &= b^{-1}\beta_1((k,a), x^l)b = \gamma_2((l,b), \beta_1((k,a), \pi_2(l,b))). \end{aligned}$$

This proves that the conditions (1)-(3) are fulfilled. It remains to show that the protocol steps 1.-2. of the Diffie-Hellman KEP are specializations of the protocol steps 1.-4. of the general AAG-KEP.

1. Alice generates an element $(k,a) \in \mathbb{Z} \times A$ such that $\pi_1(k,a) = x^k \in \langle x \rangle = S_A$, and Bob chooses $(l,b) \in \mathbb{Z} \times B$ s.t. $\pi_2(l,b) = x^l \in \langle x \rangle = S_B$.
2. Alice computes the element $\beta_1((k,a),x) = a^{-1}x^k a = y_A$ and publicly announces this element. This element is her public key. Analogously Bob computes the element $\beta_2((l,b),x) = b^{-1}x^l b = y_B$ and publishes this element.
3. Knowing that $\pi_1(k,a) = x^k$, Alice computes from Bob's public key
$$\beta_2((l,b), \pi_1(k,a)) = \beta_2((l,b), x^k) = b^{-1}(x^k)^l b = (b^{-1}x^l b)^k = y_B^k.$$
And Bob, knowing $\pi_2(l,b) = x$, computes from Alice's public key
$$\beta_1((k,a), \pi_2(l,b)) = \beta_1((k,a), x^l) = a^{-1}(x^l)^k a = (a^{-1}x^k a)^l = y_A^l.$$
4. Alice computes
$$K_A = \gamma_1((k,a), \beta_2((l,b), \pi_1(k,a))) = a^{-1}\beta_2((l,b), \pi_1(k,a))a = a^{-1}y_B^k a,$$
and symmetrically Bob computes
$$K_B = \gamma_2((l,b), \beta_1((k,a), \pi_2(l,b))) = b^{-1}\beta_1((k,a), \pi_2(l,b))b = a^{-1}y_A^l b.$$

Let us recall and emphasize that also here in step 3 the homomorphy property (1) is used in a nontrivial way. For example, Alice knowing $\pi_1(k,a) = x^k = \underbrace{x \cdots x}_{k \text{ times}} =: w_k(x)$ can compute

$$\begin{aligned} y_B^k &= (\beta_2((l,b),x))^k = (b^{-1}x^l b)^k = w_k(b^{-1}x^l b) = \underbrace{b^{-1}x^l b \cdots b^{-1}x^l b}_{k \text{ times}} \\ &\overset{(1)}{=} b^{-1}(\underbrace{x \cdots x}_{k \text{ times}})^l b = b^{-1}(w_k(x))^l b = b^{-1}(x^k)^l b = \beta_2((l,b), \pi_1(k,a)). \end{aligned}$$

Therefore, we may view the STR-KEP as an AAG-like KEP. □

3. Key establishment using non-associative operations

3.1. AAG scheme for magmas. Monoids are proposed as algebraic platform structures for the AAG key agreement protocol in [**AAG99**]. But the monoid structure is only used in the AAG scheme in order to guarantee that the secret key, e.g. Alice's key a, is an uniquely defined product of some given generators $\{s_1, \ldots, s_m\}$, i.e. $a = r_1 \cdot r_2 \cdots r_k$ with $r_i \in \{s_1, \ldots, s_m\}$ for all i. It is, of course, no problem to introduce brackets in this expression in order to handle nonassoziative operations. Therefore, there exists a straightforward generalization of the AAG scheme from monoids to magmas.

A *magma* (often also called *grupoid*) $(M, *)$ is a set M equipped with a binary operation $*$ on M, i.e. a function $M \times M \to M$. Note that there are no relations, which have to be satisfied by the elements of M. The notion of a magma was introduced by Bourbaki (see, e.g., [**Bo74**]).

We describe the AAG key establishment protocol in the - for our purposes - most general manner.

For $i = 1, 2$, let S_i be a sets and $(M, \bullet_i)$ and $(N, \circ_i)$ be magmas, i.e. there are two operations on the sets M, N, respectively. For $i = 1, 2$, we need functions

$$\beta_i : S_i \times M \to N, \quad \gamma_i : S_i \times N \to N, \quad \pi_i : S_i \to M$$

which satisfy the following three conditions:

(1) $\beta_1(x, \cdot) : (M, \bullet_2) \to (N, \circ_2)$ is for all $x \in S_1$ a *magma homomorphism*[4], i.e.

$$\forall x \in S_1, y, y' \in M : \quad \beta_1(x, y \bullet_2 y') = \beta_1(x, y) \circ_2 \beta_1(x, y').$$

Also $\beta_2(x, \cdot) : (M, \bullet_1) \to (N, \circ_1)$ is for all $x \in S_2$ a magma morphism, i.e.

$$\forall x \in S_2, y, y' \in M : \quad \beta_2(x, y \bullet_1 y') = \beta_2(x, y) \circ_1 \beta_2(x, y').$$

(2) It is, in general, not feasible to determine a secret $x \in S_i$ $(i = 1, 2)$ from the knowledge of

$$y_1, y_2, \ldots, y_k \in M \quad \text{and} \quad \beta_i(x, y_1), \beta_i(x, y_2), \ldots, \beta_i(x, y_k).$$

(3) For all $a \in S_1, b \in S_2 : \quad \gamma_1(a, \beta_2(b, \pi_1(a))) = \gamma_2(b, \beta_1(a, \pi_2(b)))$.

Consider an element y of a magma $(M, \bullet)$ which is an iterated product of other elements in M. Such an element can be described by a planar rooted binary tree T whose k leaves are labelled by these other elements $y_1, \ldots, y_k \in M$. We use the notation $y = T_\bullet(y_1, \ldots, y_k)$. Here the subscript $\bullet$ tells us that the grafting of subtrees of T corresponds to the operation $\bullet$.

Now, it is easy to prove by induction that any magma homomorphism $\beta : (M, \bullet) \to (N, \circ)$ satisfies

$$\beta(T_\bullet(y_1, \ldots, y_k)) = T_\circ(\beta(y_1), \ldots, \beta(y_k))$$

for all $y_1, \ldots, y_k \in M$. In particular, the magma morphisms $\beta_1(x, \cdot), \beta_2(x, \cdot)$ $(x \in S)$ fulfill this property.

Alice and Bob publicly assign sets $\{s_1, \ldots, s_m\}, \{t_1, \ldots, t_n\} \subset M$, respectively. The secret key spaces SK_A, SK_B of Alice and Bob are subsets of S_1, S_2, respectively, and they depend on these public elements. It is sufficient that β_1, β_2 fulfill condition (1) only for all $x \in SK_A, SK_B$, respectively, and that condition (3) holds for all $a \in SK_A, b \in SK_B$.

Now, Alice and Bob perform the following protocol steps.

1. Alice generates her secret key $a \in SK_A$, and Bob chooses his secret key $b \in SK_B$.
2. Alice computes the elements $\beta_1(a, t_1), \ldots, \beta_1(a, t_n) \in N$, and sends them to Bob. Analogously Bob computes the elements $\beta_2(b, s_1), \ldots, \beta_2(b, s_m) \in N$, and sends them to Alice.

[4] More on magmas and magma homomorphisms can be found, e.g. in [**Se65**, **Ge94**].

3. Alice, knowing $\pi_1(a) = T_{\bullet_1}(r_1, \ldots, r_k)$ with $r_i \in \{s_1, \ldots, s_m\}$, computes from Bob's public key
$$T_{\circ_1}(\beta_2(b, r_1), \ldots, \beta_2(b, r_k)) = \beta_2(b, T_{\bullet_1}(r_1, \ldots, r_k)) = \beta_2(b, \pi_1(a)).$$
And Bob, knowing $p_2(b) = T'_{\bullet_2}(u_1, \ldots, u_{k'})$ with $u_j \in \{t_1, \ldots, t_n\}$, computes from Alice's public key
$$T'_{\circ_2}(\beta_1(a, u_1), \ldots, \beta_1(a, u_{k'})) = \beta_1(a, T'_{\bullet_2}(u_1, \ldots, u_{k'})) = \beta_1(a, \pi_2(b)).$$
4. Alice computes $K := \gamma_1(a, \beta_2(b, \pi_1(a)))$. Bob also computes the shared key $\gamma_2(b, \beta_1(a, \pi_2(b))) \overset{(3)}{=} K$.

Note that the protocols described in section 2.1 are special instances of this general AAG like protocol for magmas.

A natural special case of this scheme is given by $M = N = S_1 = S_2$. This implies that the functions β_i, γ_i, for $i = 1, 2$, induce further binary operations on M. If additionally $\bullet_i = \circ_i$ holds for $i = 1, 2$, then M satisfies some distributive laws. This will lead to the notion of LD- and multi-LD-systems (see section 4).

Another specification of our general magma-based scheme is discussed in the next subsection.

3.2. Non-associative KEP based on simultaneous DCP.

3.2.1. *Specifications.* We consider the following specifications of the AAG scheme for magmas:

Let $G = M = N$ be a group, and set $S_1 = S_2 = G^2$. The group multiplication symbol in G will usually be omitted. The operations $\bullet_i, \circ_i$ $(i = 1, 2)$ on G are defined by
$$x \bullet_1 y = x \bullet_2 y = x \circ_1 y = x \circ_2 y \equiv x \bullet y := xy^{-1}x,$$
and the functions $\beta_1, \beta_2 : G^2 \times G \to G$ are defined by
$$\beta_1((x_1, x_2), y) = \beta_2((x_1, x_2), y) \equiv \beta((x_1, x_2), y) := x_1 y x_2.$$
$\beta(x, \cdot)$ fulfills the homomorphy condition (1), for all $x = (x_1, x_2) \in G^2$, because
$$\begin{aligned} \beta((x_1, x_2), y_1) \bullet \beta((x_1, x_2), y_2) &= (x_1 y_1 x_2) \bullet (x_1 y_2 x_2) = \\ (x_1 y_1 x_2) x_2^{-1} y_2^{-1} x_1^{-1} (x_1 y_1 x_2) &= x_1 (y_1 y_2^{-1} y_1) x_2 = \beta((x_1, x_2), y_1 \bullet y_2). \end{aligned}$$
Alice and Bob publicly assign sets $\{s_1, \ldots, s_m\}, \{t_1, \ldots, t_n\} \subset G$, respectively. The secret key spaces of Alice and Bob are $SK_A = G \times S_A$ and $SK_B = S_B \times G$, where $S_A = \langle s_1, \ldots, s_m \rangle_\bullet$ and $S_B = \langle t_1, \ldots, t_n \rangle_\bullet$ denote submagmas of $(G, \bullet)$ generated by the publicly assigned elements.

The projections $\pi_1, \pi_2 : G^2 \to G$ and the functions $\gamma_1, \gamma_2 : G^2 \times G \to G$ are defined by
$$\pi_1(x, y) = y, \ \pi_2(x, y) = x \quad \text{and} \quad \gamma_1((x_1, x_2), y) = x_1 y, \ \gamma_2((x_1, x_2), y) = y x_2.$$
These definitions satisfy condition (3), because
$$\begin{aligned} & \gamma_1(a, \beta(b, \pi_1(a))) = \gamma_1(a, \beta(b, a_r)) = \gamma_1(a, b_l a_r b_r) = a_l(b_l a_r b_r) \\ = \ & (a_l b_l a_r) b_r = \gamma_2(b, a_l b_l a_r) = \gamma_2(b, \beta(a, b_l)) = \gamma_2(b, \beta(a, \pi_2(b))) \end{aligned}$$
for all $a = (a_l, a_r), b = (b_l, b_r) \in G^2$.

We skip repeating all the protocol steps from section 3.1 with these specifications. The base problem for these non-associative scheme is discussed in the next subsubsection.

3.2.2. *A related associative scheme.* Consider the right part of Alice's key $a_r = T_\bullet(r_1, \ldots, r_k) \in S_A$ with $r_i \in \{s_1, \ldots, s_m\}$. If we view a_r as a word in the s_i's, then we observe that a_r is self-reverse and the exponent signs of a_r alternate, beginning and ending with a positive sign. For example, we have

$$(r_1 \bullet r_2) \bullet (r_3 \bullet (r_4 \bullet r_5)) = r_1 r_2^{-1} r_1 r_3^{-1} r_4 r_5^{-1} r_4 r_3^{-1} r_1 r_2^{-1} r_1.$$

While in this scheme alternating exponent signs are essential to gurantee that condition (1) holds, the self-reverse property seems to be superflous. It comes from the self-reverse property of the non-associative operation $\bullet$. Anyway, for example in order to compute $b_l a_r b_r$, Alice actually doesn't need to know a_r as a tree-word in the submagma $\langle s_1, \ldots, s|m\rangle_\bullet$. Rather it suffices to know a_r as an "alternating" word of the form $s_{i_1} s_{i_2}^{-1} s_{i_3} \cdots s_{i_{2l}}^{-1} s_{i_{2l+1}}$.

Therefore, we give up this restricted key choice and define modified (bigger) secret key spaces by $SK_A = G \times SK_A^{(r)}$ and $SK_B = SK_B^{(l)} \times G$ with

$$\begin{aligned} SK_A^{(r)} &= \{r_1 r_2^{-1} r_3 r_4^{-1} \cdots r_{2l}^{-1} r_{2l+1} \mid r_i \in \{s_1, \ldots, s_m\} \, \forall 1 \le i \le l, l \in \mathbb{N}\}, \\ SK_B^{(l)} &= \{u_1 u_2^{-1} u_3 u_4^{-1} \cdots u_{2l'}^{-1} u_{2l'+1} \mid u_j \in \{t_1, \ldots, t_n\} \, \forall 1 \le j \le l', l' \in \mathbb{N}\}. \end{aligned}$$

Then, Alice and Bob have to perform the following protocol steps.

1. Alice generates her secret key $(a_l, a_r) \in G \times SK_A^{(r)}$. Bob chooses his secret key $(b_l, b_r) \in SK_B^{(l)} \times G$.
2. Alice computes the elements $a_l t_1 a_r, \ldots, a_l t_n a_r$, and sends them to Bob. Analogously Bob computes the elements $b_l s_1 b_r, \ldots, b_l s_m b_r$, and sends them to Alice.
3. Alice, knowing $a_r = r_1 r_2^{-1} r_3 r_4^{-1} \cdots r_{2l}^{-1} r_{2l+1}$ with $r_i \in \{s_1, \ldots, s_m\}$, computes from Bob's public key

$$\begin{aligned} & (b_l r_1 b_r)(b_l r_2 b_r)^{-1}(b_l r_3 b_r) \cdots (b_l r_{2l} b_r)^{-1}(b_l r_{2l+1} b_r) \\ = \;& b_l(r_1 r_2^{-1} r_3 \cdots r_{2l}^{-1} r_{2l+1}) b_r = b_l a_r b_r. \end{aligned}$$

   Bob, knowing $b_l = u_1 u_2^{-1} u_3 u_4^{-1} \cdots u_{2l'}^{-1} u_{2l'+1}$ with $u_j \in \{t_1, \ldots, t_n\}$, computes from Alice's public key

$$\begin{aligned} & (a_l u_1 a_r)(a_l u_2 a_r)^{-1}(a_l u_3 a_r) \cdots (a_l u_{2l'} a_r)^{-1}(a_l u_{2l'+1} a_r) \\ = \;& a_l(u_1 u_2^{-1} u_3 \cdots u_{2l'}^{-1} u_{2l'+1}) a_r = a_l b_l a_r. \end{aligned}$$

4. Alice computes $K := a_l(b_l a_r b_r)$. Bob also computes the shared key $(a_l b_l a_r) b_r = K$.

It is easy to show that this scheme is a further instance of the generalized AAG scheme for monoids (section 2.1). Therefore one simply has to turn the sets $SK_A^{(r)}$ and $SK_B^{(l)}$ into monoids by introducing some "forgetful" operations as exercised, e.g., in the proof of 2.1.

In order to break this scheme an attacker obviously has to solve the following

Base Problem:

INPUT: Element pairs $(s_1, s_1'), \ldots, (s_m, s_m') \in G^2$ and $(t_1, t_1'), \ldots, (t_n, t_n') \in G^2$ with $s_i' = b_l s_i b_r \; \forall 1 \le i \le m$ and $t_j' = a_l t_j a_r \; \forall 1 \le j \le n$ for some (unknown) $a_l, b_r \in G$, $b_l \in SK_B^{(l)}$, $a_r \in SK_A^{(r)}$.

OBJECTIVE: Find $K = a_l b_l a_r b_r$.

A successful attack on Alice's secret key requires the solution of the following

n-simDP (n-Simultaneous Decomposition Problem):

INPUT: Element pairs $(t_1, t_1'), \ldots, (t_n, t_n') \in G^2$ with $t_j' = a_l t_j a_r$ $\forall 1 \leq j \leq n$ for some (unknown) $a_l \in G$, $a_r \in SK_A^{(r)}$.

OBJECTIVE: Find elements $a_l' \in G$, $a_r' \in SK_A^{(r)}$ with $a_l' t_j a_r' = t_j'$ for all $j = 1, \ldots, n$.

A solution (a_l', a_r') to this n-simDP satisfies the property $a_l' y a_r' = a_l y a_r$ for all $y \in SK_B^{(l)}$.

Analogeously, a successful attack on Bob's secret key requires the solution of the following

m-simDP (m-Simultaneous Decomposition Problem):

INPUT: Element pairs $(s_1, s_1'), \ldots, (s_m, s_m') \in G^2$ with $s_i' = b_l s_i b_r$ $\forall 1 \leq i \leq m$ for some (unknown) $b_l \in SK_B^{(l)}$, $b_r \in G$.

OBJECTIVE: Find elements $b_l' \in SK_B^{(l)}$, $b_r' \in G$ with $b_l' s_i b_r' = s_i'$ for all $i = 1, \ldots, m$.

A solution (b_l', b_r') to this m-simDP satisfies the property $b_l' x b_r' = b_l x b_r$ for all $x \in SK_A^{(r)}$.

Therefore, a solution to both problems provides the attacker with the shared secret, because

$$(a_l' b_l' a_r') b_r' = (a_l b_l' a_r) b_r' = a_l (b_l' a_r b_r') = a_l (b_l a_r b_r) = K.$$

Here the first and the last equality hold, because $b_l' \in SK_B^{(l)}$ and $a_r \in SK_A^{(r)}$, respectively. Alternatively, we can use equality chain

$$a_l' (b_l' a_r' b_r') = a_l' (b_l a_r' b_r) = (a_l' b_l a_r') b_r = (a_l b_l a_r) b_r = K,$$

where here the first and the last equality hold, because $a_r' \in SK_A^{(r)}$ and $b_l \in SK_B^{(l)}$, respectively. Further, the first equality chain shows us, that it is sufficient to find a solution $(a_l', a_r') \in G^2$ to the n-SDP and a solution $(b_l', b_r') \in SK_B^{(l)} \times G$ to the m-simDP. Analogously, the second equality chain shows us, that it is sufficient to find a solution $(a_l', a_r') \in G \times SK_A^{(r)}$ to the n-SDP and a solution $(b_l', b_r') \in G^2$ to the m-simDP.

Note that the knowledge of one secret key, e.g. Alice's key $(a_l, a_r) \in G \times SK_A^{(r)}$, is not sufficient for an attacker to obtain the shared secret K, because he needs not only a_r expressed in the generators of the group G, but rather an expression of the form

$$a_r = r_1 r_2^{-1} r_3 r_4^{-1} \cdots r_{2l}^{-1} r_{2l+1} \quad \text{with} \quad r_i \in \{s_1, \ldots, s_m\}.$$

Remark. An attacker might approach an n-simDP instance $\{(t_i, t_i')\}_{i \leq n}$ by considering the $\binom{n}{2}$-ssCSP instance $\{(t_i'(t_j')^{-1}, t_i t_j^{-1}) \mid 1 \leq i \neq j \leq n\}$ or the $\binom{n}{2}$-ssCSP instance $\{(t_i^{-1} t_j, (t_i')^{-1} t_j') \mid 1 \leq i \neq j \leq n\}$ in order to solve for a_l or a_r, respectively. For example, in the latter case, we have

$$a_r^{-1} t_i^{-1} t_j a_r = (a_r^{-1} t_i^{-1} a_l^{-1})(a_l t_j a_r) = (t_i')^{-1} t_j'.$$

Therefore, either the simultaneous (subgroup)-CSP has to be hard in G, or, if the simCSP is (at least heuristically) approachable in G, it is recommended that the sets $\{t_i^{-1} t_j \mid 1 \leq i \neq j \leq n\}$ and $\{t_i t_j^{-1} \mid 1 \leq i \neq j \leq n\}$ have large centralizers. This may be ensured by if the set $\{t_1, \ldots, t_n\}$ itself has a large centralizer, an thus also S_B.s Similarly S_A should have a large centralizer.

4. Non-associative schemes for LD-systems

4.1. LD- and multi-LD-systems.

4.1.1. *Definition.*

DEFINITION 4.1.. *An LD-system $(S, *)$ is a set S equipped with a binary operation $*$ on S which satisfies the left-selfdistributivity law*

$$x * (y * z) = (x * y) * (x * z) \quad \forall x, y, z \in S.$$

DEFINITION 4.2.. *(Section X.3. in* [**De00**]*) Let I be an index set. A multi-LD-system $(S, (*_i)_{i \in I})$ is a set S equipped with a family of binary operations $(*_i)_{i \in I}$ on S such that*

$$x *_i (y *_j z) = (x *_i y) *_j (x *_i z) \quad \forall x, y, z \in S$$

is satisfied for every i, j in I. Especially, it holds for $i = j$, i.e., $(S, *_i)$ is an LD-system. If $|I| = 2$ then we call S a bi-LD-system.

A classical example for an LD-system is given by a group G equipped with the conjugacy operation $x * y = x^{-1}yx$. We also mention the Laver tables (Chapter X in [**De00**]) as standard examples for finite monogenic LD-systems. Many examples for LD-, bi-LD- and multi-LD-systems are given in Dehornoy's monography [**De00**].

4.1.2. *f-conjugacy.* One may consider several generalizations of the conjugacy operation as candidates for natural LD-operations in groups. Consider an Ansatz like $x * y = f(x^{-1})g(y)h(x)$ for some group endomorphisms f, g, h.

PROPOSITION 4.3.. *Let G be a group, and $f, g, h \in End(G)$. Then the binary operation $x * y = f(x^{-1}) \cdot g(y) \cdot h(x)$ yields an LD-structure on G if and only if*

$$fh = f, \quad gh = hg = hf, \quad fg = gf = f^2, \quad h^2 = h. \tag{1}$$

PROOF. A straightforward computation yields

$$\begin{aligned} \alpha * (\beta * \gamma) &= f(\alpha^{-1})gf(\beta^{-1})g^2(\gamma)gh(\beta)h(\alpha), \quad \text{and} \\ (\alpha * \beta) * (\alpha * \gamma) &= fh(\alpha^{-1})fg(\beta^{-1})f^2(\alpha)gf(\alpha^{-1})g^2(\gamma)gh(\alpha)hf(\alpha^{-1})hg(\beta)h^2(\alpha). \end{aligned}$$

A comparison of both terms yields the assertion. □

The simplest solution of the system of equations (1) is $f = g$ and $h = \mathrm{id}$. This leads to the following definition.

DEFINITION 4.4.. *(*LD- OR f-CONJUGACY*) Let G be a group, and $f \in End(G)$. An ordered pair $(u, v) \in G \times G$ is called f-LD-conjugated or LD-conjugated, or simply f-conjugated, denoted by $u \longrightarrow_{*_f} v$, if $\exists c \in G$ such that $v = c *_f u = f(c^{-1}u)c$.*

Remark. For any non-trivial endomorphism f, the relation $\longrightarrow_{*_f}$ defines not an equivalence relation on G. Even the relation $\longrightarrow_*$ defined by $u \longrightarrow_* v$ iff $\exists f \in Aut(G)$ s.t. $u \longrightarrow_{*_f} v$ is not an equivalence relation. Indeed, transitivity requires the automorphisms (relation must be symmetric!) to be an idempotent endomorphism ($f^2 = f$) which implies $f = \mathrm{id}$.

Compare the notion of f-LD-conjugacy with the well known notion *f-twisted conjugacy* defined by $u \sim_f v$ (for $f \in Aut(G)$) iff $\exists c \in G$ s.t. $v = f(c^{-1})uc =: c *_f^{tw} u$, which yields indeed an equivalence relation. On the other hand, the operation $*^{tw} = *_f^{tw}$ is not LD - rather it satisfies the following "near" LD-law:

$$\alpha *^{tw} (\beta *^{tw} \gamma) = (\alpha *^{tw} \beta) *^{tw} (\alpha^f *^{tw} \gamma)$$

where α^f is short for $f(\alpha)$.

Anyway, it follows directly from the definitions that $u \longrightarrow_* v$ if and only if $f(u) \sim_f v$, i.e., any f-LD conjugacy problem reduces to a twisted conjugacy problem and vice versa. Here we have to extend the notion of twisted conjugacy from $f \in Aut(G)$ to all $f \in End(G)$.

4.1.3. *Shifted conjugacy.* Patrick Dehornoy introduced the following generalization of f-conjugacy, and he points out, that once the definition of shifted conjugacy is used, braids inevitably appear [**De00, De06**].

PROPOSITION 4.5.. (Exercise I.3.20. in [**De00**]) *Consider a group G, a homomorphism $f : G \to G$, and a fixed element $a \in G$. Then the binary operation*

$$x * y = x *_{f,a} y = f(x)^{-1} \cdot a \cdot f(y) \cdot x$$

yields an LD-structure on G if and only if $[a, f^2(x)] = 1$ for all $x \in G$, and a satisfies the relation $af(a)a = f(a)af(a)$. Hence the subgroup $H = \langle \{f^n(a) \mid n \in \mathbb{N}\} \rangle$ of G is a homomorphic image of the braid group

$$B_\infty = \langle \{\sigma_i\}_{i\geq 1} \mid \sigma_i\sigma_j = \sigma_j\sigma_i \text{ for } |i-j| \geq 2,\ \sigma_i\sigma_j\sigma_i = \sigma_j\sigma_i\sigma_j \text{ for } |i-j| = 1 \rangle$$

with infinitely many strands, i.e., up to an isomorphism, it is a quotient of B_∞.

There exists a straightforward generalization of Proposition 4.5 for multi-LD-systems:

PROPOSITION 4.6.. *Let I be an index set. Consider a group G, a family of endomorphisms $(f_i)_{i\in I}$ of G, and a set of fixed elements $\{a_i \in G \mid i \in I\}$. Then $(G, (*_i)_{i\in I})$ with*

$$x *_i y = f_i(x^{-1}) \cdot a_i \cdot f_i(y) \cdot x$$

is a multi-LD-system if and only if $f_i = f_j =: f$ for all $i \neq j$, $[a_i, f^2(x)] = 1$ for all $x \in G$, $i \in I$, and $a_i f(a_i) a_j = f(a_j) a_i f(a_i)$ for all $i, j \in I$.

PROOF. A straightforward computation gives

$$\begin{aligned} x *_i (y *_j z) &= f_i(x^{-1})a_i[f_i(f_j(y^{-1}))f_i(a_j)f_i(f_j(z))f_i(y)]x, \\ (x *_i y) *_j (x *_i z) &= [f_j(x^{-1})f_j(f_i(y^{-1}))f_j(a_i^{-1})f_j(f_i(x))]a_j[f_j(f_i(x^{-1}))\cdot \\ &\quad f_j(a_i)f_j(f_i(z))f_j(x)][f_i(x^{-1})a_i f_i(y)x]. \end{aligned}$$

A comparison of both terms yields the assertion. □

Note that this proof also contains proofs of Proposition 4.5 (setting $|I| = 1$) and of the following Corollary 4.7 (setting $G = B_\infty$, $I = \{1, 2\}$, $s = \partial$, $*_1 = *$, $*_2 = \bar{*}$, $a_1 = \sigma_1$ and $a_2 = \sigma_1^{-1}$).

Consider the injective *shift endomorphism* $\partial : B_\infty \longrightarrow B_\infty$ defined by $\sigma_i \mapsto \sigma_{i+1}$ forall $i \geq 1$.

COROLLARY 4.7.. (SHIFTED CONJUGACY, Example X.3.5. in [**De00**]) B_∞ *equipped with the* shifted conjugacy *operations $*, \bar{*}$ defined by*

$$x * y = \partial x^{-1} \cdot \sigma_1 \cdot \partial y \cdot x, \qquad x \bar{*} y = \partial x^{-1} \cdot \sigma_1^{-1} \cdot \partial y \cdot x$$

is a bi-LD-system. In particular, $(B_\infty, *)$ is an LD-system.

4.1.4. *Generalized shifted conjugacy in braid groups.* In the following we consider generalizations of the shifted conjugacy operations $*$ in B_∞. Therefore we set $f = \partial^p$ for some $p \in \mathbb{N}$, and we choose $a_i \in B_{2p}$ for all $i \in I$ such that

$$a_i \partial^p(a_i) a_j = \partial^p(a_j) a_i \partial^p(a_i) \quad \forall i, j \in I. \tag{2}$$

Since $a_i \in B_{2p}$, we have $[a_i, \partial^{2p}(x)] = 1$ for all $x \in B_\infty$. Thus the conditions of Proposition 4.6 are fulfilled, and $x *_i y = x\partial^p(y) a_i \partial^p(x^{-1})$ defines an multi-LD-structure on B_∞. For $|I| = 1$, $p = 1$ and $a = \sigma_1$, which implies $H = B_\infty$, we get Dehornoy's original definition of shifted conjugacy $*$.

It remains to give some natural solutions $\{a_i \in B_{2p} \mid i \in I\}$ of the equation set (1). Note that in case $|I| = 1$ (notation: $a_1 = a$), of course, every endomorphism f of B_∞ with $f(\sigma_1) \in B_{2p}$ provides such solution $a = f(\sigma_1)$.

Definition 4.8. *(Definition I.4.6. in* [**De00**]*) Let, for $n \geq 2$, be $\delta_n = \sigma_{n-1} \cdots \sigma_2 \sigma_1$. For $p, q \geq 1$, we set*

$$\tau_{p,q} = \delta_{p+1} \partial(\delta_{p+1}) \cdots \partial^{q-1}(\delta_{p+1}).$$

Since $a = \tau_{p,p}^{\pm 1} \in B_{2p}$ fulfills $a\partial^p(a)a = \partial^p(a) a \partial^p(a)$, it provides a lot of (multi)-LD-structures on B_∞.

Proposition 4.9. *(a) The binary operation $x *_a y = \partial^p(x^{-1}) a \partial^p(y) x$ with $a = a' \tau_{p,p} a''$ for some $a', a'' \in B_p$ yields an LD-structure on B_∞ if and only if $[a', a''] = 1$.*

(b) Let I be an index set. The binary operations $x *_i y = \partial^p(x^{-1}) a_i \partial^p(y) x$ with $a_i = a_i' \tau_{p,p} a_i''$ for some $a_i', a_i'' \in B_p$ $(i \in I)$ yields a multi-LD-structure on B_∞ if and only if $[a_i', a_j'] = [a_i', a_j''] = 1$ for all $i, j \in I$. (Note that a_i'' and a_j'' needn't commute for $i \neq j$.)

(c) The binary operations $x *_i y = \partial^p(x^{-1}) a_i \partial^p(y) x$ $(i = 1, 2)$ with $a_1 = a_1' \tau_{p,p} a_1''$, $a_2 = a_2' \tau_{p,p}^{-1} a_2''$ for some $a_1', a_1'', a_2', a_2'' \in B_p$ yields a bi-LD-structure on B_∞ if and only if $[a_1', a_1''] = [a_2', a_2''] = [a_1', a_2''] = [a_2', a_1''] = [a_1', a_2'] = 1$. (Note that a_1'' and a_2'' needn't commute.)

Another solution We see that there exist infinitely many (multi)-LD-structures on B_∞. Further examples are provided by Proposition 4.10, which, of course, admits a lot of variations and generalizations.

Proposition 4.10. *Let be $p, p_1, p_2 \in \mathbb{N}$ with $p_1 + p_2 = p$. The binary operation $x *_a y = \partial^p(x^{-1}) a \partial^p(y) x$ with*

$$a = a_1' \partial^{p_1}(a_2') \partial^{p_1}(\tau_{p_2,p}) \tau_{p,p_1}^{-1} a_1'' \partial^{p_1}(a_2'')$$

for some $a_1', a_1'' \in B_{p_1}$, $a_2', a_2'' \in B_{p_2}$ yields an LD-structure on B_∞ if and only if $[a_1', a_1''] = [a_2', a_2''] = 1$.

The proofs of Proposition 4.9 and 4.10 are straightforward computations. The reader is recommended to draw some pictures.

4.1.5. *Yet another group-based LD-system.* For a group G, $(G, \circ)$ is an LD-system with $x \circ y = xy^{-1}x$ (see e.g. [**St05**]).

Note that, contrary to the conjugacy operation $*$, for this *"symmetric decomposition" or conjugacy operation* $\circ$, the corresponding relation $\longrightarrow_\circ$ defined by $x \longrightarrow_\circ y$ iff $\exists c \in G$ such that $y = c \circ x$) is not an equivalence relation. In particular, $\longrightarrow_\circ$ is reflexive and symmetric, but not transitive.

One may consider several generalizations of this symmetric conjugacy operation $\circ$, as candidates for natural LD-operations in groups. Consider an Ansatz like $x \circ y = f(x)g(y^{-1})h(x)$ for some group endomorphisms f, g, h.

PROPOSITION 4.11.. *Let G be a group, and $f, g, h \in End(G)$. Then the binary operation $x \circ y = f(x) \cdot g(y^{-1}) \cdot h(x)$ yields an LD-structure on G if and only if*

$$f^2 = f, \quad fh = gh = fg, \quad hg = gf = hf, \quad h^2 = h. \tag{3}$$

PROOF. A straightforward computation yields

$$\begin{aligned} \alpha \circ (\beta \circ \gamma) &= f(\alpha)gh(\beta^{-1})g^2(\gamma)gf(\beta^{-1})h(\alpha), \quad \text{and} \\ (\alpha \circ \beta) \circ (\alpha \circ \gamma) &= f^2(\alpha)fg(\beta^{-1})fh(\alpha)gh(\alpha^{-1})g^2(\gamma)gf(\alpha^{-1})hf(\alpha)hg(\beta^{-1})h^2(\alpha). \end{aligned}$$

A comparison of both terms yields the assertion. □

Except for $f^2 = f = g = h = h^2$, the simplest solutions of the system of equations (3) are $f^2 = f = g$ and $h = \text{id}$, or $f = \text{id}$ and $g = h = h^2$.

COROLLARY 4.12.. *(LD- OR f-SYMMETRIC CONJUGACY) Let G be a group, and $f \in End(G)$ an endomorphism that is also a projector ($f^2 = f$). Then $(G, \circ_f)$ and $(G, \circ_f^{\text{rev}})$, defined by $x \circ_f y = f(xy^{-1})x$ and $x \circ_f^{\text{rev}} y = xf(y^{-1}x)$, are LD-systems.*

PROPOSITION 4.13.. *Let G be a group, and $f, g \in End(G)$.*
(i) Then the binary operations $\circ_f$ and $*_f$ (and $*_f^{\text{rev}}$), defined by $x \circ_f y = f(x) \cdot g(y^{-1}) \cdot h(x)$ and $x *_f y = f(x^{-1} \cdot y) \cdot h(x)$ ($x *_f^{\text{rev}} y = x \cdot f(y \cdot x^{-1})$), are distributive over $\circ$. In particular $*$ ($*^{\text{rev}}$) is distributive over $\circ$. In short, the following equations hold.

$$x *_f (y \circ z) = (x *_f y) \circ (x *_f z), \quad x \circ_f (y \circ z) = (x \circ_f y) \circ (\circ_f z) \forall x, y, z \in G.$$

(ii) The operations $\circ_f$ and $*_f$ ($*_f^{\text{rev}}$) are distributive over $\circ_g$ if and only if $f = gf = fg$.

4.2. Non-associative AAG f-commutator KEP. Now we consider the most natural special case of our general AAG scheme for magmas (see section 3.1). Let be $M = N = S$. This implies that the functions β_i, γ_i, for $i = 1, 2$, induce further binary operations on M. In particular, we introduce the notation $x *_i y = \beta_i(x, y)$. Now, the homomorphy condition (1) (in section 3.1) reads as

$$\begin{aligned} x *_1 (y \bullet_2 y') &= (x *_1 y) \circ_2 (x *_1 y') \quad \text{and} \\ x *_2 (y \bullet_1 y') &= (x *_2 y) \circ_1 (x *_2 y'). \end{aligned}$$

If $\bullet_i = \circ_i$ holds for $i = 1, 2$, then M fulfills two distributive laws. And if additionally $\circ_2 = \circ_1 = *_1 = *_2 =: *$, then $(M, *)$ is an LD-system.

We observe that LD-systems occur in a very natural special case of the general AAG scheme for magmas. Nevertheless, this does not imply that we get by that construction KEPs for all LD-systems. Indeed, in order to obtain a shared key, we have to specify the projections π_1 and binary operations γ_i which themselves depend on the specification of the LD-operation $*$. In the following we set $\pi_i = \text{id}_M$ for $i = 1, 2$.

Now, we establish a (non-associative) AAG-KEP for groups with f-conjugacy as LD-operation. Let $M = G$ be a group, $f \in End(G)$, then $(G, +)$ with $* = *_f$ (see Def. 4.4) is an LD-system according to Proposition 4.3.

Definition 4.14.. *(f-commutator) Let G be a group, and $f \in End(G)$. The* f-commutator *of an ordered pair $(u,v) \in G \times G$ is defined by*

$$[u,v]_f := u^{-1}f(v^{-1})f(u)v.$$

The *AAG f-commutator KEP* is given by the following further specifications of the general AAG scheme for magmas (section 3.1).

$$\gamma_1(u,v) = u^{-1}v, \qquad \gamma_2(u,v) = v^{-1}u.$$

Now, Alice and Bob perform the following protocol steps.

1. Alice generates her secret key a in the public submagma $S_1 = \langle s_1, \cdots, s_m\rangle_*$ of $(G,*)$, and Bob chooses his secret key $b \in S_2 = \langle t_1, \cdots, t_n\rangle_*$.
2. Alice computes the elements $a*t_1, \dots, a*t_n \in G$, and sends them to Bob. Analogously Bob computes the elements $b*s_1, \dots, b*s_m \in G$, and sends them to Alice.
3. Alice, knowing $a = T_*(r_1, \dots, r_k)$ with $r_i \in \{s_1, \dots, s_m\}$, computes from Bob's public key
$$T_*(b*r_1, \dots, b*r_k) = b*T_*(r_1, \dots, r_k)) = b*a = f(b^{-1}a)b.$$
And Bob, knowing $b = T'_*(u_1, \dots, u_{k'})$ with $u_j \in \{t_1, \dots, t_n\}$, computes from Alice's public key
$$T'_*(a*u_1, \dots, a*u_{k'}) = a*T'_*(u_1, \dots, u_{k'}) = a*b = f(a^{-1}b)a.$$
4. Alice computes $K := \gamma_1(a, b*a) = a^{-1}(b*a) = a^{-1}f(b^{-1}a)b = [a,b]_f$. Bob gets the shared key by $\gamma_2(b, a*b) = (a*b)^{-1}b = (f(a^{-1}b)a)^{-1}b \overset{(3)}{=} K$.

In order to break this scheme an attacker obviously has to solve the following base problem.

f-AAGP (*f-Commutator AAG-Problem*): Let $(G,*)$ be a group with $\alpha * \beta = f(\alpha^{-1}\beta)\alpha$ for some $f \in End(G)$. Furthermore, let $A = \langle a_1, \dots, a_k\rangle_*$ and $B = \langle b_1, \dots, b_m\rangle_*$ be two f.g. submagmas of $(G,*)$.

Input: $\{(a_i, y*a_i) \in G^2 | i = 1, \dots, k\} \cup \{(b_j, x*b_j) \in G^2 | j = 1, \dots, m\}$ with $x \in A$ and $y \in B$.

Objective: Find the f-commutator $[x,y]_f := x^{-1}f(y^{-1}x)y$.

But a successful attack on Bob's secret key requires at least the solution of the following

m-sim f-CSP (m-Simultaneous f-Conjugacy Search Problem):

Input: Pairs $(s_1, s'_1), \dots, (s_m, s'_m) \in G^2$ with $s'_i = b*s_i = f(b^{-1}s_i)b \ \forall 1 \le i \le m$ for some (unknown) $b \in G$.

Objective: Find an element $b' \in G$ with $f(b'^{-1}s_i)b' = f(b^{-1}s_i)b'$ for all $i = 1, \dots, m$.

Even if one solves that problem, one might have not found Bob's original secret b. This raises the question of how rigid solutions to the simultaneous f-CSP are. A vague indication for some kind of rigidity is the fact that $f(b'b^{-1})$ and $b'b^{-1}$ are conjugated with every $f(s_i)$ $(1 \le i \le m)$ being a valid conjugator.

However, even if an attacker finds Bob's original key b, then she still faces the following problem.

$*_f$-MSP ($*_f$-submagma Membership Search Problem):

Input: $b, t_1, \dots, t_n \in G$.

OBJECTIVE: Find an expression of b as a tree-word in the submagma $\langle t_1, \ldots, t_n\rangle_{*_f}$ (notation $b = T_{*_f}(u_1, \ldots, u_k)$ for $u_i \in \{t_j\}_{j\leq n}$), if it exists.

Another approach is to attack (additionally to Bob's secret key) also Alice's key, i.e., to solve for the n-simultaneous f-CSP-instance $\{(t_j, t'_j)\}_{j\leq n}$ with $t'_j = f(a^{-1}t_j)a$. An oracle to that problem provides an element $a' \in G$ such that $t'_j = f(a'^{-1}t_j)a'$ for all j. Then the attacker hopes that computation of the f-commutator $[a', b']_f =: K'$ might give her the shared key $K = [a, b]_f$.

Though the f-CSP seems to be particulary interesting for non-invertible endomorphism $f \in End(G)$, here we compare K' with K for the simplest case where $f \in Inn(G)$, i.e., there exists an element $p \in G$ s.t. $f(x) = p^{-1}xp$. Then it is easy to show that $b'b^{-1} =: c_1$ lies in $\bigcap_i C_G(s_ip)$, and $a'a^{-1} = c_2 \in \bigcap_j C_G(t_jp)$. A straightforward computation gives

$$K' = a'^{-1}p^{-1}b'^{-1}a'pb' = a^{-1}c_2^{-1}p^{-1}b^{-1}c_1^{-1}c_2apc_1b.$$

We conclude that $K' = K$ if $[c_1, c_2] = [c_1, ap] = [c_2, bp] = 1$. But, in general, we have $C_G(ap) \neq \bigcap_i C_G(s_ip)$ and $C_G(bp) \neq \bigcap_j C_G(t_jp)$. Therefore, even in the case of $f \in Inn(G)$, we can't hope to reduce the f-AAGP to a simultaneous subgroup CSP, as we have done it for the classical AAGP in Proposition 2.3.

Nevertheless, as in the remark at the end of section 3.2.2, one may approach an n-sim f-CSP instance $\{(t_i, t'_i)\}_{i\leq n}$ by considering the $\binom{n}{2}$-simCSP instance $\{(t_i^{-1}t_j, (t'_i)^{-1}t'_j) \mid 1 \leq i \neq j \leq n\}$ in order to solve for a. Indeed, here we have

$$a^{-1}t_i^{-1}t_ja = (a^{-1}f(t_i^{-1}a))(f(a^{-1}t_j)a_r) = (t'_i)^{-1}t'_j.$$

Therefore, either the simultaneous CSP has to be hard in G, or, if the simCSP is (at least heuristically) approachable in G, it is recommended that the sets $\{t_i^{-1}t_j \mid 1 \leq i \neq j \leq n\}$ and $\{t_it_j^{-1} \mid 1 \leq i \neq j \leq n\}$ have large centralizers. This may be ensured by if the set $\{t_1, \ldots, t_n\}$ itself has a large centralizer. Similarly $\{s_1, \ldots, s_m\}$ should have a large centralizer.

4.2.1. *An example in pure braid groups.* Here we provide a concrete suggestion for the group G and the endomorphism $f \in End(G)$. Let G be the n-strand pure braid group P_n. For some small integer $d \geq 1$, consider the epimorphism $\eta_d : P_n \longrightarrow P_{n-d}$ given by 'pulling out' (or erasing) the last d strands, i.e. the strands $n-d+1, \ldots, n$. Recall the shift map ∂, and note that $\partial^d(P_{n-d}) \leq P_n$. Now, we define the endomorphism $f : P_n \longrightarrow P_n$ by the composition $f = \partial^d \circ \eta_d$.

Instantiations (of more powerful systems) using finite matrix groups are discussed in [**KaT13b**].

4.3. Non-associative AAG shifted commutator KEP in braid groups. Here we establish a (non-associative) AAG-KEP for braid groups with shifted conjugacy as LD-operation. Recall from Corollary 4.7 that the braid group $(B_\infty, *, \bar{*})$ forms a bi-LD-system. Also recall the definition of shift endomorphism ∂.

DEFINITION 4.15.. (SHIFTED COMMUTATOR) *The* shifted commutator *of an ordered pair* $(u, v) \in B_\infty^2$ *is defined by*

$$[u, v]_{\text{sh}} := u^{-1}\partial(v^{-1})\sigma_1\partial(u)v.$$

The *AAG shifted commutator KEP* for the bi-LD-system $(B_\infty, *, \bar{*})$ is given by the following further specifications of the general AAG scheme for magmas (section 3.1).

Set $M = N = S = B_\infty$, $\pi_i = \mathrm{id}_M$, $\beta_i(x, y) =: x *_i y$, $\bullet_i = \circ_i = *_i$ for $i = 1, 2$, and

$$x *_1 y = x \bar{*} y = \partial(x^{-1})\sigma_1^{-1}\partial(y)x, \qquad x *_2 y = x * y = \partial(x^{-1})\sigma_1\partial(y)x, \quad \text{and}$$
$$\gamma_1(u, v) = u^{-1}v, \qquad \gamma_2(u, v) = v^{-1}u.$$

Now, Alice and Bob perform the following protocol steps.

1. Alice generates her secret key a in the public submagma $S_1 = \langle s_1, \cdots, s_m\rangle_{\bar{*}}$ of $(B_\infty, *, \bar{*})$, and Bob chooses his secret key $b \in S_2 = \langle t_1, \cdots, t_n\rangle_*$.
2. Alice computes the elements $a\bar{*}t_1, \ldots, a\bar{*}t_n \in G$, and sends them to Bob. Analogously Bob computes the elements $b * s_1, \ldots, b * s_m \in G$, and sends them to Alice.
3. Alice, knowing $a = T_{\bar{*}}(r_1, \ldots, r_k)$ with $r_i \in \{s_1, \ldots, s_m\}$, computes from Bob's public key
$$T_{\bar{*}}(b * r_1, \ldots, b * r_k) = b * T_{\bar{*}}(r_1, \ldots, r_k) = b * a = \partial(b^{-1})\sigma_1\partial(a)b.$$
And Bob, knowing $b = T'_*(u_1, \ldots, u_{k'})$ with $u_j \in \{t_1, \ldots, t_n\}$, computes from Alice's public key
$$T'_*(a\bar{*}u_1, \ldots, a\bar{*}u_{k'}) = a\bar{*}T'_*(u_1, \ldots, u_{k'}) = a\bar{*}b = \partial(a^{-1})\sigma_1^{-1}\partial(b)a.$$
4. Alice computes $K := \gamma_1(a, b * a) = a^{-1}(b * a) = a^{-1}\partial(b^{-1})\sigma_1\partial(a)b = [a, b]_{\mathrm{sh}}$. Bob gets the shared key by
$$\gamma_2(b, a\bar{*}b) = (a\bar{*}b)^{-1}b = (\partial(a^{-1})\sigma_1^{-1}\partial(b)a)^{-1}b \overset{(3)}{=} K.$$

In order to break this scheme an attacker obviously has to solve the following base problem.

sh-AAGP (shifted Commutator AAG-Problem): Consider the bi-LD-system $(B_\infty, *, \bar{*})$. Let $A = \langle a_1, \ldots, a_k\rangle_{\bar{*}}$ and $B = \langle b_1, \ldots, b_m\rangle$ be two f.g. submagmas of $(B_\infty, *, \bar{*})$.

INPUT: $\{(a_i, y * a_i) \in G^2 | i = 1, \ldots, k\} \cup \{(b_j, x\bar{*}b_j) \in G^2 | j = 1, \ldots, m\}$ with $x \in A$ and $y \in B$.

OBJECTIVE: Find the shifted commutator $[x, y]_{\mathrm{sh}} := x^{-1}\partial(y^{-1})\sigma_1\partial(x)y$.

But a successful attack on Bob's secret key requires at least the solution of the following

***m*-sim sh-CSP** (m-simultaneous shifted Conjugacy Search Problem):

INPUT: Pairs $(s_1, s'_1), \ldots, (s_m, s'_m) \in G^2$ with $s'_i = b * s_i = \partial(b^{-1})\sigma_1\partial(s_i)b$ $\forall 1 \le i \le m$ for some (unknown) $b \in G$.

OBJECTIVE: Find an element $b' \in G$ with $\partial(b'^{-1})\sigma_1\partial(s_i)b' = \partial(b^{-1})\sigma_1\partial(s_i)b'$ for all $i = 1, \ldots, m$.

As in the case of f-conjugacy, one may argue that finding b is not sufficient, since the attacker still faces a submagma MSP for $(B_\infty, *, \bar{*})$. Furthermore, as for $*_f$, one may show that solving two simultaneous sh-CSP's (for Alice's and Bob's private keys) does in general not reduce the sh-AAGP to a simultaneous subgroup CSP, as for the classical AAGP.

Remark. Note that we actually do not need a bi-LD-system, like $(B_\infty, *, \bar{*})$, in order to build a AAG shifted commutator KEP. Indeed, two LD-operations, namely $x * y = \partial(x^{-1})\sigma_1\partial(y)x$ and its reverse $x *^{\mathrm{rev}} y = x\partial(y)\sigma_1\partial(x^{-1})$, suffice. Here $(B_\infty, *, *^{\mathrm{rev}})$ is not a bi-LD-system.

Alice and Bob choose $a \in \langle s_1, \ldots, s_m \rangle_*$ and $b \in \langle t_1, \ldots, t_n \rangle_{*^{\rm rev}}$, and send $\{a^{-1} *^{\rm rev} t_j\}_{j \le n}$ and $\{b^{-1} * s_i\}_{i \le m}$, respectively. Then they may compute

$$K_A = a^{-1}(b^{-1} * a) = a^{-1}\partial(b)\sigma_1\partial(a)b^{-1} = [a, b^{-1}]_{\rm sh} = (a^{-1} *^{\rm rev} b)b^{-1} = K_B.$$

Analogeously, one may build an AAG f-commutator KEP using $*_f$ and its reverse operation.

Non-simultaneity. Analogeous to the remarks in sections 3.2.2 and 4.2, an attacker might approach an m-sim shCP instance $\{(s_i, s_i' = b * s_i)\}_{i \le m}$ by considering the $\binom{m}{2}$-simCSP instance $\{(\partial(s_i^{-1}s_j), (s_i')^{-1}s_j') \mid 1 \le i \ne j \le m\}$ in order to solve for b. Indeed, here we have

$$b^{-1}\partial(s_i^{-1}s_j)b = (b^{-1}\partial(s_i^{-1})\sigma_1^{-1}\partial(b))(\partial(b^{-1})\sigma_1\partial(s_j)b) = (s_i')^{-1}s_j'.$$

Therefore, either it is recommended that the set $\{s_i^{-1}s_j \mid 1 \le i \ne j \le m\}$ (and analogeously $\{t_i^{-1}t_j \mid 1 \le i \ne j \le n\}$) has large centralizer. This may be ensured by if the sets $\{s_1, \ldots, s_m\}$ and $\{t_1, \ldots, t_n\}$ itself have a large centralizer.

Another strategy is to *abandon simultaneity*, i.e, to consider the critical case $m = n = 1$. Note that for shifted conjugacy (and its generalizations) we have opportunity to abandon simultaneity because only here the submagmas $\langle s \rangle_*$, $\langle s \rangle_{\bar{*}}$ generated by one element are nontrivial.

Generalized shifted conjugacy. It is straightforward to construct non-associative KEPs using generalized shifted conjugacy operations. We leave this to the reader.

5. Generalizations, further work and open problems

5.1. AAG-schemes over non-associative and non-commutative algebras. It is possible to generalize the AAG-KEP for magmas from section 3.1 in several ways. One generalization is very simple - just replace the magmas $(M, \bullet_1, \bullet_2)$ and $(N, \circ_1, \circ_2)$ by $(M, \{\bullet_{1,i}\}_{i \in I_1}, \{\bullet_{2,i}\}_{i \in I_2})$ and $(N, \{\circ_{1,i}\}_{i \in I_1}, \{\circ_{2,i}\}_{i \in I_2})$ for some index sets I_1, I_2, i.e. we introduce further binary operations. In particular, in the special case given by $M = N = S_1 = S_2$ and $\pi_1 = \pi_2 = {\rm id}_M$, Alice chooses her secret key a as an element from the submagma $\langle s_1, \ldots, s_m \rangle_{\{\bullet_{1,i}\}_{i \in I_1}}$.

To describe an element of such a submagma it is not sufficient to know the planar rooted binary tree T (providing the bracket structure) and the leaf elements $r_1, \ldots, r_k \in \{s_1, \ldots, s_m\}$, but we also need to assign binary operations (from the set $\{\bullet_{1,i}\}_{i \in I_1}$) to the internal nodes of the tree T.

For example,

In the following we write $T_{\{\bullet_{1,i}\}_{i \in I_1}}$, and we assume that T is then a planar rooted binary tree accompanied with such an assignment of its internal nodes.

Here we have to modify condition (1) from section 3.1 in the obvious way:

(1') $\beta_1(x, \cdot) : (M, \bullet_{2,i}) \to (N, \circ_{2,i})$ is for all $x \in S_1$, $i \in I_2$ a magma homomorphism, i.e.

$$\forall x \in S_1,\ y, y' \in M,\ i \in I_2: \quad \beta_1(x, y \bullet_{2,i} y') = \beta_1(x, y) \circ_{2,i} \beta_1(x, y').$$

Also $\beta_2(x, \cdot) : (M, \bullet_{1,i}) \to (N, \circ_{1,i})$ is for all $x \in S_2$, $i \in I_1$ a magma morphism, i.e.

$$\forall x \in S_2,\ y, y' \in M,\ i \in I_1: \quad \beta_2(x, y \bullet_{1,i} y') = \beta_2(x, y) \circ_{1,i} \beta_2(x, y').$$

If β_1, β_2 are defined by a binary operation from a bi- or multi-LD-system, then condtion (1') is satisfied by construction. Now one may build KEPs with this obvious modification. One example is the AAG shifted commutator KEP for the bi-LD-system $(B_\infty, *, \bar{*})$. Indeed, there Alice and Bob may have choosen their secret keys from $\langle s_1, \cdots, s_m \rangle_{*,\bar{*}}$ of $(B_\infty, *, \bar{*})$ and $\langle t_1, \cdots, t_n \rangle_{*,\bar{*}}$, respectively.

Recall that bi- and multi-LD-systems fulfill more homomorphic properties (i.e. distributive laws) than is necessary to build a KEP. As an example, consider the group ring $\mathbb{Z}G$. Recall that $(G, *_f)$ is an LD-system for any $f \in End(G)$. By construction, $(\mathbb{Z}G, *_f, +)$ is a non-commutative and non-associative algebra. It is straightforward to build a non-associative KEP over $\mathbb{Z}G$ analogous to the non-associative AAG f-commutator KEP. The only modification is that we choose the secret keys $a \in \langle s_1, \cdots, s_m \rangle_{*_f,+}$ and $b \in \langle t_1, \cdots, t_n \rangle_{*_f,+}$ for $s_1, \cdots, s_m, t_1, \cdots, t_n \in \mathbb{Z}G$.

Analogoulsly, it is straightforward to build a non-associative KEP over the non-associative bialgebra $(\mathbb{Z}B_\infty, \bar{*}, +)$.

Furthermore, one could consider non-commutative (but associative) special cases of these KEPs over non-associative algebras, if one restricts the secret keys a, b (or more precisely the projection $\pi_1(a), \pi_2(b)$) to $\langle s_1, \cdots, s_m \rangle_+$ and $b \in \langle t_1, \cdots, t_n \rangle_+$, respectively.

5.2. Open problems and further work.

- The AAG-KEP for magmas (see section 3.1) describes a general framework for building non-associative key establishment protocols. Our main examples are provided by LD-operations (f-conjugacy in groups and shifted conjugacy in braid groups). Recall also the systems based on (simultaneous) symmetric DP employing the non-associative operation given by $x \bullet y = xy^{-1}x$.

  Find other interesting instances of the general AAG-KEP for magmas (see section 3.1).
- How rigid are the solutions to the f-conjugacy problem in pure braid groups and the shifted conjugacy problem in braid groups? Note that, contrary to the f-conjugacy problem in pure braid groups, there exists a solution to the shifted conjugacy problem in braid groups [**KLT09**].
- Investigate heuristic attacks, especially length-based attacks [**HT02**], [**GK+05**], on the submagma MSP for non-associative LD-operations $*$ in braid groups. Of particular interest is here the non-simultaneous case $m = 1$ which emerges only for non-associative operations. I.e. consider the submagma MSP for the submagma $\langle s_1 \rangle_*$ generated by only one element.
- Develop other primitives like signature and authentication schemes in non-associative cryptography.

  Here we concentrated on KEPs which are usually the hardest to construct. Note that, using hash functions, it is easy to build public key encryption schemes from KEPs.
- For infinite groups, like braid groups, there are limitations on the depths of the trees describing a submagma element . Consider for example f-conjugacy in an infinte group G where $f \in End(G)$ satisfies $|f(x)| \leq |x|$ for all $x \in G$. Denote by $|\cdot| = |\cdot|_X$ the word length over some given generating set X of G. We conclude that

  $$|x *_f y| \leq |f(x^{-1})| + |f(y)| + |x| \leq 2|x| + |y| \leq 3\max\{|x|, |y|\}.$$

  Now, consider the following two extreme cases of trees with k leaves defining the bracket structure of a magma element in $\langle s_1, \ldots, s_m \rangle_*$ $(* = *_f)$.

The *left comb* ($r_j \in \{s_i\}_{i \leq m}$ for all $j = 1, \ldots k$)

$$LC(r_1, \ldots, r_k) := r_1 * (r_2 * (r_3 * \cdots r_{k-2} * (r_{k-1} * r_k) \cdots)),$$

and the *right comb*

$$RC(r_1, \ldots, r_k) := ((\cdots (r_1 * r_2) * r_3 * \cdots * r_{k-2}) * r_{k-1}) * r_k.$$

If $|s_i| \leq l_0$ for all $i = 1, \ldots, m$, then one may show by induction that

$$|LC(r_1, \ldots, r_k)| \leq (2k-1)l_0, \quad |RC(r_1, \ldots, r_k)| \leq (2^k - 1)l_0.$$

I.e. we can only prove an exponential (in k) upper bound on the word length of a magma element of tree depth $k-1$. But for left combs we have a linear upper bound. In practice, one may consider as keys either only elements of small tree depth, or we choose such elements whose bracket structure defining trees have a small "distance" from a left comb.

Define a proper notion of "distance" of planar rooted binary trees, and investigate how the word length growth for trees with "small distance" from the left comb LC. Determine a method how such trees can be generated efficiently.

- Recently B. Tsaban developed a deterministic polynomial time attack on the AAG commutator KEP in linear groups [**Ts15**] which also applies to several other non-commutative schemes. In short, Tsaban's *linear centralizer attack* exploits the fact that in classical AAG-KEP the shared key is the commutator $K = a^{-1}b^{-1}ab$. So, if we find solutions (up to centralizer elements) inside the centralizer of the centralizer of, say S_A, then these centralizer elements cancel and we recover K, even if these solutions were only in the linear matrix group in which we embed our linear group. But for KEPs with shared key $K = a_l b_l a_r b_r$, or K being an f-commutator in groups or a shifted commutator in braid groups, these centralizer elements would not cancel. Therefore, we conclude that, in its present state the linear centralizer attack does not apply to most of the non-associative schemes presented in this paper.

  Can the linear centralizer attack be improved to make it work against these KEPs?
- A further conceptual simplification of the linear centralizer attack is the *algebraic span attack* [**BKT14**]. Can the algebraic span attack be applied to make it work against these KEPs?

Acknowledgements

This work is an extension of a part of my PhD thesis. Therefore, I wish to thank my supervisor L. Gerritzen for his kind support, encouragement, constant interest and steadfast patience. In particular, his great interest in non-associatve algebraic structures as well as public key cryptography formed the scientific environment that made me bring these subjects together.

I am greatly indebted to P. Dehornoy who introduced an authentication scheme based on his notion of shifted conjugacy [**De06**]. This in the first place inspired me to come up with a KEP based on shifted conjugacy and in the course of this work to invent non-associative cryptography.

I thank M. Teicher and B. Tsaban who were my hosts at Bar-Ilan University, Israel. I acknowledge financial support by the Minerva Foundation of Germany.

This paper was written up during my stay at the MPIM Bonn, Jan-March 2012, and mainly finished during my postdoctoral stay at UQ, Brisbane. For the latter I acknowledge support by the Australian Research Council (project DP110101104).

For valuable and stimulating discussions I thank L. Gerritzen, R. Holtkamp and R. Avanzi at Ruhr-University Bochum, M. Kreuzer and G. Rosenberger in Dortmund, B. Tsaban, D. Goldfeld, B. Kunyavskij and R. Cohen at BIU, A. Myasnikov, A. Ushakov and G. Zapata at CRM, D. Grigoriev and D. Tieudjo at MPIM. For further discussions during conferences I thank J. Gonzalez-Meneses, P. Bellingeri, V. Gebhardt, E. and S.J. Lee. Particularly, I thank B. Tsaban for continuing discussions over the last years.

References

[AAG99] I. Anshel, M. Anshel, and D. Goldfeld, *An algebraic method for public-key cryptography*, Math. Res. Lett. **6** (1999), no. 3-4, 287–291, DOI 10.4310/MRL.1999.v6.n3.a3. MR1713130

[AAG03] I. Anshel, M. Anshel, and D. Goldfeld, *Non-abelian key agreement protocols*, Discrete Appl. Math. **130** (2003), no. 1, 3–12, DOI 10.1016/S0166-218X(02)00585-1. The 2000 Com^{2}MaC Workshop on Cryptography (Pohang). MR2008401

[BKT14] A. Ben-Zvi and A. Kalka and B. Tsaban, *Cryptanalysis via algebraic spans*, Cryptology ePrint Archive: `eprint.iacr.org/2014/041` .

[BCP97] W. Bosma, J. Cannon, and C. Playoust, *The Magma algebra system. I. The user language*, J. Symbolic Comput. **24** (1997), no. 3-4, 235–265, DOI 10.1006/jsco.1996.0125. Computational algebra and number theory (London, 1993). MR1484478

[Bo74] N. Bourbaki, *Elements of Mathematics: Algebra I*, Hermann (1974).

[CDW07] Z. Cao, X. Dong and L.Wang, *New Public Key Cryptosystems Using Polynomials over Non-commutative Rings*, (2007).

[CK+01] J. C. Cha, K. H. Ko, S. J. Lee, J. W. Han and J. H. Cheon, *An efficie nt implementation of braid groups*, Advances in Cryptology - ASIA-CRYPT 2001, LNCS **2248**, Springer (2001).

[De00] P. Dehornoy, *Braids and self-distributivity*, Progress in Mathematics, vol. 192, Birkhäuser Verlag, Basel, 2000. MR1778150

[De06] P. Dehornoy, *Using shifted conjugacy in braid-based cryptography*, Algebraic methods in cryptography, Contemp. Math., vol. 418, Amer. Math. Soc., Providence, RI, 2006, pp. 65–73, DOI 10.1090/conm/418/07946. MR2389289

[DK74] J. Dénes and A. D. Keedwell, *Latin squares and their applications*, Academic Press, New York-London, 1974. MR0351850

[DK91] J. Dénes and A. D. Keedwell, *Latin squares*, Annals of Discrete Mathematics, vol. 46, North-Holland Publishing Co., Amsterdam, 1991. New developments in the theory and applications; With contributions by G. B. Belyavskaya, A. E. Brouwer, T. Evans, K. Heinrich, C. C. Lindner and D. A. Preece; With a foreword by Paul Erdős. MR1096296

[DK92] J. Dénes and A. D. Keedwell, *A new authentication scheme based on Latin squares*, Discrete Math. **106/107** (1992), 157–161, DOI 10.1016/0012-365X(92)90543-O. A collection of contributions in honour of Jack van Lint. MR1181910

[DK02] J. Dénes and A. D. Keedwell, *Some applications of non-associative algebraic systems in cryptology*, Pure Math. Appl. **12** (2001), no. 2, 147–195. MR1905126

[DH76] W. Diffie and M. E. Hellman, *New directions in cryptography*, IEEE Trans. Information Theory **IT-22** (1976), no. 6, 644–654. MR0437208

[El85] T. ElGamal, *A public key cryptosystem and a signature scheme based on discrete logarithms*, IEEE Trans. Inform. Theory **31** (1985), no. 4, 469–472, DOI 10.1109/TIT.1985.1057074. MR798552

[Ge94] L. Gerritzen, *Grundbegriffe der Algebra* (German, with German summary), Lehrbuch Mathematik. [Mathematics Textbook], Friedr. Vieweg & Sohn, Braunschweig, 1994. Eine Einführung unter Berücksichtigung funktorieller Aspekte. [An introduction taking into consideration functorial aspects]. MR1269105

[GK+05] D. Garber, S. Kaplan, M. Teicher, B. Tsaban, and U. Vishne, *Probabilistic solutions of equations in the braid group*, Adv. in Appl. Math. **35** (2005), no. 3, 323–334, DOI 10.1016/j.aam.2005.03.002. MR2164922

[GMK08] D. Gligoroski, S. Markovski, and S. J. Knapskog, *Public key block cipher based on multivariate quadratic quasigroups*, Cryptology ePrint Archive, Report 2008/320.

[GS10] O. Grošek and M. Sýs, *Isotopy of Latin squares in cryptography*, Tatra Mt. Math. Publ. **45** (2010), 27–36. MR2680892

[HT02] J. Hughes and A. Tannenbaum, *Length-based attacks for certain group based encryption rewriting systems*, Workshop SECI02 SEcurité de la Communication sur Internet, Tunis (2002).

[Ka07] A. Kalka, *Representations of braid groups and braid-based cryptography*, PhD thesis, Ruhr-Universität Bochum (2007).
`www-brs.ub.ruhr-uni-bochum.de/netahtml/HSS/Diss/KalkaArkadiusG/`

[KLT09] A. Kalka, E. Liberman, and M. Teicher, *A note on the shifted conjugacy problem in braid groups*, Groups Complex. Cryptol. **1** (2009), no. 2, 227–230, DOI 10.1515/GCC.2009.227. MR2598991

[KaT13a] A. Kalka and M. Teicher, *Non-associative key establishment for left distributive systems*, Groups Complex. Cryptol. **5** (2013), no. 2, 169–191. MR3245105

[KaT13b] A. Kalka and M. Teicher *Iterated LD-Problem in non-associative key establishment*, submitted to GCC, arXiv: **abs/1312.6791**

[KaT16] A. Kalka and M. Teicher, *Non-associative key establishment protocols and their implementation*, Algebra and Computer Science, Contemp. Math. 677, Amer. Math. Soc., Providence, RI, 2016.

[KL+00] K. H. Ko, S. J. Lee, J. H. Cheon, J. W. Han, J.-s. Kang, and C. Park, *New public-key cryptosystem using braid groups*, Advances in cryptology—CRYPTO 2000 (Santa Barbara, CA), Lecture Notes in Comput. Sci., vol. 1880, Springer, Berlin, 2000, pp. 166–183, DOI 10.1007/3-540-44598-6_10. MR1850042

[Ko87] N. Koblitz, *Elliptic curve cryptosystems*, Math. Comp. **48** (1987), no. 177, 203–209, DOI 10.2307/2007884. MR866109

[KM99] C. Kościelny and G. L. Mullen, *A quasigroup-based public-key cryptosystem*, Int. J. Appl. Math. Comput. Sci. **9** (1999), no. 4, 955–963. MR1736680

[Mi85] V. S. Miller, *Use of elliptic curves in cryptography*, Advances in cryptology—CRYPTO '85 (Santa Barbara, Calif., 1985), Lecture Notes in Comput. Sci., vol. 218, Springer, Berlin, 1986, pp. 417–426, DOI 10.1007/3-540-39799-X_31. MR851432

[MZ12] E. Malekian, Ali Zakerolhosseini, *A non-associative lattice-based public key cryptosystem*, Security and Communication Networks, Volume **5**, Issue 2, pages 145-163, February 2012.

[PZ03] J. Proos and C. Zalka, *Shor's discrete logarithm quantum algorithm for elliptic curves*, Quantum Inf. Comput. **3** (2003), no. 4, 317–344. MR2003569

[Ra79] M. O. Rabin, *Digitized signatures and public-key functions as intractable as factorization*, MIT Laboratory for Computer Science Technical Report, LCS/TR-212 (1979), Currently available from: `www.lcs.mit.edu/publications/pubs/pdf/MIT-LCS-TR-212.pdf`

[RSA78] R. L. Rivest, A. Shamir, and L. Adleman, *A method for obtaining digital signatures and public-key cryptosystems*, Comm. ACM **21** (1978), no. 2, 120–126, DOI 10.1145/359340.359342. MR700103

[Se65] J.-P. Serre, *Lie algebras and Lie groups*, Lectures given at Harvard University, vol. 1964, W. A. Benjamin, Inc., New York-Amsterdam, 1965. MR0218496

[Shc09] V. A. Shcherbacov, *Quasigroups in cryptology*, Comput. Sci. J. Moldova **17** (2009), no. 2, 193–228. MR2666933

[Shc12] V.A. Shcherbacov, *Quasigroup based crypto-algorithms*, `arXiv:1110.6591v1`, 2012.

[Sh97] P. W. Shor, *Polynomial-time algorithms for prime factorization and discrete logarithms on a quantum computer*, SIAM J. Comput. **26** (1997), no. 5, 1484–1509, DOI 10.1137/S0097539795293172. MR1471990

[SU06] V. Shpilrain and A. Ushakov, *The conjugacy search problem in public key cryptography: unnecessary and insufficient*, Appl. Algebra Engrg. Comm. Comput. **17** (2006), no. 3-4, 285–289, DOI 10.1007/s00200-006-0009-6. MR2233787

[SZ06] V. Shpilrain and G. Zapata, *Combinatorial group theory and public key cryptography*, Appl. Algebra Engrg. Comm. Comput. **17** (2006), no. 3-4, 291–302, DOI 10.1007/s00200-006-0006-9. MR2233788

[St05] D. Stanovský, *Left symmetric left distributive operations on a group*, Algebra Universalis **54** (2005), no. 1, 97–103, DOI 10.1007/s00012-005-1926-0. MR2217967

[STR07] E. Sakalauskas, P. Tvarijonas, and A. Raulynaitis, *Key agreement protocol (KAP) using conjugacy and discrete logarithm problems in group representation level* (English, with English and Lithuanian summaries), Informatica (Vilnius) **18** (2007), no. 1, 115–124. MR2311150

[Ts15] B. Tsaban, *Polynomial-time solutions of computational problems in noncommutative-algebraic cryptography*, J. Cryptology **28** (2015), no. 3, 601–622, DOI 10.1007/s00145-013-9170-9. MR3355794

[WM85] N. R. Wagner and M. R. Magyarik, *A public-key cryptosystem based on the word problem*, Advances in cryptology (Santa Barbara, Calif., 1984), Lecture Notes in Comput. Sci., vol. 196, Springer, Berlin, 1985, pp. 19–36, DOI 10.1007/3-540-39568-7_3. MR820010

Department of Mathematics, Bar-Ilan University, Ramat Gan 52900, Israel
E-mail address: arkadius.kalka@rub.de

Contemporary Mathematics
Volume **677**, 2016
http://dx.doi.org/10.1090/conm/677/13624

Non-associative key establishment protocols and their implementation

Arkadius Kalka and Mina Teicher

ABSTRACT. We provide implementation details for non-associative key establishment protocols. In particular, we describe the implementation of non-associative key establishment protocols for all left self-distributive and all mutually left distributive systems.

1. Introduction

Currently public-key cryptography relies mainly on a few number-theoretic problems which remain still unbroken. Nevertheless, after the advent of quantum computers, systems like RSA, Diffie-Hellman and ECC will be broken easily [**10**]. One approach to develop new cryptographic primitives for the post quantum computer era became known as *non-commutative cryptography* where the commutative groups and rings involved in number-theoretic problems are replaced by non-commutative structures, and we consider computational problems therein [**1**]. The scope of non-commutative cryptography was broadened in [**6**, **7**] as we go beyond non-commutative, associative binary operations. We utilize non-associative binary operations, i.e. magmas, thus hoping to establish *non-associative public-key cryptography.* Here we focus on key establishment protocols (KEPs) as cryptographic primitives, because they are the most important and the hardest to construct. Most suggestions for non-commutative KEPs rely on variants of the conjugacy problem, while relevant base problem for our non-associative schemes are harder than the conjugacy problem. The seminal Anshel-Anshel-Goldfeld (AAG) KEP for monoids and groups [**5**] was generalized to a general AAG-KEP for magmas in [**6**, **7**] where left self-distributive (LD) systems (and their generalizations) naturally emerge as possible non-associative platform structures. Non-associative KEPs for all LD-, multi-LD-, and other left distributive systems were introduced in [**8**, **9**].

Outline. Though it may be a bit repetitive, we split the paper, for the convenience of the reader, in three main sections, each describing non-associative KEPs of increasing complexity. Each of these main sections is subdivided in subsections on preliminaries, protocol description, and a final subsection that contains pseudocode of the actual implementations. Section 2 deals with the simple KEP scheme for all LD-systems, introduced in [**9**]. Section 3 describes an improved scheme which

2010 *Mathematics Subject Classification.* Primary 20N02, 20F36.
Key words and phrases. Non-commutative cryptography, key establishment protocol, magma (grupoid), left selfdistributive system, braid group.

©2016 American Mathematical Society

allows for iterated left-multiplication in the given LD-system [**8**]. And section 4 is about a KEP that works for all mutually left distributive systems (also allowing iterated left-multiplication) [**8**]. We close with a brief security discussion and complexity analysis in section 5.

2. Non-associative KEPs for left self-distributive systems

2.1. Preliminaries.

DEFINITION 2.1. *(1) A* left self-distributive (LD) system $(S, *)$ *is a set S equipped with a binary operation $*$ on S which satisfies the* left self-distributivity law

$$x * (y * z) = (x * y) * (x * z) \quad \text{for all } x, y, z \in S.$$

Since the operation $*$ is (in general) non-associative, an element $y \in L$ that is the $*$-product of l elements $r_1, \dots, r_l \in L$ can be described by a planar rooted binary tree T whose l leaves are labelled by these elements $r_1, \dots, r_l$. In this case we use the notation $y = T_*(r_1, \dots, r_l)$.

By induction over the tree depth, it is easy to show that, for all elements $e, e_1, \dots, e_l \in (L, *)$ and all planar rooted binary trees T with l leaves, the following equations hold.

$$e * T_*(e_1, \dots, e_l) = T_*(e * e_1, \dots, e * e_l), \tag{2.1}$$

$$e * T_*(e_1, \dots, e_l) = T_*(e * e_1, \dots, e * e_l). \tag{2.2}$$

We list some examples of LD-systems. More details can be found in [**3, 4, 7–9**].

Conjugacy. A classical example of an LD-system is $(G, *)$ where G is a group equipped with the conjugacy operation $x * y = x^{-1}yx$ (or $x *^{\text{rev}} y = xyx^{-1}$).

LD-conjugacy. Let G be a group, and $f \in End(G)$. Set $x *_f y = f(x^{-1}y)x$, then $(G, *_f)$ is an LD-system.

Shifted conjugacy. Consider the braid group on infinitely many strands

$$B_\infty = \langle \{\sigma_i\}_{i \geq 1} \mid \sigma_i\sigma_j = \sigma_j\sigma_i \text{ for } |i-j| \geq 2,\ \sigma_i\sigma_j\sigma_i = \sigma_j\sigma_i\sigma_j \text{ for } |i-j| = 1 \rangle$$

where inside σ_i the $(i+1)$-th strand crosses over the i-th strand. The *shift map* $\partial : B_\infty \longrightarrow B_\infty$ defined by $\sigma_i \mapsto \sigma_{i+1}$ for all $i \geq 1$ is an injective endomorphism. Then B_∞ equipped with the shifted conjugacy operation $*$, defined by $x * y = \partial x^{-1} \cdot \sigma_1 \cdot \partial y \cdot x$, is an LD-system.

Symmetric conjugacy. For a group G, there exists yet another LD-operation. $(G, \circ)$ is an LD-system with $x \circ y = xy^{-1}x$.

f-symmetric conjugacy. Let G be a group, and $f \in End(G)$ an endomorphism that is also a projector ($f^2 = f$). Then $(G, \circ_f)$, defined by $x \circ_f y = f(xy^{-1})x$ is an LD-system.

2.2. Protocol. Here we describe a KEP that works for any LD-system $(L, *)$ (see [**9**]).

Let $s_1, \dots, s_{m_A}, t_1, \dots, t_{m_B} \in L$ be some public elements which define the two public sub-LD-systems $S_A = \langle s_1, \cdots, s_{m_A} \rangle_*$ and $S_B = \langle t_1, \cdots, t_{m_B} \rangle_*$. Alice and Bob perform the following protocol steps.

Protocol 1: KEY ESTABLISHMENT FOR ANY LD-SYSTEM $(L, *)$.

1: Alice generates her secret key $(a_0, a) \in S_A \times L$, and Bob chooses his secret key $b \in S_B$.

2: Alice computes the elements $a * t_1, \dots, a * t_{m_B}, p_0 = a * a_0 \in L$, and sends them to Bob. Bob computes $b * s_1, \dots, b * s_{m_A} \in L$, and sends them to Alice.

3: Alice, knowing $a_0 = T_*(r_1, \dots, r_l)$ with $r_i \in \{s_1, \dots, s_{m_A}\}$, computes from the received message

$$T_*(b * r_1, \dots, b * r_l) = b * T_*(r_1, \dots, r_l) = b * a_0.$$

And Bob, knowing $b = T'_*(u_1, \dots, u_{l'})$ with $u_j \in \{t_1, \dots, t_{m_B}\}$, computes from his received message

$$T'_*(a * u_1, \dots, a * u_{l'}) = a * T'_*(u_1, \dots, u_l) = a * b.$$

4: Alice computes $K_A = a * (b * a_0)$. Bob gets the shared key by

$$K_B := (a * b) * p_0 = (a * b) * (a * a_0) \overset{(LD)}{=} K_A.$$

This protocol is an asymmetric modification of the Anshel-Anshel-Goldfeld protocols for magmas introduced in [**6**, **7**].

FIGURE 1. PROTOCOL 1: KEY ESTABLISHMENT FOR ANY LD-SYSTEM

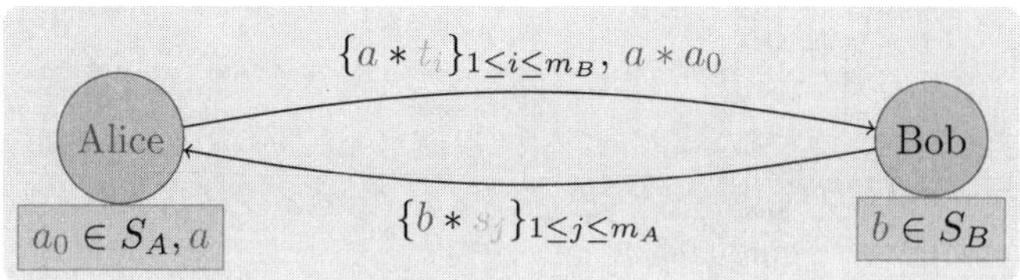

2.3. Implementation. *Planar rooted binary trees.* We need some efficient way to encode the planar rooted binary tree which determines the bracket structure of an element given as product of other elements. Let PBT_n denote the set of planar rooted binary trees (also known as *full binary trees*) with n internal nodes (and $n+1$ leaves), then $|PBT_n| = Cat(n)$ where $Cat(n) = \frac{1}{n+1}\binom{2n}{n}$ denotes the n-th Catalan number. There exists a rich variety of other Catalan sets with well understood bijections between them, e.g., diagonal avoiding paths (also known as mountain ranges), polygon triangulations, Dyck words, planar rooted trees (not only binary) and non-crossing partitions. We use the following succinct representation for Catalan sets taken from [**2**]. Denote $[n] = \{1, \dots, n\}$. To each $T \in PBT_n$ we associate a vector (array) $T \in [n]^n$ such that $T[i] \leq T[j]$ for $i < j$ and $T[i] \leq i$ for all $i \in [n]$. By abuse of notation we call the set of such vectors in $[n]^n$ also PBT_n. The function EVALUATE TREE evaluates the product of $e_1, \dots, e_{n+1}$ in $(L, *)$ where the bracket structure is given by the tree T.

Algorithm 1: EvaluateTree

Input: $(T, (e_1, \ldots, e_{n+1})) \in PBT_n \times L^{n+1}$.
Output: $e = T_*(e_1, \ldots, e_{n+1})$

Initialize $Seq := (e_1, \ldots, e_{n+1})$;
for $j := n$ **to** *1* **by** *-1* **do**
  $pos := T[j]$;
  $Seq[pos] := Seq[pos] * Seq[pos+1]$;
  Remove($\tilde{~}Seq, pos+1$); Remove($\tilde{~}T, pos$);
end
return $Seq[1]$;

Protocol implementation. Given $m_a, m_B \in \mathbb{N}$, Alice and Bob first choose random vectors $\mathcal{G}_A = (s_1, \ldots, s_{m_A}) \in L^{m_A}$ and $\mathcal{G}_B = (t_1, \ldots, t_{m_B}) \in L^{m_B}$ which determine the public submagmas $S_A = \langle \mathcal{G}_A \rangle_{O_A}$ and $S_B = \langle \mathcal{G}_B \rangle_{O_B}$, respectively. We fix some distributions on L, so that we may generate random elements from this set (according to this distribution). Given some $n_{\min} \le n_{\max}$, Alice and Bob generate their secret keys as described in the following algorithms.

Algorithm 2: GENERATE ALICE'S PRIVATE KEY

Input: $\mathcal{G}_A, n_{\min}, n_{\max} \in L^{m_A} \times \mathbb{N} \times \mathbb{N}$.
Output: $(Ia_0, Ta_0, a_0, a) \in [m_A]^{n_{a_0}+1} \times PBT_{n_{a_0}} \times L \times L$

$n_{a_0} \leftarrow$ Random($\{n_{\min}, \ldots, n_{\max}\}$);
$Ia_0 \leftarrow$ Random($[m_A]^{n_{a_0}}$);
for $i := 1$ **to** $n_{a_0} + 1$ **do**
  $Seqa_0[i] := \mathcal{G}_A[Ia_0[i]]$;
end
$Ta_0 \leftarrow$ Random($PBT_{n_{a_0}}$);
$a_0 :=$ EvaluateTree($Ta_0, Seqa_0$);
$a \leftarrow$ Random(L^{k_A});
return (Ia_0, Ta_0, a_0, a);

Algorithm 3: GENERATE BOB'S PRIVATE KEY

Input: $\mathcal{G}_B, n_{\min}, n_{\max} \in L^{m_B} \times \mathbb{N} \times \mathbb{N}$.
Output: $(Ib, Tb, b) \in [m_B]^{n_b+1} \times PBT_{n_b} \times L$

$n_b \leftarrow$ Random($\{n_{\min}, \ldots, n_{\max}\}$);
$Ib \leftarrow$ Random($[m_B]^{n_b}$);
for $i := 1$ **to** $n_b + 1$ **do**
  $Seqb[i] := \mathcal{G}_A[Ia_0[i]]$;
end
$Tb \leftarrow$ Random(PBT_{n_b});
$b :=$ EvaluateTree($Tb, Seqb$);
return (Ib, Tb, b);

Then Alice and Bob generate their public and shared keys as described in the following algorithms.

Algorithm 4: GENERATE ALICE'S PUBLIC KEY

Input: $(a, a_0, \mathcal{G}_B) \in L^{k_A} \times L \times L^{m_B}$.
Output: $(p_A, p_0) \in L^{m_B} \times L$
for $k := 1$ **to** m_B **do**
  $p_A[k] := a * \mathcal{G}_B[k]$;
end
$p_0 := a * p_0$;
return (p_A, p_0);

Algorithm 5: GENERATE BOB'S PUBLIC KEY

Input: $(b, \mathcal{G}_A) \in L \times L^{m_A}$.
Output: $p_B \in L^{m_A}$
for $k := 1$ **to** m_A **do**
  $p_B[k] := b * \mathcal{G}_A[k]$;
end
return p_B;

Algorithm 6: GENERATE ALICE'S SHARED KEY

Input: $(Ia_0, Ta_0, a, p_B) \in [m_A]^{n_{a_0}+1} \times PBT_{n_{a_0}} \times L \times L^{m_A}$.
Output: $K_A \in L$
for $i := 1$ **to** $n_{a_0} + 1$ **do**
  $Seqba_0 := (p_B[Ia_0[i]])$;
end
$K_A :=$ EvaluateTree$(Ta_0, Seqba_0)$;
$K_A := a * K_A$;
return K_A;

Algorithm 7: GENERATE BOB'S SHARED KEY

Input: $(Ib, Tb, b, p_A, p_0) \in [m_B]^{n_b+1} \times PBT_{n_b} \times L \times L^{m_B} \times L$.
Output: $K_B \in L$
for $i := 1$ **to** $n_b + 1$ **do**
  $Seqab := (p_A[Ib[i]])$;
end
$K_B :=$ EvaluateTree$(Tb[k], Seqab)$;
$K_B := K_B * p_0$;
return K_B;

3. Iterated protocol for LD-systems

3.1. Preliminaries.

PROPOSITION 3.1. [8] *Let $(L, *)$ be an LD-system and $k \in \mathbb{N}$. Then, for all $x = (x_1, \ldots, x_k) \in L^k$, the iterated left multiplication map*

$$\phi_x : y \mapsto x_k * (x_{k-1} * \cdots * (x_2 * (x_1 * y)) \cdots)$$

defines a magma endomorphism of L.

In particular, the following equations hold for all $k, l \in \mathbb{N}$, $x \in L^k$, $e_1, \ldots, e_l \in L$ and all planar rooted binary trees T with l leaves.

$$\phi_x(T_*(e_1, \ldots, e_l)) = T_*(\phi_x(e_1), \ldots, \phi_x(e_l)).$$

3.2. Protocol. Now, we are going to describe an improved KEP that applies to any LDsystem $(L, *)$. This protocol is an iterated version of Protocol 1 in [**9**] and an asymmetric modification of the Anshel-Anshel-Goldfeld protocols for magmas introduced in [**6**, **7**]. We have two subsets of public elements $\{s_1, \cdots, s_m\}$ and $\{t_1, \cdots, t_n\}$ of L. Also, recall that $S_A = \langle s_1, \cdots, s_m\rangle_*$ and $S_B = \langle t_1, \cdots, t_n\rangle_*$. Alice and Bob perform the following protocol steps.

Protocol 2: KEY ESTABLISHMENT FOR ANY LD-SYSTEM $(L, *)$.

1: Alice generates her secret key $(a_0, a_1, \ldots, a_{k_A}) \in S_A \times L^{k_A}$, and Bob chooses his secret key $b \in S_B^{k_B}$. In particular, Alice's and Bob's secret magma morphisms α and β are given by

$$\begin{aligned} \alpha(y) &= a_{k_A} * (a_{k_A-1} * \cdots * (a_2 * (a_1 * y)) \cdots) \quad \text{and} \\ \beta(y) &= b_{k_B} * (b_{k_B-1} * \cdots * (b_2 * (b_1 * y)) \cdots), \end{aligned}$$

respectively.

2: Alice computes the elements $(\alpha(t_i))_{1 \le i \le n} \in L^n, p_0 = \alpha(a_0) \in L$, and sends them to Bob. Bob computes the vector $(\beta(s_j))_{1 \le j \le m} \in L^m$, and sends it to Alice.

3: Alice, knowing $a_0 = T_*(r_1, \ldots, r_l)$ with $r_i \in \{s_1, \ldots, s_m\}$, computes from the received message

$$T_*(\beta(r_1), \ldots, \beta(r_l)) = \beta(T_*(r_1, \ldots, r_l)) = \beta(a_0).$$

And Bob, knowing for all $1 \le j \le k_B$, $b_j = T_*^{(j)}(u_{j,1}, \ldots, u_{j,l_j})$ with $u_{j,i} \in \{t_1, \ldots, t_n\} \forall i \le l_j$ for some $l_j \in \mathbb{N}$, computes from his received message for all $1 \le j \le k_B$

$$T_*^{(j)}(\alpha(u_{j,1}), \ldots, \alpha(u_{j,l_j})) = \alpha(T_*^{(j)}(u_{j,1}, \ldots, u_{j,l_j}) = \alpha(b_j).$$

4: Alice computes $K_A = \alpha(\beta(a_0))$. Bob gets the shared key by

$$K_B := \alpha(b_{k_B}) * (\alpha(b_{k_B-1}) * (\cdots (\alpha(b_2) * (\alpha(b_1) * p_0)) \cdots)) \overset{(LD)}{=} K_A.$$

FIGURE 2. PROTOCOL 2: KEY ESTABLISHMENT FOR ANY LD-SYSTEM - ITERATED VERSION.

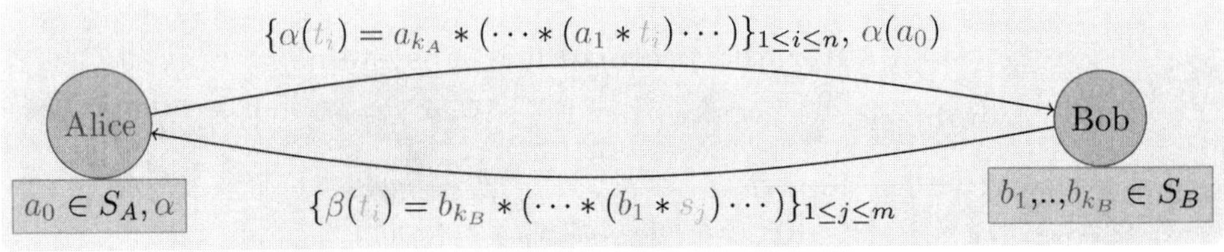

3.3. Implementation. $m_A, b_B \in \mathbb{N}$ and $S_A, S_B \leq (L, *)$ are given as in the previous section. Given some $n_{\min} \leq n_{\max}$ and $k_{\min} \leq k_{\max} \in \mathbb{N}$, Alice and Bob generate their secret keys as described in the following algorithms.

Algorithm 8: GENERATE ALICE'S PRIVATE KEY

Input: $\mathcal{G}_A, n_{\min}, n_{\max}, k_{\min}, k_{\max} \in L^{m_A} \times \mathbb{N}^4$.
Output: $(Ia_0, Ta_0, a_0, a) \in [m_A]^{n_{a_0}+1} \times PBT_{n_{a_0}} \times L \times L^{k_A}$

$n_{a_0} \leftarrow \mathsf{Random}(\{n_{\min}, \ldots, n_{\max}\})$;
$k_A \leftarrow \mathsf{Random}(\{k_{\min}, \ldots, k_{\max}\})$;
$Ia_0 \leftarrow \mathsf{Random}([m_A]^{n_{a_0}})$;
for $i := 1$ **to** $n_{a_0} + 1$ **do**
  | $Seqa_0[i] := \mathcal{G}_A[Ia_0[i]]$;
end
$Ta_0 \leftarrow \mathsf{Random}(PBT_{n_{a_0}})$;
$a_0 := \mathsf{EvaluateTree}(Ta_0, Seqa_0)$;
$a \leftarrow \mathsf{Random}(L^{k_A})$; **return** (Ia_0, Ta_0, a_0, a);

Algorithm 9: GENERATE BOB'S PRIVATE KEY

Input: $\mathcal{G}_B, n_{\min}, n_{\max}, k_{\min}, k_{\max} \in L^{m_B} \times \mathbb{N}^4$.
Output: $(Ib, Tb, b) \in ([m_B]^{n_b+1})^{k_B} \times (PBT_{n_b})^{k_B} \times L^{k_B}$

$n_b \leftarrow \mathsf{Random}(\{n_{\min}, \ldots, n_{\max}\})$;
$k_B \leftarrow \mathsf{Random}(\{k_{\min}, \ldots, k_{\max}\})$;
for $k := 1$ **to** k_B **do**
  | $Ib[k] \leftarrow \mathsf{Random}([m_B]^{n_b})$;
  | **for** $i := 1$ **to** $n_b + 1$ **do**
  | | $Seqb[k][i] := \mathcal{G}_A[Ia_0[i]]$;
  | **end**
  | $Tb[k] \leftarrow \mathsf{Random}(PBT_{n_b})$;
  | $b[k] := \mathsf{EvaluateTree}(Tb[k], Seqb[k])$;
end
return (Ib, Tb, b);

Then Alice and Bob generate their public and shared keys as described in the following algorithms.

Algorithm 10: GENERATE ALICE'S PUBLIC KEY

Input: $(a, a_0, \mathcal{G}_B) \in L^{k_A} \times L \times L^{m_B}$.
Output: $(p_A, p_0) \in L^{m_B} \times L$

for $k := 1$ **to** m_B **do**
  $p_A[k] := \mathcal{G}_B[k]$;
  for $i := 1$ **to** k_A **do**
    $p_A[k] := a[i] * p_A[k]$;
  end
end
$p_0 := a_0$;
for $i := 1$ **to** k_A **do**
  $p_0 := a[i] * p_0$;
end
return (p_A, p_0);

Algorithm 11: GENERATE BOB'S PUBLIC KEY

Input: $(b, \mathcal{G}_A) \in L^{k_B} \times L^{m_A}$.
Output: $p_B \in L^{m_A}$

for $k := 1$ **to** m_A **do**
  $p_B[k] := \mathcal{G}_A[k]$;
  for $i := 1$ **to** k_B **do**
    $p_B[k] := b[i] * p_B[k]$;
  end
end
return p_B;

Algorithm 12: GENERATE ALICE'S SHARED KEY

Input: $(Ia_0, Ta_0, a, p_B) \in [m_A]^{n_{a_0}+1} \times PBT_{n_{a_0}} \times L^{k_A} \times L^{m_A}$.
Output: $K_A \in L$

$K_A := \mathsf{EvaluateTree}(Ta_0, (p_B[Ia_0[i]])_{i \leq n_{a_0}+1})$;
for $k := 1$ **to** k_A **do**
  $K_A := a[k] * K_A$;
end
return K_A;

Algorithm 13: GENERATE BOB'S SHARED KEY

Input: $(Ib, Tb, b, p_A, p_0) \in ([m_B]^{n_b+1})^{k_B} \times (PBT_{n_b})^{k_B} \times L^{k_B} \times L^{m_B} \times L$.
Output: $K_A \in L$

Initialize $lfactors := []$; $K_B := p_0$;
for $k := 1$ **to** k_B **do**
  $lfactors[k] := \mathsf{EvaluateTree}(Tb[k], (p_A[Ib[k][i]])_{i \leq n_b+1})$;
  $K_B := lfactors[k] * K_B$;
end
return K_B;

4. Non-associative KEPs for mutually left distributive systems

4.1. Preliminaries.

DEFINITION 4.1. *(1) Let I be an index set. A* multi-LD-system $(S, (*_i)_{i\in I})$ *is a set S equipped with a family of binary operations $(*_i)_{i\in I}$ on S such that*

$$x *_i (y *_j z) = (x *_i y) *_j (x *_i z) \quad \text{for all } x, y, z \in S$$

is satisfied for every i, j in I. Especially, it holds for $i = j$, i.e., $(S, *_i)$ is an LD-system. If $|I| = 2$ then we call S a bi-LD-system.

(2) A mutually left distributive system $(S, *_a, *_b)$ *is a set S equipped with two binary operations $*_a, *_b$ on S such that*

$$x *_a (y *_b z) = (x *_a y) *_b (x *_a z) \quad x *_b (y *_a z) = (x *_b y) *_a (x *_b z) \quad \text{for all } x, y, z \in S.$$

More vaguely, we will also use the terms *partial multi-LD-system* and *mutually left distributive system* if the laws of a multi-LD-system are only fulfilled for special subsets of S or if only some of these (left) distributive laws are satisfied. A mutually left distributive system $(L, *_a, *_b)$ is only a partial bi-LD-system. The left selfdistributivity laws need not hold, i.e., $(L, *_a)$ and $(L, *_b)$ are in general not LD-systems. We list some examples. More details can be found in [**3, 4, 7–9**].

Shifted conjugacy. The braid group on infinitely many strands B_∞ equipped with the shifted conjugacy operations $*$, $\bar{*}$ defined by $x * y = \partial x^{-1} \cdot \sigma_1 \cdot \partial y \cdot x$ and $x \bar{*} y = \partial x^{-1} \cdot \sigma_1^{-1} \cdot \partial y \cdot x$ is a bi-LD-system.

Generalized shifted conjugacy in braid groups. Let, for $n \geq 2$, $\delta_n = \sigma_{n-1} \cdots \sigma_2 \sigma_1$. For $p, q \geq 1$, we set $\tau_{p,q} = \delta_{p+1} \partial(\delta_{p+1}) \cdots \partial^{q-1}(\delta_{p+1})$.

PROPOSITION 4.2. [**8, 9**] *$(B_\infty, *_1, *_2)$ with binary operations*

$$x *_i y = \partial^p(x^{-1}) a_i \partial^p(y) x$$

$(i = 1, 2)$ with $a_1 = a_1' \tau_{p,p}^{\pm 1} a_1''$, $a_2 = a_2' \tau_{p,p}^{\pm 1} a_2''$ for some $a_1', a_1'', a_2', a_2'' \in B_p$ is a mutually left distributive system if and only if $[a_1', a_2''] = [a_2', a_1''] = [a_1', a_2'] = 1$. (Note that $[a_1', a_1'']$, $[a_2', a_2'']$ and $[a_1'', a_2'']$ may be nontrivial. If, in addition $[a_1', a_1''] = [a_2', a_2''] = 1$ holds, then $(B_\infty, *_1, *_2)$ is a bi-LD-system.)

We are going to describe a KEP that works for all mutually left distributive systems, in particular all bi-LD-systems (and all LD-systems). Consider a set L equipped with a pool of binary operations $O_A \cup O_B$ (O_A and O_B non-empty) such that the operations in O_A are distributive over those in O_B and vice versa, i.e. the following holds for all $x, y, z \in L$, $*_\alpha \in O_A$ and $*_\beta \in O_B$.

$$x *_\alpha (y *_\beta z) = (x *_\alpha y) *_\beta (x *_\alpha z), \text{ and} \tag{4.1}$$

$$x *_\beta (y *_\alpha z) = (x *_\beta y) *_\alpha (x *_\beta z). \tag{4.2}$$

Then $(L, *_\alpha, *_\beta)$ is a mutually left distributive system for all $(*_\alpha, *_\beta) \in O_A \times O_B$. Note that, if $O_A \cap O_B \neq \emptyset$, then $(L, O_A \cap O_B)$ is a multi-LD-system.

Let $s_1, \ldots, s_m, t_1, \ldots, t_n \in L$ be some public elements. We denote $S_A = \langle s_1, \cdots, s_m \rangle_{O_A}$ and $S_B = \langle t_1, \cdots, t_n \rangle_{O_B}$. For example, an element y of S_A can be described by a planar rooted binary tree T whose k leaves are labelled by these other elements $r_1, \ldots, r_k$ with $r_i \in \{s_i\}_{i \leq m}$. Here the tree contains further information, namely to each internal vertex we assign a binary operation $*_i \in O_A$. We use the notation $y = T_{O_A}(r_1, \ldots, r_k)$. The subscript O_A tells us that the grafting of subtrees of T corresponds to the operation $*_i \in O_A$. Consider, for example,

FIGURE 3. $(s_3 *_{\alpha_1} ((s_3 *_{\alpha_4} (s_1 *_{\alpha_1} s_2)) *_{\alpha_2} s_1)) *_{\alpha_1} ((s_2 *_{\alpha_2} s_3) *_{\alpha_3} s_2) \in S_A$

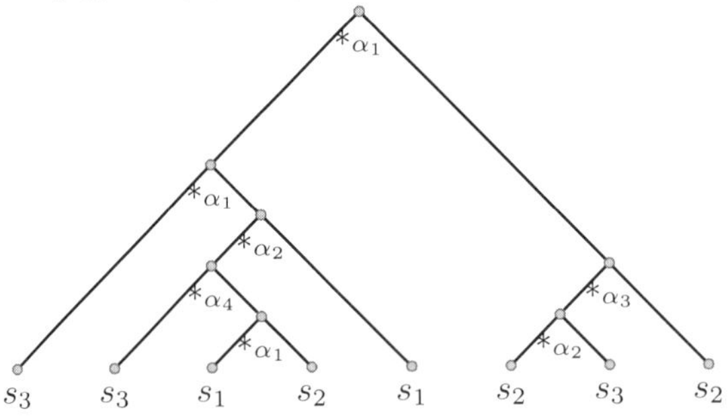

the element $y = (s_3 *_{\alpha_1} ((s_3 *_{\alpha_4} (s_1 *_{\alpha_1} s_2)) *_{\alpha_2} s_1)) *_{\alpha_1} ((s_2 *_{\alpha_2} s_3) *_{\alpha_3} s_2)$. The corresponding labelled planar rooted binary tree T is displayed in the Figure 1.

Let $*_\alpha \in O_A$ and $*_\beta \in O_B$. By induction over the tree depth, it is easy to show that, for all elements $e, e_1, \dots, e_l \in (L, O_A \cup O_B)$ and all planar rooted binary trees T with l leaves, the following equations hold.

$$e *_\alpha T_{O_B}(e_1, \dots, e_l) = T_{O_B}(e *_\alpha e_1, \dots, e *_\alpha e_l), \tag{4.3}$$
$$e *_\beta T_{O_A}(e_1, \dots, e_l) = T_{O_A}(e *_\beta e_1, \dots, e *_\beta e_l). \tag{4.4}$$

PROPOSITION 4.3. *(See Proposition 4.1 in* [**8**].*) Consider $(L, O_A \cup O_B)$ such that $(L, *_A, *_B)$ is a mutually left distributive system for all $(*_A, *_B) \in O_A \times O_B$, and let $k \in \mathbb{N}$. Then, for all $x = (x_1, \dots, x_k) \in L^k$, $o_A = (*_{A_1}, \dots, *_{A_k}) \in O_A^k$, and $o_B = (*_{B_1}, \dots, *_{B_k}) \in O_B^k$, the iterated left multiplication maps*

$$\phi_{(x,o_A)}: \quad y \mapsto x_k *_{A_k} (x_{k-1} *_{A_{k-1}} \cdots *_{A_3} (x_2 *_{A_2} (x_1 *_{A_1} y)) \cdots) \text{ and}$$
$$\phi_{(x,o_B)}: \quad y \mapsto x_k *_{B_k} (x_{k-1} *_{B_{k-1}} \cdots *_{B_3} (x_2 *_{B_2} (x_1 *_{B_1} y)) \cdots)$$

define a magma endomorphisms of (L, O_B) and (L, O_A), respectively[1].

In particular, the following equations hold for all $k, l \in \mathbb{N}$, $a, b \in L^k$, $o_A \in O_A^k$, $o_B \in O_B^k$, $e_1, \dots, e_l \in L$ and all planar rooted binary trees T with l leaves.

$$\phi_{(a,o_A)}(T_{O_B}(e_1, \dots, e_l)) = T_{O_B}(\phi_{(a,o_A)}(e_1), \dots, \phi_{(a,o_A)}(e_l)), \tag{4.5}$$
$$\phi_{(b,o_B)}(T_{O_A}(e_1, \dots, e_l)) = T_{O_A}(\phi_{(b,o_B)}(e_1), \dots, \phi_{(b,o_B)}(e_l)) \tag{4.6}$$

4.2. Protocol. Now, we are going to describe a KEP that applies to any system $(L, O_A \cup O_B)$ as described above. We have two subsets of public elements $\{s_1, \cdots, s_m\}$ and $\{t_1, \cdots, t_n\}$ of L. Also, recall that $S_A = \langle s_1, \cdots, s_m \rangle_{O_A}$ and $S_B = \langle t_1, \cdots, t_n \rangle_{O_B}$. Alice and Bob perform the following protocol steps.

Protocol 3: KEY ESTABLISHMENT FOR THE PARTIAL MULTI-LD-SYSTEM $(L, O_A \cup O_B)$.

1: Alice generates her secret key $(a_0, a, o_A) \in S_A \times L^{k_A} \times O_A^{k_A}$, and Bob chooses his secret key $(b, o_B) \in S_B^{k_B} \times O_B^{k_B}$. Denote $o_A = (*_{A_1}, \dots, *_{A_{k_A}})$

[1] Let $(L, *)$ be a set equipped with a binary operation. A *magma endomorphism* ϕ is a map from L into L such that $\phi(y_1 * y_2) = \phi(y_1) * \phi(y_2)$ for all $y_1, y_2 \in L$.

FIGURE 4. KEP FOR THE PARTIAL MULTI-LD-SYSTEM $(L, O_A \cup O_B)$.

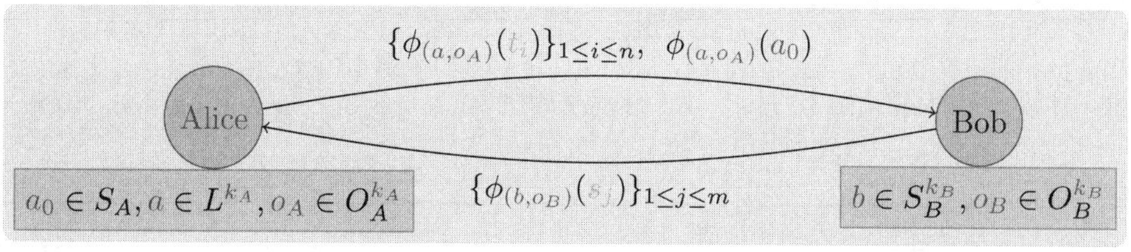

and $o_B = (*_{B_1}, \dots, *_{B_{k_B}})$, then Alice's and Bob's secret magma morphisms α and β are given by

$$\begin{aligned} \alpha(y) &= a_{k_A} *_{A_{k_A}} (a_{k_A-1} *_{A_{k_A-1}} \cdots *_{A_3} (a_2 *_{A_2} (a_1 *_{A_1} y)) \cdots) \quad \text{and} \\ \beta(y) &= b_{k_B} *_{B_{k_B}} (b_{k_B-1} *_{B_{k_B-1}} \cdots *_{B_3} (b_2 *_{B_2} (b_1 *_{B_1} y)) \cdots), \end{aligned}$$

respectively.

2: Alice computes the elements $(\alpha(t_i))_{1\le i\le n} \in L^n, p_0 = \alpha(a_0) \in L$, and sends them to Bob. Bob computes the vector $(\beta(s_j))_{1\le j\le m} \in L^m$, and sends it to Alice.

3: Alice, knowing $a_0 = T_{O_A}(r_1, \dots, r_l)$ with $r_i \in \{s_1, \dots, s_m\}$, computes from the received message

$$T_{O_A}(\beta(r_1), \dots, \beta(r_l)) = \beta(T_{O_A}(r_1, \dots, r_l)) = \beta(a_0).$$

And Bob, knowing for all $1 \le j \le k_B$, $b_j = T_{O_B}^{(j)}(u_{j,1}, \dots, u_{j,l_j})$ with $u_{j,i} \in \{t_1, \dots, t_n\} \forall i \le l_j$ for some $l_j \in \mathbb{N}$, computes from his received message for all $1 \le j \le k_B$

$$T_{O_B}^{(j)}(\alpha(u_{j,1}), \dots, \alpha(u_{j,l_j})) = \alpha(T_{O_B}^{(j)}(u_{j,1}, \dots, u_{j,l_j}) = \alpha(b_j).$$

4: Alice computes $K_A = \alpha(\beta(a_0))$. Bob gets the shared key by

$$K_B := \alpha(b_{k_B}) *_{B_{k_B}} (\alpha(b_{k_B-1}) *_{B_{k_B-1}} (\cdots (\alpha(b_2) *_{B_2} (\alpha(b_1) *_{B_1} p_0)) \cdots)) \overset{\alpha \text{ homo}}{=} K_A.$$

Here the operation vectors $o_A \in O_A^{k_A}$ and $o_B \in O_B^{k_B}$ are part of Alice's and Bob's private keys. Also explicit expressions of $a_0 \in S_A$ and all $b_i \in S_B$ as treewords $T, T^{(i)}$ (for all $1 \le i \le k_B$) are also parts of the private keys - though we did not mention it explicitly in step 1 of the protocols. But here T_{O_A} and T'_{O_B} also contain all the information about the operations (in O_A or O_B, respectively) at the internal vertices of $T, T^{(1)}, \dots, T^{(k_B)}$.

4.3. Implementation. *Planar rooted binary trees.* Let L be a magma and O be a set of binary operations on L. Given a vector of operations $o = (*_{o[1]}, \ldots, *_{o[n]}) \in O^n$ and a sequence of leave elements $(e_1, \ldots, e_{n+1}) \in L^{n+1}$, then the modified function EVALUATE TREE (see below) evaluates the product of $e_1, \ldots, e_{n+1}$ where the bracket structure is given by the tree T and the operations on the internal vertices of T are given by o.

Algorithm 14: EvaluateTree

Input: $(T, o, (e_1, \ldots, e_{n+1})) \in PBT_n \times O^n \times L^{n+1}$.
Output: $e = T_o(e_1, \ldots, e_{n+1})$

Initialize $Seq := (e_1, \ldots, e_{n+1})$;
for $j := n$ **to** *1* **by** *-1* **do**
    $pos := T[j]$;
    $Seq[pos] := Seq[pos] *_{o[pos]} Seq[pos+1]$;
    Remove($\tilde{~}Seq, pos+1$); Remove($\tilde{~}T, pos$); Remove($\tilde{~}o, pos$);
end
return $Seq[1]$;

For example, the tree in Figure 1 is given by $T = [1, 1, 2, 2, 3, 6, 6]$ and $o = (*_{\alpha_2}, *_{\alpha_3}, *_{\alpha_1}, *_{\alpha_4}, *_{\alpha_2}, *_{\alpha_1}, *_{\alpha_1})$.

Protocol implementation. Now, let (L, O_A, O_B) be as described in the KEP. We fix some distributions on L, O_A and O_B, so that we may generate random elements from these sets (according to these distributions). Given $m_a, m_B \in \mathbb{N}$, Alice and Bob first choose random vectors $\mathcal{G}_A = (s_1, \ldots, s_{m_A}) \in L^{m_A}$ and $\mathcal{G}_B = (t_1, \ldots, t_{m_B}) \in L^{m_B}$ which determine the public submagmas $S_A = \langle \mathcal{G}_A \rangle_{O_A}$ and $S_B = \langle \mathcal{G}_B \rangle_{O_B}$, respectively. Given some $n_{\min} \leq n_{\max}$ and $k_{\min} \leq k_{\max} \in \mathbb{N}$, Alice and Bob generate their secret keys as described in the following algorithms.

Algorithm 15: GENERATE ALICE'S PRIVATE KEY

Input: $\mathcal{G}_A, n_{\min}, n_{\max}, k_{\min}, k_{\max} \in L^{m_A} \times \mathbb{N}^4$.
Output: $(Ia_0, Ta_0, oa_0, a_0, a, oA) \in$
$[m_A]^{n_{a_0}+1} \times PBT_{n_{a_0}} \times O_A^{n_{a_0}} \times L \times L^{k_A} \times O_A^{k_A}$

$n_{a_0} \leftarrow \mathsf{Random}(\{n_{\min}, \ldots, n_{\max}\);$
$k_A \leftarrow \mathsf{Random}(\{k_{\min}, \ldots, k_{\max}\);$
$Ia_0 \leftarrow \mathsf{Random}([m_A]^{n_{a_0}});$
for $i := 1$ **to** $n_{a_0} + 1$ **do**
  $Seqa_0[i] := \mathcal{G}_A[Ia_0[i]];$
end
$Ta_0 \leftarrow \mathsf{Random}(PBT_{n_{a_0}});$ $oa_0 \leftarrow \mathsf{Random}(O_A^{n_{a_0}});$
$a_0 := \mathsf{EvaluateTree}(Ta_0, oa_0, Seqa_0);$
$a \leftarrow \mathsf{Random}(L^{k_A});$ $oA \leftarrow \mathsf{Random}(O_A^{k_A});$
return $(Ia_0, Ta_0, oa_0, a_0, a, oA);$

Algorithm 16: GENERATE BOB'S PRIVATE KEY

Input: $\mathcal{G}_B, n_{\min}, n_{\max}, k_{\min}, k_{\max} \in L^{m_B} \times \mathbb{N}^4$.
Output: $(Ib, Tb, ob, b, oB) \in$
$([m_B]^{n_b+1})^{k_B} \times (PBT_{n_b})^{k_B} \times (O_A^{n_b})^{k_B} \times L^{k_B} \times O_B^{k_B}$

for $k := 1$ **to** k_B **do**
  $n_b \leftarrow \mathsf{Random}(\{n_{\min}, \ldots, n_{\max}\);$
  $k_B \leftarrow \mathsf{Random}(\{k_{\min}, \ldots, k_{\max}\);$
  $Ib[k] \leftarrow \mathsf{Random}([m_B]^{n_b});$
  for $i := 1$ **to** $n_b + 1$ **do**
    $Seqb[k][i] := \mathcal{G}_A[Ia_0[i]];$
  end
  $Tb[k] \leftarrow \mathsf{Random}(PBT_{n_b});$ $ob \leftarrow \mathsf{Random}(O_B^{n_b});$
  $b[k] := \mathsf{EvaluateTree}(Tb[k], ob[k], Seqb[k]);$
end
$oB \leftarrow \mathsf{Random}(O_B^{k_B});$
return $(Ib, Tb, ob, b, oB);$

Then Alice and Bob generate their public and shared keys as described in the following algorithms.

Algorithm 17: Generate Alice's public key

Input: $(a, oA, a_0, \mathcal{G}_B) \in L^{k_A} \times O_A^{k_A} \times L \times L^{m_B}$.
Output: $(p_A, p_0) \in L^{m_B} \times L$

for $k := 1$ **to** m_B **do**
  $p_A[k] := \mathcal{G}_B[k]$;
  for $i := 1$ **to** k_A **do**
    $p_A[k] := a[i] *_{oA[i]} p_A[k]$;
  end
end
$p_0 := a_0$;
for $i := 1$ **to** k_A **do**
  $p_0 := a[i] *_{oA[i]} p_0$;
end
return (p_A, p_0);

Algorithm 18: Generate Bob's public key

Input: $(b, oB, \mathcal{G}_A) \in L^{k_B} \times O_B^{k_B} \times L^{m_A}$.
Output: $p_B \in L^{m_A}$

for $k := 1$ **to** m_A **do**
  $p_B[k] := \mathcal{G}_A[k]$;
  for $i := 1$ **to** k_B **do**
    $p_B[k] := b[i] *_{oB[i]} p_B[k]$;
  end
end
return p_B;

Algorithm 19: Generate Alice's shared key

Input: $(Ia_0, Ta_0, oa_0, a, oA, p_B) \in$
$[m_A]^{n_{a_0}+1} \times PBT_{n_{a_0}} \times O_A^{n_{a_0}} \times L^{k_A} \times O_A^{k_A} \times L^{m_A}$.
Output: $K_A \in L$

$K_A := \mathsf{EvaluateTree}(Ta_0, oa_0, (p_B[Ia_0[i]])_{i \leq n_{a_0}+1})$;
for $k := 1$ **to** k_A **do**
  $K_A := a[k] *_{oA[k]} K_A$;
end
return K_A;

Algorithm 20: GENERATE BOB'S SHARED KEY

Input: $(Ib, Tb, ob, b, oB, p_A, p_0) \in$
$([m_B]^{n_b+1})^{k_B} \times (PBT_{n_b})^{k_B} \times (O_A^{n_b})^{k_B} \times L^{k_B} \times O_B^{k_B} \times L^{m_B} \times L.$

Output: $K_A \in L$

Initialize $lfactors := []$; $K_B := p_0$;
for $k := 1$ **to** k_B **do**
  $lfactors[k] :=$ EvaluateTree$(Tb[k], ob[k], (p_A[Ib[k][i]])_{i \le n_b+1})$;
  $K_B := lfactors[k] *_{oB[k]} K_B$;
end
return K_B;

5. Security and complexity

We close with a brief security discussion and complexity analysis of the iterated Protocols 3.2 and 4.2.

5.1. Security discussion. In the case of an LD-system $(L, *)$, it is shown in [**8**] that it suffices to solve the so-called (generalized) iterated LD problem to break KEP 3.2. Obviously, the hardness of that problem depends on the platform LD-system. Braid groups (and their finite quotients) and matrix groups as natural platform LD-structures were discussed in [**6–9**]. For example, in the case of shifted conjugacy in braid and symmetric groups the so-called subgroup conjugacy coset problem appears as a relevant base problem [**8**]. For f-conjugacy and f-symmetric conjugacy as platform LD-operation in groups, the relevant base problems are the iterated f-conjugacy and f-symmetric conjugacy problem, respectively. These problem are pretty non-standard, but whenever they are comparable to the conjugacy problem, they are harder.

Efficient instantiations of these KEPs can only be achieved if L is finite. Efficient instantiations in symmetric groups and matrix groups are discussed in [**8**] where we also give concrete parameter suggestions. For example, an efficient instantiation of the f-symmetric LD-system uses a finite matrix group over the ring of N-truncated polynomials $R = \mathbb{F}_p[X]/(X^N - 1)$ (like NTRUEncrypt) for $N = p - 1$ with an evaluating homomorphism $X \mapsto a$ for some fixed $a \in \mathbb{F}_p^*$ as projector map $(f^2 = f)$.

5.2. Complexity analysis. For a finite platform LD-system L (or a finite group G) we discuss the complexity of the implementation for Protocols 3.2 and 4.2. Protocol 2.2 is a special case for $k_A = k_B = 1$. Let M be the time complexity for multiplication in L (or G). Then the complexity for evaluating and generating a random tree $T \in PBT_n$ are $O(nM)$ and $O(n \log n)$, respectively. Furthermore, let R be the complexity for generating a random element in L (or G). Then the following table displays the complexities for generating private, public and shared keys.

$O(\cdot)$	Alice	Bob
priv.	$n_{a_0}(m_A + \log n_{a_0} + M + R) + k_A R$	$k_B n_b(m_B + \log n_b + M + R)$
publ.	$(m_A + k_A)M$	$m_B k_B M$
sh.	$(n_{a_0} + k_A)M$	$k_B n_b M$

The complexities are different for Alice and Bob since the KEP is asymmetric[2]. And the KEP is asymmetric since we use the left distributive laws for key establishment which are also asymmetric.

Conclusion. We have provided implementation details for all non-associative KEPs given in [**8**, **9**]. Together with the concrete instantiations provided in [**8**], we hope this will encourage cryptanalytic examination of these new and innovative non-associative KEPs.

References

[1] A. Myasnikov, V. Shpilrain, and A. Ushakov, *Non-commutative cryptography and complexity of group-theoretic problems*, Mathematical Surveys and Monographs, vol. 177, American Mathematical Society, Providence, RI, 2011. With an appendix by Natalia Mosina. MR2850384

[2] M. Črepinšek and L. Mernik, *An efficient representation for solving Catalan number related problems*, Int. J. Pure Appl. Math. **56** (2009), no. 4, 589–604. MR2583785

[3] P. Dehornoy, *Braids and self-distributivity*, Progress in Mathematics, vol. 192, Birkhäuser Verlag, Basel, 2000. MR1778150

[4] P. Dehornoy, *Using shifted conjugacy in braid-based cryptography*, Algebraic methods in cryptography, Contemp. Math., vol. 418, Amer. Math. Soc., Providence, RI, 2006, pp. 65–73, DOI 10.1090/conm/418/07946. MR2389289

[5] I. Anshel, M. Anshel, and D. Goldfeld, *An algebraic method for public-key cryptography*, Math. Res. Lett. **6** (1999), no. 3-4, 287–291, DOI 10.4310/MRL.1999.v6.n3.a3. MR1713130

[6] A. Kalka, *Linear representations of braid groups and braid-based cryptography*, Ph.D. thesis, Ruhr-Universität Bochum, 2007.

[7] ———, *Non-associative public-key cryptography*, Algebra and Computer Science, Contemp. Math. 677, Amer. Math. Soc., Providence, RI, 2016.

[8] A. Kalka and M. Teicher, *Iterated LD-Problem in non-associative key establishment*, arXiv **abs/1312.6791** (2013).

[9] A. Kalka and M. Teicher, *Non-associative key establishment for left distributive systems*, Groups Complex. Cryptol. **5** (2013), no. 2, 169–191. MR3245105

[10] P. W. Shor, *Polynomial-time algorithms for prime factorization and discrete logarithms on a quantum computer*, SIAM J. Comput. **26** (1997), no. 5, 1484–1509, DOI 10.1137/S0097539795293172. MR1471990

Department of Mathematics, Bar-Ilan University, Ramat Gan 52900, Israel
E-mail address: arkadius.kalka@rub.de

Department of Mathematics, Bar-Ilan University, Ramat Gan 52900, Israel
E-mail address: teicher@math.biu.ac.il

[2]Therefore, it is recommended that k_B is considerably smaller than k_A.

Contemporary Mathematics
Volume **677**, 2016
http://dx.doi.org/10.1090/conm/677/13625

Knapsack and subset sum problems in nilpotent, polycyclic, and co-context-free groups

Daniel König, Markus Lohrey, and Georg Zetzsche

1. Introduction

In their paper [**22**], Myasnikov, Nikolaev, and Ushakov started the investigation of classical discrete integer optimization problems in general non-commutative groups. Among other problems, they introduced for a finitely generated (f.g.) group G the *knapsack problem* and the *subset sum problem*. The input for the knapsack problem is a sequence of group elements $g_1, \ldots, g_k, g \in G$ and it is asked whether there exists a solution $(x_1, \ldots, x_k) \in \mathbb{N}^k$ of the equation $g_1^{x_1} \cdots g_k^{x_k} = g$. For the subset sum problem one restricts the solution to $\{0,1\}^k$. For the particular case $G = \mathbb{Z}$ (where the additive notation $x_1 \cdot g_1 + \cdots + x_k \cdot g_k = g$ is usually preferred) these problems are NP-complete if the numbers $g_1, \ldots, g_k, g$ are encoded in binary representation. For subset sum this is shown in Karp's classical paper [**13**]. The statement for knapsack (in the above version) can be found in [**10**].

In [**22**] the authors encode elements of the finitely generated group G by words over the group generators and their inverses. For $G = \mathbb{Z}$ this representation corresponds to the unary encoding of integers. It is known that for unary encoded integers, knapsack and subset sum over $\mathbb{Z}$ can both be solved in polynomial time, and the precise complexity is DLOGTIME-uniform TC^0 [**6**], which is a very small complexity class that roughly speaking captures the complexity of multiplying binary coded integers. In [**22**], Myasnikov et al. proved the following new results:

- Subset sum and knapsack can be solved in polynomial time for every hyperbolic group.
- Subset sum for a virtually nilpotent group (i.e. a finite extension of a nilpotent group) can be solved in polynomial time.
- For the following groups, subset sum is NP-complete (whereas the word problem can be solved in polynomial time): free metabelian non-abelian groups of finite rank, the wreath product $\mathbb{Z} \wr \mathbb{Z}$, Thompson's group F, and the Baumslag-Solitar group $\mathrm{BS}(1,2)$.

In this paper, we continue the investigation of knapsack and subset sum for arbitrary groups. We prove the following results,where as in [**22**] group elements are

2010 *Mathematics Subject Classification.* 20F10, 20F18, 68Q17, 68Q45.

©2016 American Mathematical Society

represented by finite words over the group generators and their inverses:

- For every virtually nilpotent group, subset sum belongs to NL (nondeterministic logspace).
- There is a polycyclic group with an NP-complete subset sum problem.
- There is a nilpotent group of class two for which knapsack is undecidable. This nilpotent group is a direct product of sufficiently many copies of the discrete Heisenberg group $H_3(\mathbb{Z})$. In [**18**], the second author proved that there exists a nilpotent group (of large class) for which knapsack is undecidable. Here we improve this result to class two and at the same time simplify the construction from [**18**]. As a byproduct of our construction, we show that there exists a fixed nilpotent group of class two together with four finitely generated abelian subgroups G_1, G_2, G_3, G_4 such that membership in the product $G_1G_2G_3G_4$ is undecidable. This contrasts the known fact that membership in a product of two subgroups of a polycyclic group is decidable [**16**].
- The knapsack problem for the discrete Heisenberg group $H_3(\mathbb{Z})$ is decidable. In particular, together with the previous point it follows that decidability of knapsack is not preserved under direct products.
- The class of groups with a decidable knapsack problem is closed under finite extensions.
- The knapsack problem is decidable for every co-context-free group. Recall that a group is co-context-free if the complement of the word problem is a context-free language [**11**].

2. Nilpotent and polycyclic groups

Let A be a square matrix of dimension d over some commutative ring R. With $A[i,j]$ we denote the entry of A in row i and column j. The matrix A is called *triangular* if $A[i,j] = 0$ whenever $i > j$, i.e., all entries below the main diagonal are 0. A *unitriangular matrix* is a triangular matrix A such that $A[i,i] = 1$ for all $1 \leq i \leq d$, i.e., all entries on the main diagonal are 1. We denote the set of unitriangular matrices of dimension d over the ring R by $\mathrm{UT}_d(R)$. It is well known that for every commutative ring R, the set $\mathrm{UT}_d(R)$ is a group (with respect to matrix multiplication).

An *n-step solvable group* G is a group G that possesses a subnormal series $G = G_n \rhd G_{n-1} \rhd G_{n-2} \rhd \cdots \rhd G_1 \rhd G_0 = 1$ (i.e., G_i is a normal subgroup of G_{i+1} for all $0 \leq i \leq n-1$) such that every quotient G_{i+1}/G_i is abelian ($0 \leq i \leq n-1$). If every quotient G_{i+1}/G_i is cyclic, then G is called *polycyclic*. The number of $0 \leq i \leq n-1$ such that $G_{i+1}/G_i \cong \mathbb{Z}$ is called the *Hirsch length* of G; it does not depend on the chosen subnormal series. If $G_{i+1}/G_i \cong \mathbb{Z}$ for all $0 \leq i \leq n-1$ then G is called *strongly polycyclic*. The following characterizations of the class of polycyclic groups are known:

- A group is polycyclic if and only if it is solvable and every subgroup is finitely generated.
- A group is polycyclic if and only if it is a solvable group of integer matrices; this is a famous result by Auslander and Swan [**2, 24**]. In particular, every polycyclic group is linear, i.e., can be embedded into a matrix group over some field.

For a group G its *lower central series* is the series $G = G_0 \rhd G_1 \rhd G_2 \rhd \cdots$ of subgroups, where $G_{i+1} = [G_i, G]$, which is the subgroup generated by all commutators $[g, h]$ with $g \in G_i$ and $h \in G$. Indeed, G_{i+1} is a normal subgroup of G_i. The group G is *nilpotent of class c*, if $G_c = 1$. Every f.g. nilpotent group is polycyclic, and every group $\mathrm{UT}_d(\mathbb{Z})$ $(d \geq 1)$ is f.g. nilpotent of class $d - 1$.

The group $\mathrm{UT}_3(\mathbb{Z})$ is also denoted by $H_3(\mathbb{Z})$ and called the *discrete Heisenberg group*. Thus, $H_3(\mathbb{Z})$ is the group of all (3×3)-matrices of the form

$$\begin{pmatrix} 1 & a & c \\ 0 & 1 & b \\ 0 & 0 & 1 \end{pmatrix}$$

for $a, b, c, \in \mathbb{Z}$. The center $Z(H_3(R))$ of this group consists of all matrices of the form

$$\begin{pmatrix} 1 & 0 & c \\ 0 & 1 & 0 \\ 0 & 0 & 1 \end{pmatrix}$$

for $c \in \mathbb{Z}$. The group $H_3(\mathbb{Z})$ is nilpotent of class two (it is in fact the free nilpotent group of class two and rank two). In other words, every commutator $ABA^{-1}B^{-1}$ $(A, B \in H_3(\mathbb{Z}))$ belongs to the center $Z(H_3(\mathbb{Z}))$. The identity (3×3)-matrix will be denoted by Id_3. Clearly, a direct product of copies of $H_3(\mathbb{Z})$ and $\mathbb{Z}$ is also nilpotent of class two.

We need the following results about nilpotent groups:

THEOREM 2.1 ([**12**, Theorem 17.2.2]). *Every f.g. nilpotent group G has a torsion-free normal subgroup H of finite index (which is also f.g. nilpotent).*

THEOREM 2.2 ([**12**, Theorem 17.2.5]). *For every torsion-free f.g nilpotent group G there exists $d \geq 1$ such that G can be embedded into $\mathrm{UT}_d(\mathbb{Z})$.*

A group is *virtually nilpotent* if it has a nilpotent subgroup of finite index.

3. Subset sum and knapsack problems in groups

Let G be a f.g. group, and fix an arbitrary finite generating set Σ for G. In this paper, we consider the following computational problems for G, where elements of G are represented by finite words over $\Sigma \cup \Sigma^{-1}$:

- Subset sum problem for G (briefly **SSP**(G)): Given $g_1, \ldots, g_k, g \in G$, decide whether there exist $\varepsilon_1, \ldots, \varepsilon_k \in \{0, 1\}$ such that $g = g_1^{\varepsilon_1} \cdots g_k^{\varepsilon_k}$.
- Knapsack problem for G (briefly **KP**(G)): Given $g_1, \ldots, g_k, g \in G$, decide whether there exist natural numbers $e_1, \ldots, e_k \geq 0$ such that we have $g = g_1^{e_1} \cdots g_k^{e_k}$.

These problems were studied for general f.g. groups in [**7**, **22**], where among others the following results were shown:

- The subset sum problem for every f.g. virtually nilpotent group can be solved in polynomial time [**22**].
- The subset sum problem and the knapsack problem for every hyperbolic group can be solved in polynomial time [**22**].
- The knapsack problem can be solved in polynomial time in any free product of hyperbolic groups and finitely generated abelian groups [**7**].

- The subset sum problem for the following groups is NP-complete: $\mathbb{Z} \wr \mathbb{Z}$, free metabelian (but non-abelian) groups of finite rank, and Thompson's group F [**22**].

There is a variant of knapsack, where we ask whether for given $g_1, \ldots, g_k, g \in G$, there exist *integers* $e_1, \ldots, e_k \in \mathbb{Z}$ such that $g = g_1^{e_1} \cdots g_k^{e_k}$, i.e., whether g belongs belongs to the product of cyclic groups $\langle g_1 \rangle \cdots \langle g_k \rangle$. This second version is reducible to the above version with exponents from $\mathbb{N}$: Simply replace $g_i^{e_i}$ (where e_i ranges over $\mathbb{Z}$) with $g_i^{c_i}(g_i^{-1})^{d_i}$ (where c_i and d_i range over $\mathbb{N}$). We will prove undecidability results for the ostensibly easier version with integer exponents, whereas decidability results will be shown for the harder version with non-negative exponents.

4. Subset sum problems in nilpotent groups

In this section, we show that the subset sum problem for a finitely generated virtually nilpotent group belongs to nondeterministic logspace (NL). This is the class of all problems that can be solved on a *nondeterministic* Turing-machine with a working tape of length $O(\log n)$, where n is the length of the input, see e.g. [**1**] for details. Actually, we consider a problem more general than the subset sum problem: the membership problem for acyclic finite automaton, which was also studied in [**7**].

Recall that a *finite (nondeterministic) automaton* over a finite alphabet Σ is a tuple $\mathcal{A} = (Q, \Delta, q_0, F)$, where

- Q is a finite set of states,
- $\Delta \subseteq Q \times \Sigma^* \times Q$ is a finite set of transitions,
- $q_0 \in Q$ is the initial state, and
- $F \subseteq Q$ is the set of final states.

If the directed graph $(Q, \{(p,q) \mid \exists w \in \Sigma^* : (p,w,q) \in \Delta\})$ has no directed cycle, then the finite automaton $\mathcal{A}$ is *acyclic*. Given a word $w \in \Sigma^*$, an *accepting run for* w is a sequence of transitions $(q_0, w_1, q_1), (q_1, w_2, q_2), \ldots, (q_{n-1}, w_n, q_n) \in \Delta$ such that $w = w_1 w_2 \cdots w_n$ and $q_n \in F$. The language $L(A) \subseteq \Sigma^*$ is the set of all words over Σ that have an accepting run. By splitting transitions, one can compute in logspace from a finite automaton $\mathcal{A}$ an automaton $\mathcal{B}$ such that $L(\mathcal{A}) = L(\mathcal{B})$ and all transitions of $\mathcal{B}$ are from $Q \times (\Sigma \cup \{\varepsilon\}) \times Q$. Moreover, $\mathcal{B}$ is acyclic if $\mathcal{A}$ is acyclic.

Let G be a finitely generated group, and let Σ be a finite group generating set for G. Hence, $\Sigma \cup \Sigma^{-1}$ generates G as a monoid and there is a canonical homomorphism $h\colon (\Sigma \cup \Sigma^{-1})^* \to G$. For a finite automaton $\mathcal{A}$ over $\Sigma \cup \Sigma^{-1}$ and a word $x \in (\Sigma \cup \Sigma^{-1})^*$ we also write $x \in_G L(\mathcal{A})$ for $h(x) \in h(L(\mathcal{A}))$. The *acyclic rational subset membership problem for* G (briefly **ARatMP**(G)) is the following computational problem:

Input: An acyclic finite automaton $\mathcal{A}$ over $\Sigma \cup \Sigma^{-1}$ and a word $x \in (\Sigma \cup \Sigma^{-1})^*$.
Question: Does $x \in_G L(\mathcal{A})$ hold?

Clearly, **SSP**(G) is logspace reducible to **ARatMP**(G).

THEOREM 4.1. *For every $d \geq 1$ and every f.g. subgroup G of* $\mathsf{UT}_d(\mathbb{Z})$, *the problem* **ARatMP**(G) *belongs to* NL.

PROOF. It suffices to prove the theorem for $G = \mathsf{UT}_d(\mathbb{Z})$. Let $\mathcal{A}$ be a finite automaton with n states, whose transitions are labelled with generator matrices of $\mathsf{UT}_d(\mathbb{Z})$ or the identity matrix. We nondeterministically guess a path of length at most n from the initial state of $\mathcal{A}$ to a final state of $\mathcal{A}$ and thereby multiply the matrices along the path. We only store the current state of $\mathcal{A}$, the product of the matrices seen so far, and the length of the path travelled so far (so that after n steps we can stop). The state of the automaton as well as the length of the path need $O(\log n)$ bits. Hence, we only have to show that the product matrix can be stored in logarithmic space. For this, it suffices to show that the matrix entries are bounded polynomially in n. Then, the binary coding of the matrix needs only $O(\log n)$ many bits (note that the matrix dimension d is a constant). For this, we can use the following simple result (see [**17**, Proposition 4.18] for a proof), which only holds for unitriangular matrices: For a $(d \times d)$-matrix $M = (a_{i,j})_{1 \le i,j \le d}$ over $\mathbb{Z}$ let $|M| = \sum_{i=1}^{d} \sum_{j=1}^{d} |a_{i,j}|$. Let $M_1, \ldots, M_n \in \mathsf{UT}_d(\mathbb{Z})$, $n \ge 2d$, and let $m = \max\{|M_i| \mid 1 \le i \le n\}$. For the product of these matrices we have

$$|M_1 M_2 \cdots M_n| \le d + (d-1)\binom{n}{d-1} d^{2(d-2)} m^{d-1}.$$

In our situation, the matrices M_i are from a fixed set (generators and the identity matrix). Hence, m and also d are constants. Hence, the above bound is polynomial in n, which means that every entry of the product $M_1 M_2 \cdots M_n$ can be stored with $O(\log n)$ bits. □

THEOREM 4.2. *Let H be a finite index subgroup of the f.g. group G (hence, H is f.g. too). Then* **ARatMP**(G) *is logspace-reducible to* **ARatMP**(H).

PROOF. Let G and H be as in the statement of the theorem. Let Γ (resp., Σ) be a finite generating set for G (resp., H). Let $Hg_0, Hg_1, \ldots, Hg_n$ be a list of all right cosets of H, where $g_0 = 1$.

Let $\mathcal{A} = (Q, \Delta, q_0, F)$ be an acyclic finite automaton over the alphabet $\Gamma \cup \Gamma^{-1}$ and let $x \in (\Gamma \cup \Gamma^{-1})^*$. We can assume that $\Delta \subseteq Q \times (\Gamma \cup \Gamma^{-1} \cup \{\varepsilon\}) \times Q$. Assume that $x = yg_s$ in G, where $y \in (\Sigma \cup \Sigma^{-1})^*$. We can compute the word y and the coset representative g_s in logspace as follows: Let $x = a_1 a_2 \cdots a_m$. We store an index $i \in \{0, \ldots, n\}$, which is initially set to 0. Then, for $1 \le j \le m$ we do the following: If $g_i a_j = wg_k$ for $w \in (\Sigma \cup \Sigma^{-1})^*$, then we append the word w at the output tape and we set $i := k$. At the end, the word y is written on the output tape and the final index i is s such that $x = yg_s$.

We now construct a new acyclic automaton $\mathcal{B}$ over the alphabet $\Sigma \cup \Sigma^{-1}$ as follows:

- The state set is $Q \times \{g_0, g_1 \ldots, g_n\}$.
- Assume that $(p, a, q) \in \Delta$ is a transition of $\mathcal{A}$ ($a \in \Gamma \cup \Gamma^{-1} \cup \{\varepsilon\}$) and let $i \in \{0, 1, \ldots, n\}$. Assume that $g_i a = wg_j$ in G, where $w \in (\Sigma \cup \Sigma^{-1})^*$. Then, we add the transition $(\langle p, g_i \rangle, w, \langle q, g_j \rangle)$ to $\mathcal{B}$.
- The initial state of $\mathcal{B}$ is $\langle q_0, g_0 \rangle$.
- The set of final states of $\mathcal{B}$ is $F \times \{g_s\}$.

From the construction, we get $x \in_G L(\mathcal{A})$ if and only if $y \in_H L(\mathcal{B})$. □

THEOREM 4.3. *Let G be finitely generated virtually nilpotent. Then, the problem* **ARatMP**(G) *is* NL*-complete.*

PROOF. Hardness for NL follows immediately from the NL-hardness of the graph reachability problem for acyclic directed graphs. For the membership in NL let G be finitely generated virtually nilpotent. By Theorem 2.1 and 2.2, G has a f.g. finite index subgroup H such that H is isomorphic to a subgroup of $\mathsf{UT}_d(\mathbb{Z})$.

W.l.o.g we assume that H is a subgroup of $\mathsf{UT}_d(\mathbb{Z})$. Membership in NL follows from Theorem 4.1 and Theorem 4.2. □

By Theorem 4.3, the subset sum problem for a finitely generated virtually nilpotent belongs to NL. It is open whether this upper bound can be improved further. In particular, it is open whether the subset sum problem for the Heisenberg group $H_3(\mathbb{Z})$ can be solved in deterministic logspace. Recall from the introduction that subset sum for $\mathbb{Z}$ (and unary encoded numbers) belongs to $\mathsf{DLOGTIME}$-uniform TC^0 which is a subclass of deterministic logspace. This result generalizes easily to any f.g. abelian group.

5. Subset sum in polycyclic groups

We show in this section that there exists a polycyclic group with an NP-complete subset sum problem, which is in sharp contrast to nilpotent groups (assuming $\mathsf{NL} \neq \mathsf{NP}$). Let us start with a specific example of a polycyclic group. Consider the two matrices

$$g_a = \begin{pmatrix} a & 0 \\ 0 & 1 \end{pmatrix} \text{ and } h = \begin{pmatrix} 1 & 1 \\ 0 & 1 \end{pmatrix},$$

where $a \in \mathbb{R}$, $a \geq 2$. Let $G_a = \langle g_a, h \rangle \leq \mathsf{GL}_2(\mathbb{R})$. Let us remark that, for instance, the group G_2 is not polycyclic, see e.g. [**25**, p. 56]. On the other hand, we have:

Proposition 5.1 (cf. [**14**]). *The group $G_{1+\sqrt{2}}$ is polycyclic.*

Theorem 5.2. **SSP**$(G_{1+\sqrt{2}})$ *is* NP*-complete.*

Proof. Let $\alpha = 1+\sqrt{2}$. We follow the standard proof for the NP-completeness of subset sum for binary encoded integers. However, we will work with real numbers of the form

$$x = \sum_{i=0}^{n} x_i \cdot \alpha^{3i},$$

where the x_i are natural numbers with $0 \leq x_i \leq 5$. The numbers x_i are uniquely determined by x in the following sense:

Claim 1: If

$$\sum_{i=0}^{n} x_i \cdot \alpha^{3i} = \sum_{i=0}^{m} y_i \cdot \alpha^{3i} \tag{5.1}$$

with $x_0, \ldots x_n, y_0, \ldots, y_m \in \{0, 1, \ldots, 5\}$ and $x_n \neq 0 \neq y_m$, then $n = m$ and $x_i = y_i$ for all $0 \leq i \leq n$.

Proof of Claim 1. Assume that the conclusion of the claim fails. Then, by canceling α-powers with highest exponent, we obtain from (5.1) an identity of the form

$$\sum_{i=0}^{n} x_i \cdot \alpha^{3i} = \sum_{i=0}^{m} y_i \cdot \alpha^{3i}$$

where $n > m$, $x_0, \ldots x_n, y_0, \ldots, y_m \in \{0, 1, \ldots, 5\}$ and $x_n \neq 0$. In order to lead this to a contradiction, it suffices to show

$$\alpha^{3n} > \sum_{i=0}^{n-1} 5 \cdot \alpha^{3i}.$$

Indeed, we have

$$\sum_{i=0}^{n-1} 5 \cdot \alpha^{3i} < \sum_{i=0}^{n-1} (\alpha^{3i} + \alpha^{3i+1} + \alpha^{3i+2}) = \sum_{i=0}^{3n-1} \alpha^i = \frac{\alpha^{3n} - 1}{\alpha - 1} < \alpha^{3n}.$$

Let us now take a 3CNF-formula $C = \bigwedge_{i=1}^m C_i$, where $C_i = (z_{i,1} \vee z_{i,2} \vee z_{i,3})$. Every $z_{i,j}$ is a literal, i.e., a Boolean variable or a negated Boolean variable. Let $x_1, \ldots, x_n$ be the Boolean variables appearing in C.

We now define numbers $u_1, \ldots, u_{2n+2m}$, and t as follows, where $1 \le i \le n$ and $1 \le j \le m$:

$$\begin{aligned} u_{2i-1} &= \alpha^{3i-3} + \sum_{x_i \in C_k} \alpha^{3n+3k-3} \\ u_{2i} &= \alpha^{3i-3} + \sum_{\overline{x}_i \in C_k} \alpha^{3n+3k-3} \\ u_{2n+2j-1} &= u_{2n+2j} = \alpha^{3n+3j-3} \\ t &= \sum_{i=1}^{n} \alpha^{3i-3} + \sum_{k=1}^{m} 3 \cdot \alpha^{3n+3k-3} \end{aligned}$$

Claim 2: C is satisfiable if and only if there exists a subset $I \subseteq \{1, \ldots, 2n + 2m\}$ such that $\sum_{k \in I} u_k = t$.

Proof of Claim 2. Suppose C is satisfiable and let $\varphi\colon \{x_1, \ldots, x_n\} \to \{0, 1\}$ be a satisfying assignment for C. We define $\varphi(\overline{x}_i) = 1 - \varphi(x_i)$. For every clause $C_j = (z_{j,1} \vee z_{j,2} \vee z_{j,3})$ let $\gamma_j = |\{k \in \{1, 2, 3\} \mid \varphi(z_{j,k}) = 1\}|$ be the number of literals in C_j that are true under φ. Thus, we have $1 \le \gamma_j \le 3$.

We define the set I as follows, where $1 \le i \le n$ and $1 \le j \le m$:

- $2i - 1 \in I$ iff $\varphi(x_i) = 1$
- $2i \in I$ iff $\varphi(x_i) = 0$
- If $\gamma_j = 3$, then $2n + 2j - 1 \notin I$ and $2n + 2j \notin I$.
- If $\gamma_j = 2$, then $2n + 2j - 1 \in I$ and $2n + 2j \notin I$.
- If $\gamma_j = 1$, then $2n + 2j - 1 \in I$ and $2n + 2j \in I$.

With this set I we have indeed $\sum_{k \in I} u_k = t$.

For the other direction, let $I \subseteq \{1, \ldots, 2n + 2m\}$ such that $\sum_{k \in I} u_k = t$. Note that in the sum $\sum_{k \in I} u_k$ no power α^{3k} can appear more than 5 times (a power $\alpha^{3n+3j-3}$ with $1 \le j \le m$ can appear at most 5 times, since it appears in at most 3 of the numbers $u_1, \ldots, u_{2n}$ and in 2 of the numbers $u_{2n+1}, \ldots, u_{2n+2m}$). This allows to use Claim 1. A comparison of t and $\sum_{k \in I} u_k$ shows that either $2i - 1 \in I$ or $2i \in I$. We define the assignment $\varphi\colon \{x_1, \ldots, x_n\} \to \{0, 1\}$ as follows:

- $\varphi(x_i) = 1$ iff $2i - 1 \in I$
- $\varphi(x_i) = 0$ iff $2i \in I$

As above, let γ_j be the number of literals in C_j that are true under φ. Moreover, let $\delta_j = |I \cap \{2n + 2j - 1, 2n + 2j\}|$ for $1 \le j \le m$. We get

$$\sum_{k \in I} u_k = \sum_{i=1}^{n} \alpha^{3i-3} + \sum_{j=1}^{m} (\gamma_j + \delta_j) \cdot \alpha^{3n+3j-3} = t = \sum_{i=1}^{n} \alpha^{3i-3} + \sum_{j=1}^{m} 3 \cdot \alpha^{3n+3j-3}.$$

Since $\delta_j \in \{0, 1, 2\}$ we must have $\gamma_j \ge 1$ for all $1 \le j \le m$. This shows that φ satisfies C.

We now map each of the numbers $u_1, \dots, u_{2n+2m}, t$ to a word over the generators g_α, h (and their inverses) of the polycyclic group G_α. First, for $i \geq 0$ let us define

$$w_i = g_\alpha^i h g_\alpha^{-i}$$

In the group G_α we have

$$w_i = \begin{pmatrix} 1 & \alpha^i \\ 0 & 1 \end{pmatrix}$$

Finally, take a number $Y = \sum_{i=0}^n y_i \cdot \alpha^i$. We define the word

$$w_Y = \prod_{i=0}^n w_i^{y_i}.$$

In the group G_α we have

$$w_Y = \begin{pmatrix} 1 & Y \\ 0 & 1 \end{pmatrix}.$$

The words $w_{u_1}, \dots, w_{u_{2n+2m}}, w_t$ can be computed in polynomial time (even in logspace) from the 3CNF-formula C. Moreover, the construction implies that C is satisfiable iff there exists a subset $I \subseteq \{1, \dots, 2n+2m\}$ such that $\sum_{k \in I} u_k = t$ iff there are $\varepsilon_1, \dots, \varepsilon_{2n+2m} \in \{0,1\}$ such that $w_{u_1}^{\varepsilon_1} \cdots w_{u_{2n+2m}}^{\varepsilon_{2n+2m}} = w_t$ in the group G_α. □

6. Knapsack problems in nilpotent groups

The goal of this section is to prove that the knapsack problem is undecidable for a direct product of sufficiently many copies of $H_3(\mathbb{Z})$, which is nilpotent of class two.

6.1. Exponential expressions. Let $\mathcal{X}$ be a countably infinite set of variables. An *exponential expression* E over a group G is a formal product of the form

$$E = g_1^{x_1} g_2^{x_2} \cdots g_l^{x_l}$$

with $x_1, \dots, x_l \in \mathcal{X}$ and $g_1, \dots, g_l \in G$. We do not assume that $x_i \neq x_j$ for $i \neq j$. The group elements $g_1, \dots, g_l$ will also be called the *base elements* of E. The *length* of E is l. Let $\mathrm{Var}(E) = \{x_1, \dots, x_l\}$ be the set of variables that appear in E. For a finite set X with $\mathrm{Var}(E) \subseteq X \subseteq \mathcal{X}$ and $h \in G$, the set of X*-solutions* of the equation $E = h$ is the set of mappings

$$S_X(E = h) = \{\nu \colon X \to \mathbb{Z} \mid g_1^{\nu(x_1)} g_2^{\nu(x_2)} \cdots g_l^{\nu(x_l)} = h \text{ in } G\}.$$

Note that not every variable from X has to appear as an exponent in E. We moreover set $S(E = h) = S_{\mathrm{Var}(E)}(E = h)$.

For every $1 \leq i \leq n$ consider an exponential expression E_i over a group G_i. Then we can define the exponential expression $E = \prod_{i=1}^n E_i$ over the product group $G = \prod_{i=1}^n G_i$. It is defined by replacing in E_i every occurrence of a base element $g \in G_i$ with the corresponding element

$$(\underbrace{1, \dots, 1}_{i-1 \text{ many}}, g, \underbrace{1, \dots, 1}_{n-i \text{ many}}) \in G$$

and taking the concatenation of the resulting exponential expressions. With this definition, the following lemma is obvious.

LEMMA 6.1. *For $1 \le i \le n$ let E_i be an exponential expression over a group G_i. Let $h_i \in G_i$ for $1 \le i \le n$. Let $X = \bigcup_{i=1}^n \mathrm{Var}(E_i)$. Then for the exponential expression $E = \prod_{i=1}^n E_i$ and the element $h = (h_1, \dots, h_n) \in \prod_{i=1}^n G_i$ we have:*

$$S_X(E = h) = \bigcap_{i=1}^{n} S_X(E_i = h_i).$$

PROPOSITION 6.2. *There are fixed constants $d, e \in \mathbb{N}$ and a fixed exponential expression E over $G = H_3(\mathbb{Z})^d \times \mathbb{Z}^e$ such that the following problem is undecidable:*

Input: A element $h \in G$.
Question: Does $S(E = h) \neq \emptyset$ hold?

PROOF. Let $P(x_1, \dots, x_n) \in \mathbb{Z}[x_1, \dots, x_n]$ be a fixed polynomial such that the following question is undecidable:

Input: A number $a \in \mathbb{N}$.
Question: Is there a tuple $(z_1, \dots, z_n) \in \mathbb{Z}^n$ such that $P(z_1, \dots, z_n) = a$.

By Matiyasevich's proof for the unsolvability of Hilbert's 10th problem, we know that such a polynomial exists, see [**20**] for details. By introducing additional variables, we can construct from the polynomial $P(x_1, \dots, x_n)$ a system $\mathcal{S}$ of equations of the form $x \cdot y = z$, $x + y = z$, $x = c$ (for $c \in \mathbb{Z}$) such that the equation $P(x_1, \dots, x_n) = a$ has a solution in $\mathbb{Z}$ if and only if the system of equations $\mathcal{S}_a := \mathcal{S} \cup \{x_0 = a\}$ has a solution in $\mathbb{Z}$. Here x_0 is a distinguished variable of $\mathcal{S}$. Let X be the set of variables that occur in $\mathcal{S}_a$.

Take an integer $a \in \mathbb{Z}$ (the input for our reduction). Assume that $\mathcal{S}_a$ contains d many equations of the form $x \cdot y = z$ and e many equations of the form $x + y = z$ or $x = c$. Enumerate all equations as $\mathcal{E}_1, \dots, \mathcal{E}_{d+e}$, where $\mathcal{E}_1, \dots, \mathcal{E}_d$ are all equations of the form $x \cdot y = z$. Let $G_i = H_3(\mathbb{Z})$ for $1 \le i \le d$, $G_i = \mathbb{Z}$ for $d+1 \le i \le d+e$, and $G = \prod_{i=1}^{d+e} G_i$. We define for every $1 \le i \le d+e$ an element $h_i \in G_i$ and an exponential expression E_i over G_i as follows:

Case 1. $\mathcal{E}_i = (x \cdot y = z)$ and thus $G_i = H_3(\mathbb{Z})$. Then, we set $h_i = \mathrm{Id}_3$ (the identity matrix) and

$$E_i = \begin{pmatrix} 1 & 0 & 0 \\ 0 & 1 & 1 \\ 0 & 0 & 1 \end{pmatrix}^x \begin{pmatrix} 1 & 1 & 0 \\ 0 & 1 & 0 \\ 0 & 0 & 1 \end{pmatrix}^y \begin{pmatrix} 1 & 0 & 0 \\ 0 & 1 & -1 \\ 0 & 0 & 1 \end{pmatrix}^x \begin{pmatrix} 1 & -1 & 0 \\ 0 & 1 & 0 \\ 0 & 0 & 1 \end{pmatrix}^y \begin{pmatrix} 1 & 0 & 1 \\ 0 & 1 & 0 \\ 0 & 0 & 1 \end{pmatrix}^z.$$

Note that this product evaluates to the matrix

$$\begin{pmatrix} 1 & 0 & z - xy \\ 0 & 1 & 0 \\ 0 & 0 & 1 \end{pmatrix}.$$

Hence, a mapping $\nu \colon X \to \mathbb{Z}$ is a solution of $E_i = h_i$ if and only if $\nu(x) \cdot \nu(y) = \nu(z)$.

Case 2. $\mathcal{E}_i = (x + y = z)$ and thus $G_i = \mathbb{Z}$. Then, we set $h_i = 0$ and $E_i = x + y - z$ (written in additive form, or, written multiplicatively, $E_i = g^x g^y g^{-z}$, where g is a generator of $\mathbb{Z}$). Then, a mapping $\nu \colon X \to \mathbb{Z}$ is a solution of $E_i = h_i$ if and only if $\nu(x) + \nu(y) = \nu(z)$.

Case 3. $\mathcal{E}_i = (x = c)$ (this includes the distinguished equation $x_0 = a$) and thus $G_i = \mathbb{Z}$. Then, $h_i = c$ and $E_i = x$ (or, written multiplicatively, $E_i = g^x$). Then, a mapping $\nu \colon X \to \mathbb{Z}$ is a solution of $E_i = h_i$ if and only if $\nu(x) = c$.

Let $E = \prod_{i=1}^{d} E_i$ and $h = (h_1, \ldots, h_d) \in G$. By Lemma 6.1, a mapping $\nu \colon X \to \mathbb{Z}$ is a solution of $E = h$ if and only if ν is a solution of the system $\mathcal{S}_a$. Also note that $h \in G$ depends on the input integer a, but the exponential expression E only depends on the fixed polynomial $P(x_1, \ldots, x_n)$. □

REMARK 6.3. The fixed exponential expression E from Proposition 6.2 has the following property that will be exploited in the next section: We can write E as $E'_1 E'_2 \cdots E'_m$ such that every exponential expression E'_i has length at most four and every base element g from E'_i commutes with every base element h from E'_j whenever $i \neq j$. For this, note that the last matrix in the exponential expression from Case 1 is central in $H_3(\mathbb{Z})$ and hence central in the direct product G. The first four matrices in the exponential expression from Case 1 yield an expression E'_i of length four.

6.2. Undecidability of knapsack for nilpotent groups of class two. Let $E = g_1^{x_1} g_2^{x_2} \cdots g_l^{x_l}$ be an exponential expression over the f.g. group G and let $X = \mathrm{Var}(E)$. Consider the group $G \times \mathbb{Z}^l$. For $1 \leq i \leq l$ let $e_i \in \mathbb{Z}^l$ be the i-th unit vector from $\mathbb{Z}^l$. For every $x \in X$ define

$$e_x = \sum_{1 \leq i \leq l, x_i = x} e_i \in \mathbb{Z}^l \quad \text{and} \quad k_x = (1, e_x) \in G \times \mathbb{Z}^l.$$

Note that k_x is central in $G \times \mathbb{Z}^l$. Moreover, for $1 \leq i \leq l$ let

$$k_i = (g_i, -e_i) \in G \times \mathbb{Z}^l.$$

Then, for a given group element $h \in G$, we have $S(E = h) \neq \emptyset$ if and only if

$$(h, 0) \in \prod_{x \in X} \langle k_x \rangle \prod_{i=1}^{l} \langle k_i \rangle.$$

By applying the above construction to the fixed exponential expression E over the fixed group $G = H_3(\mathbb{Z})^d \times \mathbb{Z}^e$ from Proposition 6.2, we obtain (note that $\mathbb{Z} \leq H_3(\mathbb{Z})$):

THEOREM 6.4. *There exist a fixed constant c and a fixed list $g_1, \ldots, g_\lambda \in H_3(\mathbb{Z})^c$ of group elements such that membership in the product $\prod_{i=1}^{\lambda} \langle g_i \rangle$ is undecidable.*

In particular, we have:

THEOREM 6.5. *There exists a fixed constant c such that* $\mathbf{KP}(H_3(\mathbb{Z})^c)$ *is undecidable.*

Finally, from the construction in the previous section, we also obtain the following undecidability result.

THEOREM 6.6. *There exist a fixed constant c and a fixed list of four abelian subgroups $G_1, G_2, G_3, G_4 \leq H_3(\mathbb{Z})^c$ such that membership in the product $G_1 G_2 G_3 G_4$ is undecidable.*

PROOF. Recall from Remark 6.3 that the exponential expression from Proposition 6.2 can be written as $E'_1 E'_2 \cdots E'_m$ such that every E'_i has length at most four, and every base element g from E'_i commutes with every base element h from E'_j whenever $i \neq j$. The above construction implies that the sequence of group elements $g_1, g_2, \ldots, g_\lambda$ from Theorem 6.4 can be split into blocks $B_1, B_2, \ldots, B_\mu$ of length at most four such that every group element from block B_i commutes with every group element from B_j whenever $i \neq j$. This allows to rearrange the product of cyclic groups $\prod_{i=1}^{\lambda} \langle g_i \rangle$ as a product of four abelian subgroups G_1, G_2, G_3, G_4, where G_i is generated by all group elements that are at the i-th position in their block. $\square$

REMARK 6.7. In contrast to Theorem 6.6, it was shown in [**16**] that a product of two subgroups of a polycyclic group is closed in the profinite topology. Since polycyclic groups are finitely presented, it follows that membership in a product of two subgroups of a polycyclic group is decidable. This leaves open whether membership in a product of three subgroups of a polycyclic (or nilpotent) group is decidable.

Let us finally prove that the knapsack problem for the discrete Heisenberg group $H_3(\mathbb{Z})$ is decidable.

THEOREM 6.8. *For every* $e \geq 0$, $\mathbf{KP}(H_3(\mathbb{Z}) \times \mathbb{Z}^e)$ *is decidable.*

PROOF. Let us first show the result for $H_3(\mathbb{Z})$. Take matrices $A, A_1, \ldots, A_l$ from $H_3(\mathbb{Z})$ and let

$$A = \begin{pmatrix} 1 & a & c \\ 0 & 1 & b \\ 0 & 0 & 1 \end{pmatrix} \quad \text{and} \quad A_i = \begin{pmatrix} 1 & a_i & c_i \\ 0 & 1 & b_i \\ 0 & 0 & 1 \end{pmatrix}$$

A straightforward induction over n shows that

$$A_i^n = \begin{pmatrix} 1 & a_i \cdot n & c_i \cdot n + a_i b_i \frac{(n-1)n}{2} \\ 0 & 1 & b_i \cdot n \\ 0 & 0 & 1 \end{pmatrix}$$

Hence, there is a solution $(x_1, \ldots, x_l) \in \mathbb{N}^l$ of $A = A_1^{x_1} \cdots A_l^{x_l}$ if and only if the following system of three Diophantine equations has a solution over $\mathbb{N}$:

$$a = \sum_{i=1}^{l} a_i \cdot x_i$$

$$b = \sum_{i=1}^{l} b_i \cdot x_i$$

$$c = \sum_{i=1}^{l} c_i \cdot x_i + \sum_{i=1}^{l} a_i b_i \frac{(x_i - 1)x_i}{2} + \sum_{1 \leq i < j \leq l} a_i b_j x_i x_j$$

This is a Diophantine system with a single quadratic equation and two linear equations. By [**4**], a system consisting of a single quadratic Diophantine equation together with an arbitrary number of linear equations can be reduced to a single quadratic Diophantine equation, which has the same solutions over $\mathbb{Z}$. By [**9**], one can decide whether this quadratic Diophantine equation has a solution over $\mathbb{N}$.

Finally, the above proof also works for the group $H_3(\mathbb{Z}) \times \mathbb{Z}^e$, since we only get additional linear equations. $\square$

COROLLARY 6.9. *The class of f.g. groups with a decidable knapsack problem is not closed under direct products.*

PROOF. This follows directly from Theorem 6.5 and 6.8. $\square$

7. Knapsack problems for finite extensions

We show that in contrast to direct products, decidability of the knapsack problem is preserved under finite extensions. For this, it will be convenient to consider a slightly extended version of the knapsack problem, which we will prove equivalent (with respect to polynomial time reducibility) to the knapsack problem. The *generalized knapsack problem* (briefly $\mathbf{GKP}(G)$) is the following decision problem: Given $f_0, \dots, f_k \in G$ and $g_1, \dots, g_k \in G$, decide whether

$$f_0 g_1^{n_1} f_1 g_2^{n_2} f_2 \cdots g_k^{n_k} f_k = 1 \tag{7.1}$$

for some $n_1, \dots, n_k \in \mathbb{N}$. An *instance* of the generalized knapsack problem is therefore a tuple $(f_0, g_1, f_1, \dots, g_k, f_k)$ with $f_0, \dots, f_k \in G$ and $g_1, \dots, g_k \in G$. If (7.1) holds, we call the tuple $(n_1, \dots, n_k)$ a *solution*. If two instances have the same set of solutions, we call them *equivalent*.

PROPOSITION 7.1. $\mathbf{KP}(G)$ *and* $\mathbf{GKP}(G)$ *are inter-reducible in polynomial time.*

PROOF. Since $g_1^{n_1} \cdots g_k^{n_k} = g$ if and only if $g^{-1} g_1^{n_1} \cdots g_k^{n_k} = 1$, $\mathbf{KP}(G)$ clearly reduces to $\mathbf{GKP}(G)$ in polynomial time.

Let us reduce $\mathbf{GKP}(G)$ to $\mathbf{KP}(G)$. Let $(f_0, g_1, f_1, \dots, g_k, f_k)$ be an instance of $\mathbf{GKP}(G)$. Observe that since $g_i^{n_i} f_i = f_i (f_i^{-1} g_i f_i)^{n_i}$, if we replace f_{i-1}, g_i, and f_i with $f_{i-1} f_i$, $f_i^{-1} g_i f_i$ and 1, respectively, we obtain an equivalent instance in which $f_i = 1$. By repeating this step k times, starting with f_k, we arrive at an instance with $f_1 = \cdots = f_k = 1$. Then, clearly, $f_0 g_1^{n_1} \cdots g_k^{n_k} = 1$ is equivalent to $g_1^{n_1} \cdots g_k^{n_k} = f_0^{-1}$. □

From now on, let G be finitely generated and H be a finite index subgroup of G, which is therefore finitely generated too. Furthermore, let $R \subseteq G$ be a finite set of representatives of right cosets of H in G. Then for each $g \in G$, there is a unique $\rho(g) \in R$ such that $g \in H\rho(g)$. Also recall from the proof of Theorem 4.2 that from a given element $g \in G$ we can compute effectively a decomposition $g = hr$ with $h \in H$ and $r \in R$. This fact will be implicitly used throughout this section.

LEMMA 7.2. *Let* $g_1, g_2 \in G$ *and* $\rho(g_1 g_2) = \rho(g_1)$. *We can compute* $h_1, h_2 \in H$ *and* $r \in R$ *such that* $g_1 g_2^t = h_1 h_2^t r$ *for every* $t \geq 0$.

PROOF. Since $\rho(g_1 g_2) = \rho(g_1)$, we can write $g_1 = h_1 r$ and $g_1 g_2 = h_{12} r$ for $h_1, h_{12} \in H$ and $r \in R$. Moreover, we can find $h_2 \in H$ and $r_2 \in R$ with $r g_2 = h_2 r_2$. Then

$$h_{12} r = g_1 g_2 = h_1 r g_2 = h_1 h_2 r_2$$

and hence $r_2 = r$. This means $r g_2 = h_2 r$ and thus $r g_2^t = h_2^t r$ and

$$g_1 g_2^t = h_1 r g_2^t = h_1 h_2^t r.$$

□

THEOREM 7.3. *Let* H *be a finite-index subgroup of a finitely generated group* G. *Then* $\mathbf{KP}(G)$ *is decidable if and only if* $\mathbf{KP}(H)$ *is decidable.*

PROOF. Since the "only if" direction is trivial, it remains to prove the "if" direction. According to 7.1, it suffices to show that if $\mathbf{GKP}(H)$ is decidable, then $\mathbf{GKP}(G)$ is decidable.

We say that an instance $I = (f_0, g_1, f_1, \ldots, g_k, f_k)$ of $\mathbf{GKP}(G)$ is *j-pure* if $f_0, g_1, \ldots, f_{j-1}, g_j \in H$. In particular, every instance is 0-pure. We call an instance *pure* if it is k-pure. If an instance is j-pure, but not $(j+1)$-pure, then $k-j$ is its *impurity*.

First, we prove the following claim by induction on the impurity of I: For every instance $I = (f_0, g_1, f_1, \ldots, g_k, f_k)$ of $\mathbf{GKP}(G)$, we can construct finitely many pure instances of $\mathbf{GKP}(G)$ such that the solution set of I is the union of affine images of their solution sets.

Suppose I is j-pure but not $(j+1)$-pure. Write $f_j = hr$ for $h \in H$ and $r \in R$. Since R is finite, there are $m, \ell \in \mathbb{N}$ with $\ell \geq 1$ and $\rho(rg_{j+1}^m) = \rho(rg_{j+1}^{m+\ell})$. We use 7.2 to find $h_1, h_2 \in H$ and $r' \in R$ such that $rg_{j+1}^{m+t\ell} = h_1 h_2^t r'$ for all $t \geq 0$. In particular

$$f_j g_{j+1}^{m+t\ell} = hrg_{j+1}^{m+t\ell} = hh_1h_2^t r'.$$

We can also find for each $0 \leq s < m$ elements $\hat{h}_s \in H$ and $\hat{r}_s \in R$ with $rg_{j+1}^s = \hat{h}_s \hat{r}_s$. Finally, we can find for each $0 \leq s < \ell$ a decomposition $r' g_{j+1}^s = \bar{h}_s \bar{r}_s$ with $\bar{h}_s \in H$, $\bar{r}_s \in R$. With these notations, we get

$$f_j g_{j+1}^n = \begin{cases} h\hat{h}_n\hat{r}_n & \text{for } n < m \text{ and} \\ hh_1h_2^t\bar{h}_s\bar{r}_s & \text{for } n = m + t\ell + s \text{ with } t \geq 0 \text{ and } 0 \leq s < \ell. \end{cases}$$

We therefore construct two types of instances. The first type consists of the instances

$$(f_0, g_1, f_1, \ldots, g_{j-1}, f_{j-1}, g_j, h\hat{h}_s\hat{r}_s f_{j+1}, g_{j+2}, f_{j+2}, \ldots, g_k, f_k), \tag{7.2}$$

for $0 \leq s < m$. The second type consists of instances

$$(f_0, g_1, f_1, \ldots, g_{j-1}, f_{j-1}, g_j, hh_1, h_2, \bar{h}_s\bar{r}_s f_{j+1}, g_{j+2}, f_{j+2}, \ldots, g_k, f_k) \tag{7.3}$$

for each $0 \leq s < \ell$. Observe that I has a solution if and only if one of these new instances has one. Furthermore, each of these new instances has lower impurity than I: In (7.2), we reduced the number of elements after g_j and in (7.3), both elements hh_1 and h_2 belong to H. Hence, the induction hypothesis yields the desired finite set of instances. This proves our claim.

Let us now prove the 7.3. Given an instance I of $\mathbf{GKP}(G)$, we construct pure instances $I_1, \ldots, I_m$ of $\mathbf{GKP}(G)$ such that I has a solution if and only if one of $I_1, \ldots, I_m$ has one. Since I_i is pure, if $I_i = (f_0, g_1, f_1, \ldots, g_k, f_k)$, then $f_0, g_1, \ldots, f_{k-1}, g_k \in H$, but f_k may not be in H. However, the equation

$$f_0 g_1^{n_1} f_1 \cdots g_k^{n_k} f_k = 1$$

can only have a solution if $f_k \in H$. Moreover, if $f_k \in H$, then I is in fact an instance of $\mathbf{GKP}(H)$. Since we can decide whether $f_k \in H$, we can pick from $I_1, \ldots, I_m$ those that are instances of $\mathbf{GKP}(H)$. This means, from I we have constructed finitely many instances of $\mathbf{GKP}(H)$ such that I has a solution if and only if one of the new instances has one. This proves the 7.3. □

8. Knapsack problems for co-context-free groups

In this 8, we exhibit another class of groups with a decidable knapsack problem, namely co-context-free groups, which we introduce first.

A *language* is a subset of a free monoid X^*, where X is an *alphabet*, i.e. a finite set of abstract symbols. A *context-free grammar* is a tuple $\Gamma = (N, T, P, S)$, where

- N and T are disjoint alphabets, their members are called *nonterminals* and *terminals*, respectively,
- $P \subseteq N \times (N \cup T)^*$ is a finite set of *productions*,
- $S \in N$ is the *start symbol*.

A production $(A, w) \in P$ is also denoted $A \to w$. In a context-free grammar, the productions allow us to rewrite words. Specifically, for $u, v \in (N \cup T)^*$, we write $u \Rightarrow_\Gamma v$ if there are $x, y \in (N \cup T)^*$ such that $u = xAy$ and $v = xwy$ for some production $A \to w$ in P. Furthermore, $\Rightarrow_\Gamma^*$ denotes the reflexive transitive closure of $\Rightarrow_\Gamma$. The language *generated by* Γ is then defined as

$$L(\Gamma) = \{w \in T^* \mid S \Rightarrow_\Gamma^* w\}.$$

A language is called *context-free* if it is generated by some context-free grammar.

Let Σ be a finite generating set of the group G and let $h\colon (\Sigma \cup \Sigma^{-1})^* \to G$ be the canonical monoid homomorphism. The *word problem* and the *co-word problem (with respect to* $\Sigma \cup \Sigma^{-1}$*)* of G are the languages

$$\{w \in (\Sigma \cup \Sigma^{-1})^* \mid h(w) = 1\} \text{ and}$$
$$\{w \in (\Sigma \cup \Sigma^{-1})^* \mid h(w) \neq 1\},$$

respectively. Since it does not depend on the chosen generating set whether the word problem or the co-word problem are context-free [**11**], we may define a group G to be *(co-)context-free* if its (co-)word problem is a context-free language. Co-context-free groups were introduced by Holt, Rees, Röver, and Thomas [**11**] and shown to significantly extend the class of context-free groups (which are, by a well-known result of Muller and Schupp and Dunwoody, precisely the virtually free groups [**5**, **21**]): The class of co-context-free groups is closed under taking direct products, taking restricted standard wreath products with a context-free top-group, passing to finitely generated subgroups and finite index overgroups. Furthermore, Lehnert and Schweitzer [**15**] have shown that the Higman-Thompson groups are co-context-free as well. Here, we show the following.

THEOREM 8.1. *Every co-context-free group has a decidable knapsack problem.*

Note that this means in particular that the wreath product $\mathbb{Z} \wr \mathbb{Z}$ has a decidable knapsack problem, which is in contrast to the fact that this group has an undecidable submonoid membership problem [**19**].

PROOF OF 8.1. Let W be the co-word problem of G with respect to $\Sigma \cup \Sigma^{-1}$ and let W be context-free.

We will need some terminology. A language $L \subseteq X^*$ is called *regular* if it can be obtained from the empty set and the singletons $\{x\}$, $x \in X$, by the operations

- *union*, which turns $K \subseteq X^*$ and $M \subseteq X^*$ into $K \cup M$,
- *concatenation*, which turns $K, M \subseteq X^*$ into $\{uv \mid u \in K,\ v \in M\}$, and
- *iteration*, which maps $M \subseteq X^*$ to the submonoid of X^* generated by M.

For every context-free language $L \subseteq X^*$, homomorphisms $\alpha\colon X^* \to Y^*$ and $\beta\colon Z^* \to X^*$ and regular language $K \subseteq X^*$, the languages $\alpha(L)$, $\beta^{-1}(L)$, and $L \cap K$ are context-free as well and we can effectively compute a grammar for the resulting languages [**3**].

Suppose we are given $g_1, \ldots, g_k, g$ as an instance of the knapsack problem and let these elements be written as words $w_1, \ldots, w_k, w$, respectively, over $\Sigma \cup \Sigma^{-1}$. Consider the alphabets $X = \{a_1, \ldots, a_k\}$, $Y = X \cup \{a\}$, and the homomorphisms

$\alpha\colon Y^* \to (\Sigma \cup \Sigma^{-1})^*$, with $\alpha(a_i) = w_i$ for $1 \le i \le k$ and $\alpha(a) = w^{-1}$. Here, w^{-1} is the word obtained from w by inverting the generators and then reversing the word. Furthermore, observe that the language $K = \{a_1\}^* \cdots \{a_k\}^*\{a\}$ is regular. Moreover, let $\beta\colon Y^* \to X^*$ be the homomorphism with $\beta(a_i) = a_i$ for $1 \le i \le k$ and $\beta(a) = \varepsilon$. Then, the language

$$M = \beta(\alpha^{-1}(W) \cap K) = \{a_1^{e_1} \cdots a_k^{e_k} \mid g_1^{e_1} \cdots g_k^{e_k} \neq g\}$$

is effectively context-free. Clearly, there exist $e_1, \ldots, e_k \in \mathbb{N}$ with $g_1^{e_1} \cdots g_k^{e_k} = g$ if and only if $M \neq \{a_1\}^* \cdots \{a_k\}^*$. In order to decide the latter, we will employ Parikh's Theorem.

For each $w \in X^*$, let $\Psi(w) = (e_1, \ldots, e_k)$, where e_i is the number of occurrences of a_i in w for $1 \le i \le k$. The resulting map $\Psi\colon X^* \to \mathbb{N}^k$ is called the *Parikh map*. Parikh's Theorem [**23**] states that for each context-free $L \subseteq X^*$, its *Parikh image* $\Psi(L) = \{\Psi(w) \mid w \in L\}$ is *semilinear*, meaning that it is a finite union of sets of the form

$$\{v_0 + x_1 \cdot v_1 + \cdots + x_n \cdot v_n \mid x_1, \ldots, x_n \in \mathbb{N}\},$$

where $v_0 \in \mathbb{N}^k$ and $v_1, \ldots, v_n \in \mathbb{N}^k$ are called the *base vectors* and the *period vectors*, respectively. Again, Parikh's theorem is effective, meaning that given a context-free grammar, we can compute base vectors and period vectors for its semilinear Parikh image.

Furthermore, given a semilinear set $S \subseteq \mathbb{N}^k$, its complement $\mathbb{N}^k \setminus S$ is effectively semilinear as well [**8**]. Since $M = \{a_1\}^* \cdots \{a_k\}^*$ if and only if $\Psi(M) = \mathbb{N}^k$, we can compute $\mathbb{N}^k \setminus \Psi(M)$ and check if it is non-empty. This concludes the proof of the theorem. □

References

[1] S. Arora and B. Barak, *Computational complexity*, Cambridge University Press, Cambridge, 2009. A modern approach. MR2500087

[2] L. Auslander, *On a problem of Philip Hall*, Ann. of Math. (2) **86** (1967), 112–116. MR0218454

[3] J. Berstel, *Transductions and context-free languages*, Leitfäden der Angewandten Mathematik und Mechanik [Guides to Applied Mathematics and Mechanics], vol. 38, B. G. Teubner, Stuttgart, 1979. MR549481

[4] M. Duchin, H. Liang, and M. Shapiro, *Equations in nilpotent groups*, Proc. Amer. Math. Soc. **143** (2015), no. 11, 4723–4731, DOI 10.1090/proc/12630. MR3391031

[5] M. J. Dunwoody, *The accessibility of finitely presented groups*, Invent. Math. **81** (1985), no. 3, 449–457, DOI 10.1007/BF01388581. MR807066

[6] M. Elberfeld, A. Jakoby, and T. Tantau, *Algorithmic meta theorems for circuit classes of constant and logarithmic depth*, 29th International Symposium on Theoretical Aspects of Computer Science, LIPIcs. Leibniz Int. Proc. Inform., vol. 14, Schloss Dagstuhl. Leibniz-Zent. Inform., Wadern, 2012, pp. 66–77. MR2909304

[7] E. Frenkel, A. Nikolaev, and A. Ushakov, *Knapsack problems in products of groups*, J. Symbolic Comput. **74** (2016), 96–108, DOI 10.1016/j.jsc.2015.05.006. MR3424034

[8] S. Ginsburg and E. H. Spanier, *Semigroups, Presburger formulas, and languages*, Pacific J. Math. **16** (1966), 285–296. MR0191770

[9] F. Grunewald and D. Segal, *On the integer solutions of quadratic equations*, J. Reine Angew. Math. **569** (2004), 13–45, DOI 10.1515/crll.2004.023. MR2055712

[10] C. Haase. *On the complexity of model checking counter automata*. PhD thesis, University of Oxford, St Catherine's College, 2011.

[11] D. F. Holt, S. Rees, C. E. Röver, and R. M. Thomas, *Groups with context-free co-word problem*, J. London Math. Soc. (2) **71** (2005), no. 3, 643–657, DOI 10.1112/S002461070500654X. MR2132375

[12] M. I. Kargapolov and Ju. I. Merzljakov, *Fundamentals of the theory of groups*, Graduate Texts in Mathematics, vol. 62, Springer-Verlag, New York-Berlin, 1979. Translated from the second Russian edition by Robert G. Burns. MR551207

[13] R. M. Karp, *Reducibility among combinatorial problems*, Complexity of computer computations (Proc. Sympos., IBM Thomas J. Watson Res. Center, Yorktown Heights, N.Y., 1972), Plenum, New York, 1972, pp. 85–103. MR0378476

[14] D. König and M. Lohrey. Evaluating matrix circuits. In *Proceedings of the 21st International Conference on Computing and Combinatorics, COCOON 2015*, volume 9198 of *Lecture Notes in Computer Science*, pages 235–248. Springer, 2015.

[15] J. Lehnert and P. Schweitzer, *The co-word problem for the Higman-Thompson group is context-free*, Bull. Lond. Math. Soc. **39** (2007), no. 2, 235–241, DOI 10.1112/blms/bdl043. MR2323454

[16] J. C. Lennox and J. S. Wilson, *On products of subgroups in polycyclic groups*, Arch. Math. (Basel) **33** (1979/80), no. 4, 305–309, DOI 10.1007/BF01222760. MR564284

[17] M. Lohrey, *The compressed word problem for groups*, Springer Briefs in Mathematics, Springer, New York, 2014. MR3289040

[18] M. Lohrey, *Rational subsets of unitriangular groups*, Internat. J. Algebra Comput. **25** (2015), no. 1-2, 113–121, DOI 10.1142/S0218196715400068. MR3325879

[19] M. Lohrey, B. Steinberg, and G. Zetzsche, *Rational subsets and submonoids of wreath products*, Inform. and Comput. **243** (2015), 191–204, DOI 10.1016/j.ic.2014.12.014. MR3352783

[20] Y. V. Matiyasevich, *Hilbert's tenth problem*, Foundations of Computing Series, MIT Press, Cambridge, MA, 1993. Translated from the 1993 Russian original by the author; With a foreword by Martin Davis. MR1244324

[21] D. E. Muller and P. E. Schupp, *Groups, the theory of ends, and context-free languages*, J. Comput. System Sci. **26** (1983), no. 3, 295–310, DOI 10.1016/0022-0000(83)90003-X. MR710250

[22] A. Myasnikov, A. Nikolaev, and A. Ushakov, *Knapsack problems in groups*, Math. Comp. **84** (2015), no. 292, 987–1016, DOI 10.1090/S0025-5718-2014-02880-9. MR3290972

[23] R. J. Parikh, *On context-free languages*, J. Assoc. Comput. Mach. **13** (1966), 570–581. MR0209093

[24] R. G. Swan, *Representations of polycyclic groups*, Proc. Amer. Math. Soc. **18** (1967), 573–574. MR0213442

[25] B. A. F. Wehrfritz. *Infinite Linear Groups*. Springer, 1977.

Department für Elektrotechnik und Informatik, Universität Siegen
E-mail address: koenig@eti.uni-siegen.de

Department für Elektrotechnik und Informatik, Universität Siegen
E-mail address: lohrey@eti.uni-siegen.de

Fachbereich Informatik, Technische Universität Kaiserslautern
E-mail address: zetzsche@cs.uni-kl.de

Contemporary Mathematics
Volume **677**, 2016
http://dx.doi.org/10.1090/conm/677/13626

On the Tits alternative for a class of finitely presented groups with a special focus on symbolic computations

Anja I. S. Moldenhauer, Gerhard Rosenberger, and Kristina Rosenthal

ABSTRACT. A group is said to satisfy the Tits alternative if it contains a non-abelian free subgroup or is virtually solvable, that is, it contains a solvable subgroup of finite index. This property was established by J. Tits, who proved that it is satisfied by finitely generated linear groups over a field [**38**].

In this paper, we consider groups admitting a presentation

$$\langle a, b, c \mid a^l = b^m = c^n = R_1^p(a,b) = R_2^q(a,c) = (abc)^r = 1 \rangle,$$

$2 \leq l, m, n, p, q, r$, where each $R_i(x,y)$ is a cyclically reduced word involving both x and y. Many of these groups appear in different contexts, inter alia, as fundamental groups of hyperbolic orbifolds or as factor groups of the extended modular group. To prove the Tits alternative for these groups under certain conditions concerning their presentations we build on previous work on the Tits alternative for generalized tetrahedron groups and apply Gröbner bases computations in non-commutative polynomial rings. Furthermore, the groups with $R_1(a,b) = a^\alpha b^\beta$ and $R_2(a,c) = a^\gamma c^\delta$, $1 \leq \alpha,\ \gamma < l$, $1 \leq \beta < m$, $1 \leq \delta < n$ are proved to satisfy the Tits alternative and a classification of the finite groups among these groups is presented. Additionally, we remark in which cases these groups are hyperbolic.

1. Introduction

The Tits alternative as well as the hyperbolicity are considered for an interesting class of groups. This class consists of groups having a presentation of the form

$$\langle a, b, c \mid a^l = b^m = c^n = R_1^p(a,b) = R_2^q(a,c) = (abc)^r = 1 \rangle, \tag{1.1}$$

where $2 \leq l, m, n, p, q, r$ and each $R_i(x,y)$ is a cyclically reduced word involving both x and y. Without loss of generality, we assume that each word $R_i(x,y)$ is not a proper power in the free product on x and y. Moreover, we assume that $R_1(a,b) = a^{\alpha_1}b^{\beta_1}\cdots a^{\alpha_{k_1}}b^{\beta_{k_1}}$ with $k_1 \geq 1$, $1 \leq \alpha_i < l$, $1 \leq \beta_i < m$ for $1 \leq i \leq k_1$ and $R_2(a,c) = a^{\gamma_1}c^{\delta_1}\cdots a^{\gamma_{k_2}}c^{\delta_{k_2}}$ with $k_2 \geq 1$, $1 \leq \gamma_i < l$, $1 \leq \delta_i < n$ for $1 \leq i \leq k_2$ (after conjugation if necessary). We call k_1 and k_2 the *block length* of R_1 and R_2, respectively. We say that a group G with a presentation of the form (1.1) has block length 1 if both R_1 and R_2 have block length 1. There are operations on presentations of the form (1.1) which do not change the defined groups. Two

---

2010 *Mathematics Subject Classification.* Primary 20F05; Secondary 20F65, 20H20.

Key words and phrases. Groups with periodic relations, generalized triangle groups, generalized tetrahedron groups, essential representation, trace polynomial, Fortsetzungssatz.

©2016 American Mathematical Society

presentations of the form (1.1) are called *equivalent* if one can be obtained from the other by a sequence of operations of the following type:

(1) Let $S_i(x,y)$ be a cyclically reduced conjugate of $R_i(x,y)$, $i \in \{1,2\}$. Replace the relator $R_i^k(x,y)$ by $S_i^k(x,y)$, $k \in \{p,q\}$, respectively.
(2) Replace the relator $R_i^k(x,y)$ by $S_i^k(x,y)$ for $i \in \{1,2\}$ and $k \in \{p,q\}$ where $S_i(x,y)$ is the inverse of $R_i(x,y)$.
(3) Let x be a generator of order 2, y a generator of order k and α, β coprime to k. Replace the relator $R(x,y) = (xy^{\alpha})^2$ by $S(x,y) = \left(xy^{\beta}\right)^2$.

Two groups defined by equivalent presentations are called *equivalent*, too. Obviously, two equivalent groups are isomorphic. We need this equivalence relation especially to reduce the number of cases to consider in the proofs or to describe lists of finite groups with a representation (1.1).

Groups with the presentation (1.1) appear quite often in the mathematical literature and in several contexts. First of all, these groups are factor groups for the three generator case of so called Pride groups [**29**] and of groups defined by periodic paired relations as considered by Vinberg [**41**]. Moreover, many of the groups with a presentation (1.1) have so called $M^\star$-groups as factor groups. $M^\star$-*groups* are defined as finite factor groups of the extended modular group Γ [**27**]. Γ is the projective general linear group $\mathrm{PGL}(2,\mathbb{Z})$ and provides the presentation

$$\langle x, y, z \mid x^2 = y^3 = z^2 = (xy)^2 = (xyz)^2 = 1\rangle.$$

$M^\star$-groups appear as the maximal automorphism groups of Kleinian surfaces. The groups (1.1) have often finite factor groups which are generated by three involutions, two of which commute [**4**]. These groups provide a presentation of the form

$$\langle r_1, r_2, r_3 \mid r_1^2 = r_2^2 = r_3^2 = 1, r_1r_2 = r_2r_1, r = 1, r \in R\rangle,$$

where R is a set of additional relators needed to present a finite group.

Cherkassoff and Sjerve [**4**] considered the question which of the alternating groups A_n, the symmetric groups S_n, the projective special linear groups $\mathrm{PSL}(2,q)$ and projective general linear groups $\mathrm{PGL}(2,q)$ are generated by three involutions, two of which commute. Their motivation came from the observation that such groups are amenable to the conjecture, arising from a question of Lovász, that every vertex transitive graph, with just four exceptions, has a Hamilton path. Cherkassoff and Sjerve proved that the Cayleygraph of a finite group with a presentation

$$\langle r_1, r_2, r_3 \mid r_1^2 = r_2^2 = r_3^2 = 1, r_1r_2 = r_2r_1, r = 1, r \in R\rangle$$

as above contains a Hamilton circle.

In addition, several of these group presentations (1.1) appear as fundamental groups of hyperbolic orbifolds. Many generalized triangle groups and tetrahedron groups appear as fundamental groups of threedimensional hyperbolic orbifolds (see for instance [**21**]). Since groups with a presentation (1.1) often are isomorphic to generalized tetrahedron groups or are in a canonical manner finite extensions of certain generalized triangle or tetrahedron groups, several of them also appear as fundamental groups of threedimensional (orientable or non-orientable) hyperbolic orbifolds. Last but not least groups with a presentation (1.1) are all factor groups of cocompact Fuchsian groups of genus 0 and with four cone points. Hence, they are interesting for the structure of their normal subgroups of finite index, especially in the case of arithmetic groups (see [**34**]).

We ask if a group with a presentation (1.1) satisfies the Tits alternative, that is, if it contains a non-abelian free subgroup or a solvable subgroup of finite index. To prove the Tits alternative for the groups under certain conditions regarding the exponents in the presentations of the form (1.1) several methods are applied. First, homomorphic images of the groups are considered. Second, we prove the existence of essential representations in the projective special linear group PSL $(2, \mathbb{C})$. Based on these representations, the existence of non-abelian free subgroups and thus the Tits alternative follows in many cases. Regarding the groups with block length 1, we also investigate the dimension of the group rings considered as vector spaces to prove a group is finite and thus satisfies the Tits alternative. For this purpose, Gröbner bases computations in non-commutative polynomial rings are used. The considered groups with block length 1 are proved to satisfy the Tits alternative in all cases. As a conclusion of the corresponding proof, a classification of the finite groups with block length 1 is obtained.

In addition to the Tits alternative we shortly describe some conditions under which the groups with a presentation (1.1) are hyperbolic. The hyperbolicity of groups is a property accompanied with strong results regarding the structure of the groups, for instance, hyperbolic groups admit a finite presentation and a Dehn algorithm [**3**]. More precisely, a hyperbolic group is a group admitting a Cayley graph fulfilling certain metric conditions. To prove hyperbolicity for the groups (1.1) under certain conditions we apply Small Cancellation Theory and take into account the exponents in the presentations of the respective groups.

The main purpose of this paper is to prove the Tits alternative for the groups (1.1) under certain conditions concerning their presentations as well as for those groups with block length 1. As a corollary, a classification of the finite groups with block length 1 is obtained. In addition, results regarding the hyperbolicity of the groups (1.1) are given. For the details see [**36**].

We thank the referee for some very helpful comments which helped to improve the paper. We also thank M. Kreuzer and X. Xiu for explaining how to apply Gröbner bases computations in non-commutative polynomial rings.

2. Generalized triangle and tetrahedron groups

Here we describe preliminary the definitions and results which we need.

First, we give several preliminary results about linear representations and the Tits alternative for generalized triangle and tetrahedron groups. We suppose that the group K is a generalized triangle group, that is, a group with a presentation of the form $\langle a, b \mid a^l = b^m = W^p(a,b) = 1\rangle$, where l, m, $p \geq 2$ and $W_1(a,b)$ is a cyclically reduced word involving both a and b which is not a proper power. If L is a linear group and $\rho : K \to L$ is a representation of K in L, then we call ρ *essential* if $A = \rho(a)$, $B = \rho(b)$, $\rho(W_1(a,b))$ have orders l, m, p respectively in $\rho(K)$.

THEOREM 2.1. [**2**], [**11**] *Each generalized triangle group admits an essential representation in* PSL $(2, \mathbb{C})$.

Now we suppose H to be a generalized tetrahedron group, that is, a group with a presentation of the form

$$\langle a, b, c \mid a^l = b^m = c^n = W_1^p(a,b) = W_2^q(a,c) = W_3^r(b,c) = 1\rangle,$$

where l, m, n, p, q, $r \geq 2$ and $W_i(x,y)$, $i = 1, 2, 3$, is a cyclically reduced word involving both x and y which is not a proper power. These groups are generalizations of the ordinary tetrahedron groups which admit a presentation of the form

$$\langle a, b, c \mid a^l = b^m = c^n = \left(ab^{-1}\right)^p = \left(ca^{-1}\right)^q = \left(bc^{-1}\right)^r = 1\rangle,$$

with l, m, n, p, q, $r \geq 2$. The definition of the ordinary tetrahedron groups is motivated by the symmetries of the tetrahedrons in the euclidean, hyperbolic or spherical space. If L is a linear group and $\rho : H \to L$ is a representation of H in L, then we call ρ *essential* if $A = \rho(a)$, $B = \rho(b)$, $C = \rho(c)$, $\rho(W_1(a, b))$, $\rho(W_2(a, c))$, $\rho(W_3(b, c))$ have orders l, m, n, p, q, r respectively in $\rho(H)$.

THEOREM 2.2. [**14**] *Each generalized tetrahedron group admits an essential representation in* $\mathrm{PSL}(2, \mathbb{C})$.

The next theorem is essential to prove the Tits alternative for generalized tetrahedron groups and will be adapted for groups with a presentation of the form (1.1) in this paper.

THEOREM 2.3 (Fortsetzungssatz). [**8**],[**14**] *Let H be a generalized tetrahedron group and for $G_1 = \langle a, b \mid a^l = b^m = R_1^p(a, b) = 1\rangle$ suppose that*

$$\rho_1 : G_1 \to \mathrm{PSL}(2, \mathbb{C})$$

is an essential representation with $\rho_1(a) = A$ and $\rho_1(b) = B$. Furthermore, assume that one of the following two possibilities holds:

(i) $\mathrm{tr}([A, B]) \neq 2$,
(ii) $(n, q, r) \neq (2, 2, 2)$ *and $\langle A, B\rangle$ is an infinite metabelian subgroup of* $\mathrm{PSL}(2, \mathbb{C})$.

Then ρ_1 can be extended to an essential representation $\rho : H \to \mathrm{PSL}(2, \mathbb{C})$ with $\rho(a) = A$ and $\rho(b) = B$. Moreover, if in addition $\langle A, B\rangle$ is non-elementary in case (i) *then H has a non abelian free subgroup in both cases* (i) *and* (ii).

A finitely generated subgroup of $\mathrm{PSL}(2, \mathbb{C})$ is said to be non-elementary if it is not solvable-by-finite. We have the following lemma.

LEMMA 2.4. *Let K be a finitely generated subgroup of* $\mathrm{PSL}(2, \mathbb{C})$. *Then the following are equivalent:*

(a) *K is solvable by finite.*
(b) *Any two elements of K of infinite order have at least one common fixed point in $\hat{\mathbb{C}} = \mathbb{C} \cup \{\infty\}$, considered as linear fractional transformations.*
(c) *For any two elements $A, B \in K$ of infinite order we have* $\mathrm{tr}([A, B]) = 2$ *for the commutator* $[A, B]$.

The proof is done by straightforward calculations (see for instance [**24**]). Hence, if a finitely generated subgroup K of $\mathrm{PSL}(2, \mathbb{C})$ is non-elementary then K has two elements A and B of infinite order with $\mathrm{tr}([A, B]) \neq 2$. In [**31**] and [**32**] there is an algorithmic construction of non-abelian free subgroups of $\langle A, B\rangle$, the groups generated by A and B. Hence, non-elementary subgroups of $\mathrm{PSL}(2, \mathbb{C})$ contain a non-abelian free subgroup (see also [**38**]).

There are several extensions of Theorem 2.3 showing the Tits alternative for generalized tetrahedron groups under certain conditions [**14**], [**7**]. We state some relevant results which we need directly.

THEOREM 2.5. [**10**] *Let H be a generalized tetrahedron group with*

$$(p,q,r) \neq (2,2,2).$$

Then H satisfies the Tits alternative.

The case $(p,q,r) = (2,2,2)$ is more complicated; the methods used for the last Theorem 2.5 do not work. But the Tits alternative is proved for Tsaranov generalized tetrahedron groups. A generalized tetrahedron group H is called a *Tsaranov generalized tetrahedron group* if $W_1(a,b) = a^{\alpha}b^{\beta}$, $W_2(a,c) = a^{\gamma}c^{\delta}$, $W_3(b,c) = b^{\varepsilon}c^{\zeta}$ with $1 \leq \alpha, \gamma < l$, $1 \leq \beta, \varepsilon < m$, $1 \leq \delta, \zeta < n$.

THEOREM 2.6. [**19**] *Let H be a Tsaranov generalized tetrahedron group. Then H satisfies the Tits alternative.*

In the proofs of Theorems 2.5 and 2.6 we have detailed lists for the cases where the respective H has a non-abelian free subgroup. Nevertheless there are several cases where H is solvable by finite or in fact finite. The most general situation where H always has a non-abelian free subgroup is given by Howie and Kopteva [**22**]. If we write - up to conjugacy - $W_1(a,b) = a^{\alpha_1}b^{\beta_1}\cdots a^{\alpha_{k_1}}b^{\beta_{k_1}}$ with $k_1 \geq 1$, $1 \leq \alpha_i < l$, $1 \leq \beta_i < m$ for $i = 1,\ldots,k$, and $W_2(a,c)$, $W_3(b,c)$ analogously, then we have the best general result.

THEOREM 2.7. *Let H be a generalized tetrahedron group as above.*

(1) *If $\frac{1}{pk_1} + \frac{1}{qk_2} + \frac{1}{rk_3} < 1$ then H contains a non-abelian free subgroup.*

(2) *If $\frac{1}{pk_1} + \frac{1}{qk_2} + \frac{1}{rk_3} = 1$ then H contains a non-abelian free subgroup except for the case $(l,m,n) = (2,2,2)$. If $(l,m,n) = (2,2,2)$ then H is solvable by finite.*

3. The Tits alternative for groups with a presentation (1.1)

We use as before the following notations:

$$G_1 = \langle a,b \mid a^l = b^m = R_1^p(a,b) = 1\rangle$$

and

$$G_2 = \langle a,c \mid a^l = c^n = R_2^q(a,c) = 1\rangle.$$

Homomorphic images of G_1 and G_2 are used to prove the Tits alternative under certain conditions. As a corollary of Theorem 2.5, we obtain the following result by applying Tietze transformations to transform the presentation of a considered group in the presentation of a generalized tetrahedron group.

THEOREM 3.1. *Let G be a group with a presentation of the form (1.1) with $R_1(a,b) = ab$ or, symmetrically, with $R_2(a,c) = ac$. Then G is isomorphic to a generalized tetrahedron group. Moreover, if in addition $(m,q,r) \neq (2,2,2)$ or $(l,p,r) \neq (2,2,2)$, respectively, then G satisfies the Tits alternative.*

PROOF. Let $R_1(a,b) = ab$. Then

$$G = \langle a,b,c \mid a^l = b^m = c^n = (ab)^p = R_2^q(a,c) = (abc)^r = 1\rangle.$$

Define $x = ab$. Then $b = a^{-1}x$, and we get

$$G = \langle a,x,c \mid a^l = x^p = c^n = (a^{-1}x)^m = R_2^q(a,c) = (xc)^r = 1\rangle$$

which is a generalized tetrahedron group. If $(m,q,r) \neq (2,2,2)$ then G satisfies the Tits alternative by Theorem 2.5. The case $R_2(a,c) = ac$ is analogously because $ca = cacc^{-1}$ and we may replace $R_2(a,c) = ac$ by $\tilde{R}_2(a,c) = ca$. $\square$

If here G is isomorphic to a Tsaranov generalized tetrahedron group, then G satisfies the Tits alternative by Theorem 2.6.

More types of homomorphic images appear quite often which can be used to prove the Tits alternative for generalized tetrahedron groups [**10**].

First, groups of special SN-type may appear as homomorphic images of a group with a presentation (1.1), for instance in the proof of Theorem 3.10 or in several cases in section 4. Groups of special SN-type are groups with a presentation

$$H = \langle a, b, c \mid a^{e_1} = b^{e_2} = c^{e_3} = R_1^{f_1}(a, b) = R_2^{f_2}(a, c) = 1 \rangle$$

where $e_1, e_2, e_3, f_1, f_2 \geq 2$ and $R_i(x, y)$, $i = 1, 2$, is a cyclically reduced word involving both x and y. For more details regarding groups of SN-type see [**17**]. We remark that in [**17**] the definition of a group of SN-type is more general. Hence, we call those groups of SN-type we need in this paper *special groups of SN-type.* We use the following result.

THEOREM 3.2. [**17**] *Let H be as above. If at least one of e_2, e_3, f_1, f_2 is greater than 2, then H contains a non-abelian free subgroup.*

With this result, we know that the group G contains a non-abelian free subgroup whenever we get a corresponding group of special SN-type as in Theorem 3.2 as a factor group of G.

Second, if we have a group G admitting a presentation

$$\langle a, b, c \mid a^2 = b^m = c^n = \left(ab^\beta\right)^2 = R_2^q(a, c) = (abc)^r = 1 \rangle,$$

then, for $\gcd(m, \beta) = 1$, the group G is isomorphic to the group with the presentation

$$\langle a, b, c \mid a^2 = b^m = c^n = (ab)^2 = R_2^q(a, c) = (abc)^r = 1 \rangle.$$

This group is a generalized tetrahedron group and if it has a non-abelian free subgroup, then G also has one. For $d = \gcd(m, \beta) \geq 2$ we consider the factor group

$$G^\star = \langle a, b, c \mid a^2 = b^d = c^n = R_2^q(a, c) = (abc)^r = 1 \rangle$$

of G. If $G^\star$ has a non-abelian free subgroup, then G also has one. The cases of these two types are just used but not mentioned explicitly in the following. More details can be found in [**36**].

Third, by introducing new relations, the images of the relators $R_1^p(a, b)$ and $R_2^q(a, c)$ may become superfluous, for instance in the above example with $d = gcd(m, \beta) \geq 2$ we introduced the relation $b^d = 1$, and then $(ab^\beta)^2 = 1$ follows from $a^2 = b^d = 1$, and we obtain a factor group of the group G with a presentation of the form

$$G^\star = \langle a, b, c \mid a^l = b^m = c^n = (abc)^r = 1 \rangle$$

with $l, m, n, r \geq 2$. This group is a one-relator product of cyclic groups which is SQ-universal if $(l, m, n, r) \neq (2, 2, 2, 2)$, that is, each countable group can be embedded into a quotient of the group $G^\star$ [**33**].

Now we introduce essential representations in $\mathrm{PSL}(2, \mathbb{C})$ of the groups (1.1). Let G be a group with a presentation of the form (1.1). If L is a linear group and $\rho : G \to L$ is a representation of G in L, then we call ρ *essential* if $A = \rho(a)$, $B = \rho(b)$, $C = \rho(c)$, $\rho(R_1(a, b))$, $\rho(R_2(a, c))$, $\rho(abc)$ have orders l, m, n, p, q, r respectively in $\rho(G)$. To prove the existence of essential representations we argue similar as in the case for generalized tetrahedron groups in [**14**]. In this sense the next lemma is an adapted version of [**30**, Lemma 1], and the prove is analogous.

LEMMA 3.3. *Let* $A, B, D, X \in \mathrm{SL}(2, \mathbb{C})$ *with* $AB = D$ *and* X *arbitrary. We write*

$$A = \begin{pmatrix} a_1 & a_2 \\ a_3 & a_4 \end{pmatrix}, \; B = \begin{pmatrix} b_1 & b_2 \\ b_3 & b_4 \end{pmatrix}, \; D = \begin{pmatrix} d_1 & d_2 \\ d_3 & d_4 \end{pmatrix}, \; X = \begin{pmatrix} x_1 & x_2 \\ x_3 & x_4 \end{pmatrix}.$$

Moreover, let $\vec{x}, \vec{r} \in \mathbb{C}^4$ *and* $M \in \mathbb{C}^{4\times 4}$ *be defined as*

$$\vec{x} = \begin{pmatrix} x_1 \\ x_2 \\ x_3 \\ x_4 \end{pmatrix}, \; \vec{r} = \begin{pmatrix} \mathrm{tr}(X) \\ \mathrm{tr}(AX) \\ \mathrm{tr}(BX) \\ \mathrm{tr}(ABX) \end{pmatrix}, \; M = \begin{pmatrix} 1 & 0 & 0 & 1 \\ a_1 & a_3 & a_2 & a_4 \\ b_1 & b_3 & b_2 & b_4 \\ d_1 & d_3 & d_2 & d_4 \end{pmatrix}.$$

Then $M\vec{x} = \vec{r}$ *and* $\det(M) = \mathrm{tr}([A, B]) - 2$. *Moreover, if* $\det(M) \neq 0$, *that is,* $\mathrm{tr}([A, B]) \neq 2$, *and* $\mathrm{tr}(BX)$ *is substituted by the variable* t *in* $\vec{r}$, *then*

$$\det(X) = x_1x_4 - x_2x_3 = 1$$

defines a quadratic polynomial $f(t) \in K[t]$, $K = \mathbb{Q}(a_1, a_2, a_3, a_4, b_1, b_2, b_3, b_4, \mathrm{tr}(X), \mathrm{tr}(AX), \mathrm{tr}(ABX))$. *In particular,* $\mathrm{tr}(BX)$ *is a zero of this polynomial.*

With this lemma we are able to prove the existence of essential representations for the groups with a presentation (1.1). Again, the approach is similar to that for generalized tetrahedron groups and the prove is analogous [**14**, Theorem 1].

THEOREM 3.4. *Let* G *be a group with a presentation of the form* (1.1). *Then* G *admits an essential representation in* $\mathrm{PSL}(2, \mathbb{C})$.

PROOF. Let G_1 be as above. Then G_1 is a generalized triangle group and thus admits an essential representation in $\mathrm{PSL}(2, \mathbb{C})$ by Theorem 2.1. Let A, B be the respective images of a, b. For $R_1(A, B) = A^{\alpha_1}B^{\beta_1} \cdots A^{\alpha_{k_1}}B^{\beta_{k_1}}$, $1 \leq \alpha_i < l$ and $1 \leq \beta_i < m$, the trace $\mathrm{tr}(R_1(A, B))$ is a polynomial of degree k_1 in $\mathrm{tr}(AB)$ [**17**, p. 176].

First we suppose $\mathrm{tr}([A, B]) \neq 2$. Now, analogously, $\mathrm{tr}(R_2(A, C))$ is a non-constant polynomial of degree k_2 in $\mathrm{tr}(AC)$. With Lemma 3.3 we can construct $C \in \mathrm{PSL}(2, \mathbb{C})$ such that

$$\mathrm{tr}(C) = 2\cos\frac{\pi}{n}, \mathrm{tr}(R_2(A, C)) = -2\cos\frac{\pi}{q}$$

and

$$\mathrm{tr}(ABC) = 2\cos\frac{\pi}{r}.$$

As a consequence, the map $a \mapsto A$, $b \mapsto B$, $c \mapsto C$ defines an essential representation. We remark that a matrix in $\mathrm{PSL}(2, \mathbb{C})$ has order k if it has trace $2\cos\frac{\pi}{k}$.

Now suppose $\mathrm{tr}([A, B]) = 2$. Lemma 3.3 cannot be used directly. If

$$\mathrm{tr}([A, B]) = 2$$

then A and B have at least one common fixed point and the commutator subgroup of $\langle A, B\rangle$ is abelian. Hence, $\langle A, B\rangle$ is abelian or infinite metabelian, and, without loss of generality, we may assume that $\langle A, B\rangle$ is infinite metabelian because $\mathrm{tr}(A, B)$ is the same in both cases. After a suitable conjugation we may assume A and B as

$$A = \pm\begin{pmatrix} \alpha & \gamma \\ 0 & \alpha^{-1} \end{pmatrix}, \; B = \pm\begin{pmatrix} \beta & \delta \\ 0 & \beta^{-1} \end{pmatrix}.$$

Because $\langle A, B\rangle$ is not abelian we may assume that $\gamma = 0$, $\delta \neq 0$ and thus

$$A = \pm\begin{pmatrix} \alpha & 0 \\ 0 & \alpha^{-1}\end{pmatrix} \text{ and } B = \pm\begin{pmatrix} \beta & \delta \\ 0 & \beta^{-1}\end{pmatrix}.$$

Since A and B have finite order we have further that $\alpha \neq \pm 1$ and $\beta \neq \pm 1$ for $l, m \geq 2$. Let

$$C = \pm\begin{pmatrix} c_1 & c_2 \\ c_3 & c_4\end{pmatrix}$$

be an element in $\mathrm{PSL}(2,\mathbb{C})$ with variables c_1, c_2, c_3, c_4.

We choose C such that $\mathrm{tr}(C) = 2\cos\frac{\pi}{n} = r_1$ and $\mathrm{tr}(ABC) = 2\cos\frac{\pi}{r} = r_2$ and $\mathrm{tr}(AC) = r_2$ a zero of the non-constant polynomial $\mathrm{tr}(R_2(A,C)) + 2\cos\frac{\pi}{q}$ in $\mathrm{tr}(AC)$. We first assume that $(\mathrm{tr}(C), \mathrm{tr}(AC), \mathrm{tr}(ABC)) \neq (0,0,0)$ and consider the following system of linear equations:

$$\begin{aligned} c_1 + c_4 = {} & \mathrm{tr}(C) && = r_1 \\ \alpha c_1 + \alpha^{-1} c_4 = {} & \mathrm{tr}(AC) && = r_2 \\ \alpha\beta c_1 + \alpha\delta c_3 + \alpha^{-1}\beta^{-1} c_4 = {} & \mathrm{tr}(ABC) && = r_3 \end{aligned} \tag{3.1}$$

We have

$$\det\begin{pmatrix} 1 & 0 & 1 \\ \alpha & 0 & \alpha^{-1} \\ \alpha\beta & \alpha\delta & \alpha^{-1}\beta^{-1}\end{pmatrix} = \delta\left(\alpha^2 - 1\right) \neq 0.$$

So there exists a unique solution (c_1, c_3, c_4) with

$$\begin{aligned} c_1 &= -\frac{\alpha^{-1}}{\alpha - \alpha^{-1}} r_1 + \frac{1}{\alpha - \alpha^{-1}} r_2, \\ c_3 &= \frac{\beta\alpha^{-1} - \alpha^{-1}\beta^{-1}}{\delta\left(\alpha - \alpha^{-1}\right)} r_1 + \frac{\alpha^{-2}\beta^{-1} - \beta}{\delta\left(\alpha - \alpha^{-1}\right)} r_2 + \alpha^{-1}\delta^{-1} r_3, \\ c_4 &= \frac{\alpha}{\alpha - \alpha^{-1}} r_1 - \frac{1}{\alpha - \alpha^{-1}} r_2. \end{aligned}$$

Further, with this we have $\mathrm{tr}(BC) = \beta c_1 + \delta c_3 + \beta^{-1} c_4$. For $c_3 \neq 0$, we can choose c_2 such that $c_1 c_4 - c_2 c_3 = 1$ is satisfied. Together with (c_1, c_3, c_4) we obtain the desired $C \in \mathrm{PSL}(2,\mathbb{C})$ and thus the desired essential representation. Now suppose $c_3 = 0$. For $n \neq 2$ we may replace $\mathrm{tr}(C) = 2\cos\frac{\pi}{n}$ by $\mathrm{tr}(C) = -2\cos\frac{\pi}{n}$. So we obtain $c_3 \neq 0$ and the desired $C \in \mathrm{PSL}(2,\mathbb{C})$. For $r \neq 2$ we may replace $\mathrm{tr}(ABC) = 2\cos\frac{\pi}{r}$ by $\mathrm{tr}(ABC) = -2\cos\frac{\pi}{r}$ and obtain with $c_3 \neq 0$ the desired $C \in \mathrm{PSL}(2,\mathbb{C})$. For $q \neq 2$ we obtain $c_3 \neq 0$ by replacing $\mathrm{tr}(R_2(A,C)) + 2\cos\frac{\pi}{q}$ by $\mathrm{tr}(R_2(A,C)) - 2\cos\frac{\pi}{q}$, because the polynomials $\mathrm{tr}(R_2(A,C)) + 2\cos\frac{\pi}{q}$ and $\mathrm{tr}(R_2(A,C)) - 2\cos\frac{\pi}{q}$ in $\mathrm{tr}(AC)$ have no common zero. If $q = 2$ and $\mathrm{tr}(R_2(A,C))$ has two different zeros for $\mathrm{tr}(AC)$, we may replace one zero by the other and obtain the desired $C \in \mathrm{PSL}(2,\mathbb{C})$. Therefore we can assume $n = q = r = 2$ for $(\mathrm{tr}(C), \mathrm{tr}(AC), \mathrm{tr}(ABC)) \neq (0,0,0)$. We may also start with an essential representation of $G_2 = \langle a, c \mid a^l = c^n = R_2(a,c)^q = 1\rangle$, consider the case when $(\mathrm{tr}(B), \mathrm{tr}(AB), \mathrm{tr}(ABC)) \neq (0,0,0)$, and proceed analogously. So, if we assume that

$$(\mathrm{tr}(C), \mathrm{tr}(AC), \mathrm{tr}(ABC)) \neq (0,0,0)$$

and

$$(\mathrm{tr}(B), \mathrm{tr}(AB), \mathrm{tr}(ABC)) \neq (0,0,0),$$

an essential representation of G is obtained except possibly in the case when $m = n = p = q = r = 2$, $\operatorname{tr}(R_1(A,B))$ as well as $\operatorname{tr}(R_2(A,C))$ have exactly one zero and $\langle A, B\rangle$ and $\langle A, C\rangle$ are abelian or infinite metabelian for each essential representation of G_1 and G_2, respectively, in $\mathrm{PSL}(2,\mathbb{C})$. In this case, the traces of AB, AC and ABC are completely determined. We have $\operatorname{tr}(AB) = 2\cos\left(\frac{\pi}{l} \pm \frac{\pi}{2}\right)$, $\operatorname{tr}(AC) = 2\cos\left(\frac{\pi}{l} \pm \frac{\pi}{2}\right)$ and $\operatorname{tr}(ABC) = 0$. If we now choose

$$A = \pm\begin{pmatrix} \alpha & 0 \\ 0 & \alpha^{-1} \end{pmatrix} \text{ and } B = \pm\begin{pmatrix} \beta & \delta \\ 0 & \beta^{-1} \end{pmatrix}$$

with $\alpha + \alpha^{-1} = 2\cos\frac{\pi}{l}$, $\delta \neq 0$ and $\beta + \beta^{-1} = 0$, we directly obtain $c_1c_4 = 1$ in the system of linear equations (3.1) and thus the desired $C \in \mathrm{PSL}(2,\mathbb{C})$ and the essential representation.

We finally have to consider $(\operatorname{tr}(C), \operatorname{tr}(AC), \operatorname{tr}(ABC)) = (0,0,0)$ or $(\operatorname{tr}(B), \operatorname{tr}(AB), \operatorname{tr}(ABC)) = (0,0,0)$. Without loss of generality we assume $(\operatorname{tr}(C), \operatorname{tr}(AC), \operatorname{tr}(ABC)) = (0,0,0)$. In this case, we may also assume

$$n = q = r = 2$$

because for $(n,q,r) \neq (2,2,2)$ we may reduce to the situation $(\operatorname{tr}(C), \operatorname{tr}(AC), \operatorname{tr}(ABC)) \neq (0,0,0)$. For instance, let $n = 2$, $q > 2$. We replace $\operatorname{tr}(R_2(A,C)) = -2\cos\frac{\pi}{q}$ by $\operatorname{tr}(R_2(A,C)) = 2\cos\frac{\pi}{q}$ and obtain that $\operatorname{tr}(AC) = 0$ is not a solution of the equation $\operatorname{tr}(R_2(A,C)) = 2\cos\frac{\pi}{q}$ if it is one of the equation $\operatorname{tr}(R_2(A,C)) = -2\cos\frac{\pi}{q}$. For $(\operatorname{tr}(C), \operatorname{tr}(AC), \operatorname{tr}(ABC)) = (0,0,0)$ and $n = q = r = 2$ there exists an epimorphism, which preserves the orders, of G onto the group $G^\star = \langle a,b,c \mid a^l = b^m = c^2 = R_1^p(a,b) = (ac)^2 = (abc)^2 = 1\rangle$. Here, ${G_2}^\star = \langle a, c \mid a^l = c^2 = (ac)^2 = 1\rangle$ is a dihedral group, so we may choose an essential representation of $G_2^\star$ in $\mathrm{PSL}(2,\mathbb{C})$ with $a \mapsto A$, $c \mapsto C$ and $\operatorname{tr}([A,C]) \neq 2$. With this essential representation of ${G_2}^\star$ we are able to construct a $B \in \mathrm{PSL}(2,\mathbb{C})$ with Lemma 3.3 which together with A and C gives the desired essential representation of $G^\star$ and thus defines one of G. □

REMARK 3.5. Let G be a group with a presentation (1.1). From the proof of Theorem 3.4 we see that there exists in fact an irreducible essential representation of G in $\mathrm{PSL}(2,\mathbb{C})$. Here an irreducible representation means that in the image group of G there are at least two elements which have no common fixed point, considered as linear fractional transformations. From the proof of Theorem 3.4 we also see that we find three generators x, y and z of G and a representation of G in $\mathrm{PSL}(2,\mathbb{C})$ with respective matrices X, Y and Z such that X and Y have no common fixed point. For X and Y we have three degrees of freedom with $\operatorname{tr}(X)$, $\operatorname{tr}(Y)$, $\operatorname{tr}(XY)$, $\det(X)$ and $\det(Y)$ being specified while for Z we have no degree of freedom. Hence, the dimension of the character space of G in $\mathrm{PSL}(2,\mathbb{C})$ - as an affine algebraic set - is $3 - 3 = 0$.

As a consequence of the proof we obtain the following theorem, the Fortsetzungssatz, analogously as in [**8**], [**14**] for generalized tetrahedron groups. This theorem - especially the additional results concerning the existence of non-abelian free subgroups - will be very useful to prove the Tits alternative in several cases.

THEOREM 3.6 (Fortsetzungssatz). *Let G be a group with a presentation of the form* (1.1) *and suppose $\rho_1 : G_1 \to \mathrm{PSL}(2,\mathbb{C})$ is an essential representation with*

$\rho_1(a) = A$ and $\rho_1(b) = B$. Furthermore, let one of the following two possibilities holds:

(i) $\mathrm{tr}([A,B]) \neq 2$,
(ii) $(n,q,r) \neq (2,2,2)$ *and* $\langle A,B\rangle$ *is an infinite metabelian subgroup of* $\mathrm{PSL}(2,\mathbb{C})$.

Then ρ_1 can be extended to an essential representation $\rho : G \to \mathrm{PSL}(2,\mathbb{C})$ with $\rho(a) = A$ and $\rho(b) = B$. Moreover, if in addition $\langle A,B\rangle$ is non-elementary in case (i), *then G has a non-abelian free subgroup in both cases* (i) *and* (ii).

Proof. If $\mathrm{tr}([A,B]) \neq 2$ then by the proof of Theorem 3.4 we may extend ρ_1 directly to an essential representation $\rho : G \to \mathrm{PSL}(2,\mathbb{C})$, and if $\langle A,B\rangle$ is non-elementary then $\langle A,B\rangle$, and therefore also G, has a non-abelian free subgroup. Now, let case (ii) hold, that is, $(n,q,r) \neq (2,2,2)$ and $\langle A,B\rangle$ is an infinite metabelian group. We may assume that

$$A = \begin{pmatrix} \alpha & 0 \\ 0 & \alpha^{-1} \end{pmatrix} \text{ and } B = \begin{pmatrix} \beta & \delta \\ 0 & \beta^{-1} \end{pmatrix}$$

with $\delta \neq 0$, $\alpha \neq \pm 1$ and $\beta \neq \pm 1$. Since $(n,q,r) \neq (2,2,2)$ we may construct, as in the proof of Theorem 3.4, an extension $\rho : G \to \mathrm{PSL}(2,\mathbb{C})$ of ρ_1 with

$$\rho(c) = C = \begin{pmatrix} c_1 & c_2 \\ c_3 & c_4 \end{pmatrix} \in \mathrm{PSL}(2,\mathbb{C}) \quad \text{and} \quad c_3 \neq 0.$$

But then $\langle A,B,C\rangle$ is non-elementary since C and $[A,B]$ have no common fixed point and $[A,B]$ has infinite order. Thus $\langle A,B,C\rangle$, and therefore also G, has a non-abelian free subgroup. □

Remark 3.7. If in Theorem 3.6 the subgroup $\langle A,B\rangle$ is finite and non-abelian then automatically $\mathrm{tr}([A,B]) \neq 2$ and ρ_1 can be extended to an essential representation $\rho : G \to \mathrm{PSL}(2,\mathbb{C})$. If now $\mathrm{tr}(BC)$, $C = \rho(c)$, is non-real then $\rho(G)$, and hence G, has a non-abelian free subgroup.

Lemma 3.3 can also be used in another manner to prove the Tits alternative [**10**]. For the group G with a presentation of the form (1.1) and G_1 as above let $\rho_1 : G_1 \to \mathrm{PSL}(2,\mathbb{C})$ be an essential representation with $\rho_1(a) = A$ and $\rho_1(B) = B$. Moreover, let $\langle A,B\rangle$ be finite and $\mathrm{tr}([A,B]) \neq 2$, that is, $\langle A,B\rangle$ isomorphic to D_{2n}, $(2 \leq n)$, A_4, S_4 or A_5. With Lemma 3.3 we construct $C \in \mathrm{PSL}(2,\mathbb{C})$ such that $\rho : G \to \mathrm{PSL}(2,\mathbb{C})$ with $\rho(a) = A$, $\rho(b) = B$ and $\rho(c) = C$ is an essential representation. If some possible value for $\mathrm{tr}(BC)$ is non-real, then BC is of infinite order and $\rho(G)$ is non-elementary. It follows that G contains a non-abelian free subgroup.

For G_2 we can argue symmetrically.

From the Fortsetzungssatz we get the following theorems as extensions by having a look at special cases for G_1. First, we consider the SQ-universality in analogy to [**14**, Theorem 3] for generalized tetrahedron groups.

Theorem 3.8. *Let G be a group with a presentation of the form* (1.1).

(a) *If*

$$\frac{1}{l} + \frac{1}{m} + \frac{1}{n} + \frac{1}{p} + \frac{1}{q} + \frac{1}{r} < 2,$$

then G is SQ-universal.

(b) *If*

$$\frac{1}{l}+\frac{1}{m}+\frac{1}{n}+\frac{1}{p}+\frac{1}{q}+\frac{1}{r}=2,$$

then G is infinite.

PROOF. Let ρ be an essential representation of G in PSL $(2,\mathbb{C})$. From Selberg's Theorem on finitely generated subgroups of linear groups over fields of characteristic 0 there exists a normal subgroup H of finite index in $\rho(G)$ such that $\rho(a)$ has order l, $\rho(b)$ has order m, $\rho(c)$ has order n, $\rho(R_1(a,b))$ has order p, $\rho(R_2(a,c))$ has order q and $\rho(abc)$ has order r modulo H [**37**]. Let π be the canonical epimorphism from $\rho(G)$ onto $\rho(G)/H$. The composition of maps $\phi=\pi\circ\rho$ is a representation of G onto a finite group. The rest of the proof is analogous to the proof of [**14**, Theorem 3]. □

Non-abelian free subgroups can be obtained for the groups (1.1) under other conditions as well. We proceed analogously as in [**14**, Theorem 4,5] together with the little correction in [**7**].

THEOREM 3.9. *Let G be a group with a presentation of the form* (1.1) *and assume $k_1\geq 2$ for the block length k_1 of $R_1(a,b)$ as described in the presentation* (1.1). *Further suppose that either*

(i) $m\geq 4$ *and* $p\geq 3$,
(ii) $m\geq 3$ *and* $p\geq 4$ *or*
(iii) $l\geq 3$ *and* $p\geq 3$.

Then G contains a non-abelian free subgroup.

PROOF. If G_1 admits an essential representation in PSL $(2,\mathbb{C})$ with non-elementary image, then G contains a non-abelian free subgroup by Theorem 3.6. For $k_1\geq 2$ and under the conditions (i), (ii) or (iii), such an essential representation exists except in the following cases (a) to (f) [**15**], [**7**]:

(a) $l=2$, $m=6t\geq 6$, $p=3$, $k_1=2$ and

$$G_1=\langle x,y\mid x^2=y^{6t}=\left(xy^txy^{3t}\right)^3=1\rangle.$$

(b) $l=2$, $m=4t\geq 4$, $p=4$, $k_1=2$ and

$$G_1=\langle x,y\mid x^2=y^{4t}=\left(xy^txy^{2t}\right)^4=1\rangle.$$

(c) $l=m=p=3$, $k_1=2$ and

$$G_1=\langle x,y\mid x^3=y^3=\left(xyxy^2\right)^3=1\rangle.$$

(d) $l=m=p=4$, $k_1=2$ and

$$G_1=\langle x,y\mid x^4=y^4=\left(xy^2x^2y^2\right)^4=1\rangle.$$

(e) $l=2$, $m=p=k_1=4$ and

$$G_1=\langle x,y\mid x^2=y^4=\left(xyxy^2xy^2xy^2\right)^4=1\rangle.$$

(f) $l=2$, $m=6t\geq 6$, $p=6$, $k_1=2$ and

$$G_1=\langle x,y\mid x^2=y^{6t}=\left(xy^{2t}xy^{3t}\right)^6=1\rangle.$$

Without loss of generality we may assume $x = a$ and $y = b$. First, we consider case (a). We introduce the relation $b^{3t} = 1$. The corresponding factor group $G^{\star} = \langle a, b, c \mid a^2 = b^{3t} = c^n = R_2^q(a, c) = (abc)^r = 1\rangle$ contains a non-abelian free subgroup because for $G_1^{\star} = \langle a, b \mid a^2 = b^{3t} = 1\rangle$ there exists an essential representation in $\mathrm{PSL}(2, \mathbb{C})$ with non-elementary image [**12**].

Case (b) is handled analogously. By introducing the relation $b^{2t} = 1$ we obtain the factor group $G^{\star} = \langle a, b, c \mid a^2 = b^{2t} = c^n = R_2^q(a, c) = (abc)^r = 1\rangle$. For $t > 1$ there exists an essential representation of $\langle a, b \mid a^2 = b^{2t} = 1\rangle$ in $\mathrm{PSL}(2, \mathbb{C})$ with non-elementary image [**12**]. Thus $G^{\star}$ and G contain non-abelian free subgroup of rank 2. For $t = 1$ we can choose an essential representation ρ_1 of

$$G_1 = \langle a, b \mid a^2 = b^4 = \left(abab^2\right)^4 = 1\rangle$$

in $\mathrm{PSL}(2, \mathbb{C})$ such that $\rho_1(G_1)$ is infinite metabelian. For $(n, q, r) \neq (2, 2, 2)$ then G itself contains a non-abelian free subgroup by Theorem 3.6. Now we assume $(n, q, r) = (2, 2, 2)$. G is isomorphic to

$$G' = \langle x, y, z \mid x^2 = y^4 = z^2 = \left(xyxy^2\right)^4 = (xz)^2 = (yz)^2 = 1\rangle.$$

G', and thus G, contains a non-abelian free subgroup [**14**].

In case (c) we consider an essential representation of G_1 in $\mathrm{PSL}(2, \mathbb{C})$ with $A = \rho_1(a)$ and $B = \rho_1(b)$. It is possible to choose $\rho_1(G_1)$ infinite metabelian. So G contains a non-abelian free subgroup by Theorem 3.6 if $(n, q, r) \neq (2, 2, 2)$. If $(n, q, r) = (2, 2, 2)$ we introduce the relation $(ab)^3 = 1$ and get the factor group $G^{\star} = \langle a, b, c \mid a^3 = b^3 = c^2 = (ab)^3 = R_2^2(a, c) = (abc)^2 = 1\rangle$. $G^{\star}$ is isomorphic to $G'^{\star} = \langle x, y, z \mid x^3 = y^3 = z^2 = \left(x^2y\right)^3 = R_2^2(x, z) = (yz)^2 = 1\rangle$ with $R_2(x, z) = x^{\gamma_1}z \cdots x^{\gamma_{k_2}}z$, $k_2 \geq 1$, $1 \leq \gamma_i < 3$ for $1 \leq i \leq k_2$. For $k_2 \geq 2$ this generalized tetrahedron group contains a non-abelian free subgroup [**18**], [**22**]. If $k_2 = 1$, then G is isomorphic to the generalized tetrahedron group

$$\begin{aligned} G' &= \langle x, y, z \mid x^3 = y^3 = z^2 = \left(xyxy^2\right)^3 = \left(zx^2\right)^2 = (yz)^2 = 1\rangle \\ &\cong \langle x, y, z \mid x^3 = y^3 = z^2 = \left(xyxy^2\right)^3 = (xz)^2 = (yz)^2 = 1\rangle. \end{aligned}$$

This group and thus G contains a non-abelian free subgroup [**14**].

Now consider case (d). We introduce in G the new relation $\left(ab^2a^2b^2\right)^2 = 1$ to get a factor group $G^{\star}$ of G. The respective $G_1^{\star} = \langle a, b \mid a^4 = b^4 = \left(ab^2a^2b^2\right)^2 = 1\rangle$ admits an essential representation in $\mathrm{PSL}(2, \mathbb{C})$ with non-elementary image. Thus, $G^{\star}$ and G contain a non-abelian free subgroup.

In case (e) we introduce the new relation $\left(abab^2ab^2ab^2\right)^2 = 1$ in G to get a factor group $G^{\star}$. Again, $G_1^{\star} = \langle a, b \mid a^2 = b^4 = \left(abab^2ab^2ab^2\right)^2 = 1\rangle$ admits an essential representation in $\mathrm{PSL}(2, \mathbb{C})$ with non-elementary image [**25**]. Thus, $G^{\star}$ and G contain a non-abelian free subgroup.

In case (f) we add the relation $\left(ab^{2t}ab^{3t}\right)^3 = 1$ to G and get a factor group $G^{\star}$. Again, $G_1^{\star} = \langle a, b \mid a^2 = b^{6t} = \left(ab^{2t}ab^{3t}\right)^3 = 1\rangle$ admits an essential representation in $\mathrm{PSL}(2, \mathbb{C})$ with a non-elementary image [**7**]. Thus, $G^{\star}$ and G contain a non-abelian free subgroup. □

THEOREM 3.10. *Let G be a group with a presentation of the form* (1.1) *with $k_1 \geq 1$ and $\frac{1}{l} + \frac{1}{m} + \frac{1}{p} < 1$. If either $k_1 = 1$ or $k_1 \geq 2$ and $p \geq 3$ then G contains a non-abelian free subgroup.*

PROOF. Let first $k_1 = 1$. Let ρ_1 be an essential representation of

$$G_1 = \langle a, b \mid a^l = b^m = R_1^p(a,b) = 1 \rangle$$

in $\mathrm{PSL}(2,\mathbb{C})$ with $A = \rho_1(a)$ and $B = \rho_1(b)$. Moreover, assume $\langle A, B\rangle$ is not abelian. For $p \geq 7$ or $p = 5$ there exists an essential representation of G_1 in $\mathrm{PSL}(2,\mathbb{C})$ with non-elementary image [**16**]. Thus, in this cases G contains a non-abelian free subgroup of rank 2. Now suppose $2 \leq p \leq 4$ or $p = 6$ and let

$$\rho_1 : G_1 \to \mathrm{PSL}(2,\mathbb{C})$$

be an essential representation. The image $\rho_1(G_1)$ can be chosen to be non-elementary, such that G contains a non-abelian free subgroup of rank 2, except in the following four cases [**16**]:

(a) $p = 6$, $l = 2$, $m = 6$ and $R_1(a,b) = ab^2$.
(b) $p = 6$, $l = 3$, $m = 6$ and $R_1(a,b) = ab^3$.
(c) $p = 6$, $l = 6$, $m = 6$ and $R_1(a,b) = a^2b^3$ or $R_1(a,b) = a^3b^2$.
(d) $p = 4$, $l = 4$, $m = 4$ and $R_1(a,b) = ab^2$ or $R_1(a,b) = a^2b$.

In the cases (a), (b) and (c) we add the relation $R_1^3(a,b) = 1$ to G to get the factor group $G^\star$ of G. The respective $G_1^\star = \langle a, b \mid a^l = b^m = R_1^3(a,b) = 1\rangle$ admits an essential representation in $\mathrm{PSL}(2,\mathbb{C})$ with non-elementary image [**16**]. Thus, $G^\star$ and G contain a non-abelian free subgroup.

Now we consider case (d). First we assume $(n,q,r) \neq (2,2,2)$. We can choose the image of $G_1 = \langle a, b \mid a^4 = b^4 = (ab^2)^4 = 1\rangle$ respectively

$$G_1 = \langle a, b \mid a^4 = b^4 = (a^2b)^4 = 1\rangle$$

under an essential representation in $\mathrm{PSL}(2,\mathbb{C})$ to be infinite metabelian. If $(n,q,r) \neq (2,2,2)$ then G contains a non-abelian free subgroup by Theorem 3.6.

Now we assume $(n,q,r) = (2,2,2)$ and first let $R_1(a,b) = ab^2$. By adding the relation $b^2 = 1$ to G we get the factor group

$$G^\star = \langle a, b, c \mid a^4 = b^2 = c^2 = R_2^2(a,c) = (abc)^2 = 1\rangle.$$

$G^\star$ is isomorphic to

$$\langle x, y, z \mid x^4 = (x^3y)^2 = z^2 = R_2^2(x,z) = (yz)^2 = 1\rangle.$$

This group contains a non-abelian free subgroup of rank 2 because

$$\langle x, y \mid x^4 = (x^3y)^2 = 1\rangle$$

admits an essential representation in $\mathrm{PSL}(2,\mathbb{C})$ with non-elementary image [**15**]. Thus, $G^\star$ contains a non-abelian free subgroup and so G. Now let $R_1(a,b) = a^2b$. We have

$$G = \langle a, b, c \mid a^4 = b^4 = c^2 = (a^2b)^4 = R_2^2(a,c) = (abc)^2 = 1\rangle$$

with $R_2(a,c) = a^{\alpha_1}ca^{\alpha_2}c\cdots a^{\alpha_{k_2}}c$, $k_2 \geq 1$ and $1 \leq \alpha_i < 4$ for $1 \leq i \leq k_2$. After adding the relation $a^2 = 1$ to G, there are two possibilities for $R_2(a,c)$. First, $R_2(a,c)$ may have the form $(ac)^h$ with $h \geq 0$, where $(ac)^h = 1$ for $h = 0$. Second, $R_2(a,c)$ may be equal to a or c. In the first case we get the factor group $G^\star = \langle a, b, c \mid a^2 = b^4 = c^2 = (ac)^{2h} = (abc)^2 = 1\rangle$ with $h \geq 0$. $G^\star$ is isomorphic to $\langle x, y, z \mid x^2 = y^4 = z^{2h} = (xz)^2 = (yz)^2 = 1\rangle$, $h \geq 0$. This group is of special SN-type and thus G contains a non-abelian free subgroup by Theorem 3.2. In the second case we get the factor group $G^{\star\star} = \langle a, b, c \mid a^2 = b^4 = c^2 = (abc)^2 = 1\rangle$. $G^{\star\star}$

is a one-relator product of cyclics and contains a non-abelian free subgroup [**33**]. So G contains a non-abelian free subgroup.

Now, let $k_1 \geq 2$ and $p \geq 3$. Then G contains a non-abelian free subgroup by Theorem 3.9 because $\frac{1}{l} + \frac{1}{m} + \frac{1}{p} < 1$. □

COROLLARY 3.11. *Let G be a group with a presentation* (1.1). *Assume that* $\frac{1}{l}+\frac{1}{m}+\frac{1}{p} \leq 1$ *and* $(n, q, r) \neq (2, 2, 2)$. *Then G contains a non-abelian free subgroup.*

PROOF. If $p \geq 3$ and $\frac{1}{l}+\frac{1}{m}+\frac{1}{p} < 1$ then the result follows from Theorem 3.10.

Now let $\frac{1}{l} + \frac{1}{m} + \frac{1}{p} = 1$ or $p = 2$ and $\frac{1}{l} + \frac{1}{m} + \frac{1}{2} < 1$. Let $\rho_1 : G_1 \to \mathrm{PSL}\,(2, \mathbb{C})$ be an essential representation with $\rho_1(a) = A$ and $\rho_1(b) = B$. If $\mathrm{tr}\,([A, B]) = 2$ then $\langle A, B\rangle$ is abelian or infinite metabelian, and, as in the proof of Theorem 3.2, we may assume that $\langle A, B\rangle$ is infinite metabelian, and then G has a non-abelian free subgroup by Theorem 3.6 because $(n, q, r) \neq (2, 2, 2)$. If $\mathrm{tr}\,([A, B]) \neq 2$ then $\langle A, B\rangle$ can not be finite, and hence $\langle A, B\rangle$ is non-elementary, because $\frac{1}{l} + \frac{1}{m} + \frac{1}{p} = 1$. Recall that a finite subgroup of $\mathrm{PSL}\,(2, \mathbb{C})$ either is abelian or a dihedral group or an alternating group A_4 or A_5 or a symmetric group S_4. □

REMARK 3.12. (1) The analogous results in Theorem 3.9 and 3.10 and in Corollary 3.11 hold for G_2 instead of G_1.

(2) Many of our arguments and methods fail if just $\frac{1}{l} + \frac{1}{m} + \frac{1}{p} \geq 1$ or $\frac{1}{l} + \frac{1}{n} + \frac{1}{q} \geq 1$, respectively. Here, we may show in many cases that at least G is infinite.

We consider

$$G_1 = \langle a, b \mid a^l = b^m = R_1^p(a, b) = 1\rangle.$$

Suppose $A, B \in \mathrm{PSL}\,(2, \mathbb{C})$ are matrices with $\mathrm{tr}\,(A) = 2\cos\frac{\pi}{l}$ and $\mathrm{tr}\,(B) = 2\cos\frac{\pi}{m}$. Let $\tau = \tau(t) = \mathrm{tr}\,(R_1(A, B))$ be **the trace polynomial** of $R_1(A, B)$, it is a non-constant polynomial in $t = \mathrm{tr}\,(AB)$. We define a polynomial

$$\sigma(t) = \prod_{\alpha} \left(\tau(t) - \alpha - \alpha^{-1}\right),$$

where the product is taken over all primitive rth roots α of -1 if r is even and of ± 1 if r is odd. Howie, Metaftsis and Thomas [**23**] showed that G_1 is infinite if $(l, m) \neq (2, 2)$ and if $\sigma(t)$ has a multiple root β, that is, a root of multiplicity > 1. They chose integers k coprime to l and s coprime to m such that

$$\left(\cos\frac{k\pi}{l} + \cos\frac{s\pi}{m}\right)^2 \neq \beta + 2$$

and defined

$$f(t) = \left(-t + 2\cos\frac{k\pi}{l}\right)\left(t + 2\cos\frac{s\pi}{m}\right) - 2.$$

Let $\gamma \in \mathbb{C}$ with $f(\gamma) = \beta$. Since β is a multiple root of $\sigma(t)$ then γ is a multiple root of $\sigma \circ f(t)$. Using this, they constructed a representation

$$\varphi_1 : G_1 \to PSL(2, \Lambda),\ a \mapsto A_1,\ b \mapsto B_1,$$

with

$$\Lambda = {}^{\mathbb{C}[t]}/\left({}_{(t-\gamma)^2}\right),$$

such that $\operatorname{tr}(A_1) = 2\cos\frac{k\pi}{l}$, $\operatorname{tr}(B_1) = 2\cos\frac{s\pi}{m}$ and $\varphi_1(G_1)$ is infinite.

We now apply this for a group with a presentation (1.1) with $(l,m) \neq (2,2)$. Again, we choose integers k, coprime to l, and s, coprime to m, such that

$$\left(\cos\frac{k\pi}{l} + \cos\frac{s\pi}{m}\right)^2 \neq \beta + 2.$$

Moreover, to have a suitable Fortsetzungssatz as Theorem 2.25 of [**35**], we need to choose k and s such that in addition $\cos\frac{k\pi}{l} \neq \cos\frac{s\pi}{m}$, which we can do. This is clear for $l \neq m$ and certainly to arrange if $l = m$ (see [**35**]).

Let $\sigma(t)$ and $f(t)$ be the polynomials as defined above.

THEOREM 3.13. *Let G be a group with a presentation* (1.1) *with* $(l,m) \neq (2,2)$. *Assume that $\sigma(t)$ has a multiple root β. Then G is infinite.*

PROOF. Assume first that there is an essential representation

$$\rho_1 : G_1 \to \operatorname{PSL}(2,\mathbb{C})$$

such that $\rho_1(G_1)$ is abelian or infinite metabelian. From the proof of Theorem 3.4 and 3.6 we know that G is infinite unless G has a factor group

$$\overline{G} = \langle a,b,c \mid a^l = b^m = c^2 = R_1^p(a,b) = (ca)^2 = (cab)^2 = 1\rangle,$$

which is isomorphic to

$$\langle a,b,x \mid a^l = b^m = x^2 = R_1^p(a,b) = (ax)^2 = (bx)^2 = 1\rangle,$$

and this is an infinite generalized tetrahedron group (see [**35**]). Hence, G is infinite if there is an essential representation $\rho_1 : G_1 \to \operatorname{PSL}(2,\mathbb{C})$ such that $\rho_1(G_1)$ is abelian or infinite metabelian. Now, assume there is no such representation ρ_1. We choose k and s as in the preliminary remarks. Also we choose $\gamma \in \mathbb{C}$ with

$$f(\gamma) = \left(-\gamma + 2\cos\frac{k\pi}{l}\right)\left(\gamma + 2\cos\frac{s\pi}{m}\right) - 2 = \beta.$$

Using the construction in [**23**] we get a representation $\varphi_1 : G_1 \to PSL(2,\Lambda)$, $\Lambda = \mathbb{C}[t]/((t-\gamma)^2)$, such that $\varphi_1(G_1)$ is infinite. By Theorem 2.25 of [**35**], the Fortsetzungssatz, we may extend φ_1 to a representation $\varphi : G \to PSL(2,\Lambda)$ such that $\varphi(G)$ is infinite. Hence, G is infinite. □

REMARK 3.14. The analogous result holds for G_2 instead of G_1.

COROLLARY 3.15. *Let G be a group with a presentation* (1.1) *with* $(l,m) \neq (2,2)$. *Assume that one of the following holds:*

(i) $k_1 \geq 9$ *if* $l = 2$ *or* $m = 2$,
(ii) $k_1 \geq 5$ *if* $l \geq 3$ *and* $m \geq 3$,
(iii) $k_1 \geq 5$ *if* $l \geq 4$ *or* $m \geq 4$.

Then G is infinite.

PROOF. If G_1 has an essential representation $\rho_1 : G_1 \to \operatorname{PSL}(2,\mathbb{C})$ with $\rho_1(G_1)$ abelian or infinite metabelian then G is infinite, (see the proof of Theorem 3.13). Assume now that there is no such ρ_1.

Let $\rho_1 : G_1 \to \mathrm{PSL}(2,\mathbb{C})$ an essential representation. Then $\rho_1(G_1)$ is non-elementary or finite. If $\rho_1(G_1)$ is non-elementary then G has a non-abelian free subgroup.

If $\rho_1(G_1)$ is finite then, by the given conditions, the polynomial $\sigma(t)$ must have a multiple root (see [**23**]). □

REMARK 3.16. The analogous results hold for G_2 instead of G_1.

4. The groups (1.1) with block length 1

Unfortunately, the used methods do not work in the remaining cases and we are not able to prove the Tits alternative for the groups (1.1) in general. So in the following, we focus on the groups with block length 1 and prove that these groups satisfy the Tits alternative.

THEOREM 4.1. *Let G be a group with a presentation*

$$\langle a,b,c \mid a^l = b^m = c^n = \left(a^\alpha b^\beta\right)^p = \left(a^\gamma c^\delta\right)^q = (abc)^r = 1\rangle$$

with $l,m,n,p,q,r \geq 2$, $1 \leq \alpha,\gamma < l$, $1 \leq \beta < m$ and $1 \leq \delta < n$. Then G satisfies the Tits alternative.

First, we remark that the group with the presentation

$$\langle a,b,c \mid a^l = b^m = c^n = \left(a^\alpha b^\beta\right)^p = \left(a^\gamma c^\delta\right)^q = (abc)^r = 1\rangle$$

with $l,m,n,p,q,r \geq 2$, $1 \leq \alpha,\gamma < l$, $1 \leq \beta < m$ and $1 \leq \delta < n$ is isomorphic to the group with the presentation

$$\langle a,b,c \mid a^l = b^n = c^m = \left(a^\gamma b^\delta\right)^q = \left(a^\alpha c^\beta\right)^p = (abc)^r = 1\rangle$$

by an analogous argument as in the proof of Theorem 3.1.
(Recall that

$$x^\alpha (x^{-1}y)^\beta = x^{-1}x^\alpha (yx^{-1})^\beta x.)$$

This reduces the number of cases to consider significantly. Second, from Theorem 3.10 and the above symmetry we know that G has a non-abelian free subgroup if

$$\frac{1}{l}+\frac{1}{m}+\frac{1}{p}<1 \quad \text{or} \quad \frac{1}{l}+\frac{1}{n}+\frac{1}{q}<1.$$

Hence, from now on we may assume that

$$\frac{1}{l}+\frac{1}{m}+\frac{1}{p}\geq 1 \quad \text{and} \quad \frac{1}{l}+\frac{1}{n}+\frac{1}{q}\geq 1.$$

Also we may exclude all the cases where $R_1(a,b) = ab$ or $R_2(a,c) = ac$ because here the results follow directly from Theorem 3.1 and 2.6. This handles also automatically the cases with $(l,m) = (2,2)$ or $(l,n) = (2,2)$. This can be seen as follows. Let, without loss of generality, $(l,m) = (2,2)$. Then

$$G = \langle a,b,c \mid a^2 = b^2 = c^n = (ab)^p = (ac^\beta)^q = (abc)^r = 1\rangle.$$

If we define $x = ab$, then $b = ax$ and

$$G = \langle a,x,c \mid a^2 = x^p = c^n = (ax)^2 = (ac^\beta)^q = (xc)^r = 1\rangle$$

is a Tsaranov generalized tetrahedron group, and the result follows from Theorem 2.6.

Altogether, using all the above remarks and the symmetry, we may divide the remaining cases to consider in the following seven subclasses:

Case 1: $(l,p) = (2,2)$,
Case 2: $l = 2$, $p \geq 3$ and $q \geq 3$,
Case 3: $l = 3$ and $(m,p) \neq (2,2)$,
Case 4: $l = 4$ and $(m,p) \neq (2,2)$,
Case 5: $l = 5$ and $(m,p) \neq (2,2)$,
Case 6: $l = 6$ and $(m,p) \neq (2,2)$,
Case 7: $(m,p) = (2,2)$.

In all the cases involved we had to make case-by-case considerations and to use, besides others, the results and methods described in the previous sections. In most of the cases we easily found a non-abelian free subgroup or a solvable subgroup of finite index by either considering essential representation ρ into $\mathrm{PSL}(2,\mathbb{C})$, also in a symmetric manner, or suitable subgroups and factor groups or isomorphisms to generalized tetrahedron groups. First, to show what is going on, we describe for each of these cases 1 to 7 at least one typical example for which the respective group G satisfies the Tits alternative just by applying the mentioned methodical aspects. Second, we finally come up with those situations where the described methods do not give automatically the desired result. But it turns that in each of the remaining cases the group G is finite, and, hence, also satisfies the Tits alternative.

(1) For case 1, let

$$G = \langle a,b,c \mid a^2 = b^m = c^n = (ab^\beta)^2 = (ac^\delta)^q = (abc)^r = 1\rangle.$$

If $\gcd(\beta, m) = 1$ then

$$G = \langle a,b,c \mid a^2 = b^m = c^n = (ab)^2 = (ac^\delta)^q = (abc)^r = 1\rangle,$$

because of the dihedral relation. With $x = ab$ we get $b = ax$ and

$$G = \langle a,x,c \mid a^2 = x^2 = c^n = (ax)^m = (ac^\delta)^q = (xc)^r = 1\rangle,$$

a generalized tetrahedron group. Hence, G satisfies the Tits alternative if $\gcd(\beta, m) = 1$.

Now, let $\gcd(\beta, m) = d > 1$. Then, especially, $m = 4$ or $m \geq 6$. We introduce the relation $b^d = 1$, and get the factor group

$$\overline{G} = \langle a,b,c \mid a^2 = b^d = c^n = (ac^\delta)^q = (abc)^r = 1\rangle.$$

In $\overline{G}$ let $x = ab$. Then $b = ax$ and

$$\overline{G} = \langle a,x,c \mid a^2 = c^n = (ax)^d = (ac^\delta)^q = (xc)^r = 1\rangle.$$

If $(d,q,r) \neq (2,2,2)$ the $\overline{G}$, and hence also G, has a non-abelian free subgroup by [**17**, Theorem 9.3.4 and 9.3.5]. Hence, we are left with the case $(d,q,r) = (2,2,2)$. Then

$$G = \langle a,b,c \mid a^2 = b^m = c^n = (ab^\beta)^2 = (ac^\delta)^2 = (abc)^2 = 1\rangle$$

with $\gcd(\beta, m) = 2$.
If $\gcd(\delta, n) = 1$ then, analogously as for the case $\gcd(\beta, m) = 1$, G is a generalized tetrahedron group. Now, let $\gcd(\delta, n) = f > 1$. Recall that in a dihedral group

$$\langle g,h \mid g^2 = h^m = (gh)^2 = 1\rangle$$

the relation $(gh^\delta)^2 = 1$ is a consequence of the given relations.

Therefore, if we introduce in addition the relation $c^f = 1$, we get the factor group

$$\overline{\overline{G}} = \langle a, b, c \mid a^2 = b^m = c^f = (ab)^2 = (abc)^2 = 1 \rangle.$$

Again, let $x = ab$, then $b = ax$ and

$$\overline{\overline{G}} = \langle a, x, c \mid a^2 = x^2 = c^f = (ax)^m = (xc)^2 = 1 \rangle$$

which is a group of special SN-type.

Now, $\overline{\overline{G}}$, and hence also G, contains a non-abelian free subgroup by Theorem 3.2, because $m \geq 4$.

(2) For case 2, that is, $l = 2$, $p \geq 3$ and $q \geq 3$, we consider the example

$$G = \langle a, b, c \mid a^2 = b^5 = c^5 = (ab^2)^3 = (ac^2)^3 = (abc)^2 = 1 \rangle.$$

We choose $A, B \in \mathrm{PSL}(2, \mathbb{C})$ with $\mathrm{tr}(A) = 0$, $\mathrm{tr}(B) = 2\cos\frac{\pi}{5}$ and $\mathrm{tr}(AB) = 2\cos\frac{\pi}{5} - 1$.

The map $a \mapsto A$, $b \mapsto B$ defines an essential representation $\rho_1 : G_1 \to \mathrm{PSL}(2, \mathbb{C})$, and $\langle A, B \rangle$ is isomorphic to the alternating group A_5. We want to construct $C \in \mathrm{PSL}(2, \mathbb{C})$ such that $a \mapsto A$, $b \mapsto B$ and $c \mapsto C$ defines an essential representation $\rho : G \to \mathrm{PSL}(2, \mathbb{C})$. For doing this we apply Lemma 3.3. By setting $\mathrm{tr}(C) = 2\cos\frac{\pi}{5}$ we get $\mathrm{tr}(AC) = 2\cos\frac{\pi}{5} - 1$ from the relation $(ac^2)^3 = 1$. With $\mathrm{tr}(ABC) = 0$ we obtain $\mathrm{tr}(BC) \approx 1{,}118 \pm 0{,}866\mathrm{i}$, and $\mathrm{tr}(BC)$ is non-real, especially BC has infinite order. Therefore G has a non-abelian free subgroup.

(3) For case 3, that is, $l = 3$ and $(m, p) \neq (2, 2)$, we consider the example

$$G = \langle a, b, c \mid a^3 = b^4 = c^3 = (ab)^2 = (ac^2)^2 = (abc)^r = 1 \rangle, \quad r \geq 2.$$

We choose $A, B \in \mathrm{PSL}(2, \mathbb{C})$ with $\mathrm{tr}(A) = 1$, $\mathrm{tr}(B) = \sqrt{2}$ and $\mathrm{tr}(AB) = \frac{1}{\sqrt{2}}$. The map $a \mapsto A$, $b \mapsto B$ defines an essential representation $\rho_1 : G_1 \to \mathrm{PSL}(2, \mathbb{C})$ with $\langle A, B \rangle$ non-elementary because $\mathrm{tr}([A, B]) \neq 2$ and AB has infinite order. Hence, we may extend ρ_1 to an essential representation $\rho : G \to \mathrm{PSL}(2, \mathbb{C})$ such that $\rho(G)$, and therefore G, has a non-abelian subgroup.

(4) For case 4, that is $l = 4$ and $(m, p) \neq (2, 2)$ we consider two examples.

(a) Let

$$G = \langle a, b, c \mid a^4 = b^3 = c^4 = (ab^2)^2 = (a^2c)^2 = (abc)^r = 1 \rangle, \quad r \geq 2.$$

If we introduce the relation $a^2 = 1$ then we get the factor group

$$\begin{aligned} G^\star &= \langle a, b, c \mid a^2 = b^3 = c^2 = (ab)^2 = (abc)^r = 1 \rangle \\ &\cong \langle x, y, z \mid x^2 = y^3 = z^2 = (xy)^2 = (xz)^r = 1 \rangle, \end{aligned}$$

a group of special SN-type. From Theorem 3.6 we easily get that $G^\star$ contains the modular group $\mathrm{PSL}(2, \mathbb{Z}) \cong \langle y, z \mid y^3 = z^2 \rangle$ as a subgroup. Hence, G contains a non-abelian free subgroup.

(b) Let

$$G = \langle a, b, c \mid a^4 = b^3 = c^4 = (a^3b)^2 = (ac^3)^2 = (abc)^r = 1 \rangle$$

with $r \geq 2$.

Now, instead of G_1, we consider the group

$$G_2 = \langle a, c \mid a^4 = c^4 = (ac^3)^2 = 1 \rangle.$$

We have an essential representation $\rho_2 : G_2 \to \mathrm{PSL}(2,\mathbb{C})$ with $\rho_2(G_2)$ infinite and metabelian. Since $(m,p,r) = (3,2,r) \neq (2,2,2)$, we get that G has a non-abelian free subgroup (see also Corollary 3.11).

(5) For case 5, that is, $l = 5$ and $(m,p) \neq (2,2)$, we consider the example

$$G = \langle a, b, c \mid a^5 = b^3 = c^3 = (a^2b)^2 = (a^3c)^2 = (abc)^3 = 1\rangle.$$

We choose $A, B \in \mathrm{PSL}(2,\mathbb{C})$ with $\mathrm{tr}(A) = 1 - 2\cos(\frac{\pi}{5})$, $\mathrm{tr}(B) = 1$ and $\mathrm{tr}(AB) = -2\cos(\frac{\pi}{5})$. The map $a \mapsto A$, $b \mapsto B$ defines an essential representation $\rho_1 : G_1 \to \mathrm{PSL}(2,\mathbb{C})$. We want to construct $C \in \mathrm{PSL}(2,\mathbb{C})$ such that $a \mapsto A$, $b \mapsto B$ and $c \mapsto C$ defines an essential representation $\rho : G \to \mathrm{PSL}(2,\mathbb{C})$. For doing this we apply Lemma 3.3. By setting $\mathrm{tr}(C) = 1$, $\mathrm{tr}(AC) = 1$ and $\mathrm{tr}(ABC) = 1$ we obtain $\mathrm{tr}(BC) \approx 1 \pm 1,272\mathrm{i}$, and $\mathrm{tr}(BC)$ is non-real, especially BC has infinite order. Therefore G has a non-abelian free group. □

We remark that this group in a sense is typical for case 5. Let G be one of the groups in case 5. Then we may rewrite G as a generalized tetrahedron group, we find as in the example a representation $\rho : G \to \mathrm{PSL}(2,\mathbb{C})$ such that $\rho(G)$ is non-elementary or we can not apply one of the described methods to show that the Tits alternative holds for G. One group for the latter situation is the group where we replace $r = 3$ in the above example by $r = 2$. This group with $r = 2$ is shown to be finite of order 7200 by using Gröbner bases computations in non-commutative polynomial rings (see the section on Gröbner bases computations).

(6) For case 6, that is $l = 6$ and $(m,p) \neq (2,2)$, we consider the example

$$G = \langle a, b, c \mid a^6 = b^3 = c^n = (a^2b)^2 = (a^\gamma c^\delta)^q = (abc)^r = 1\rangle$$

with $1 \leq \gamma < 6$, $1 \leq \delta < n$, $(\gamma, \delta) \neq (1,1)$, $2 \leq q$ and $2 \leq r$. We can not rewrite G as a generalized tetrahedron group in an obvious manner by applying Theorem 3.1. But if we choose $A, B \in \mathrm{PSL}(2,\mathbb{C})$ with $\mathrm{tr}(A) = \sqrt{3}$, $\mathrm{tr}(B) = 1$ and $\mathrm{tr}(AB) = \frac{1}{\sqrt{3}}$, then we get an essential representation $\rho_1 : G_1 \to \mathrm{PSL}(2,\mathbb{C})$ such that $\rho_1(G_1)$ is non-elementary. This representation we may extend to an essential representation $\rho : G \to \mathrm{PSL}(2,\mathbb{C})$ by Theorem 3.4. Hence, G contains a non-abelian free subgroup.

(7) For case 7, that is $(m,p) = (2,2)$, we consider three typical examples.

(a) Let

$$G = \langle a, b, c \mid a^{2k} = b^2 = c^2 = (a^2b)^2 = (a^2c)^2 = (abc)^2 = 1\rangle.$$

The subgroup $H = \langle a^2 \rangle$ is a normal subgroup of G. The factor group ${}^{G}/_{H}$ has a presentation

$${}^{G}/_{H} = \langle a, b, c \mid a^2 = b^2 = c^2 = (abc)^2 = 1\rangle$$

which is an infinite and solvable group because the subgroup $\langle ab, ac \rangle < {}^{G}/_{H}$ is free abelian of rank 2 and index 2 (see [**17**, proof of Theorem 7.2.1]).

Therefore G is infinite and solvable.

(b) Let

$$G = \langle a, b, c \mid a^4 = b^2 = c^3 = (a^2b)^2 = (a^2c)^2 = (abc)^r = 1\rangle \quad \text{with } r \geq 2.$$

Now, instead of G_1, we consider the group

$$G_2 = \langle a, c \mid a^4 = c^3 = (a^2c)^2 = 1 \rangle.$$

If we choose $A, B \in \mathrm{PSL}(2, \mathbb{C})$ with $\mathrm{tr}(A) = \sqrt{2}$, $\mathrm{tr}(C) = 1$ and $\mathrm{tr}(AC) = \frac{1}{\sqrt{2}}$, then we get an essential representation

$$\rho_2 : G_2 \to \mathrm{PSL}(2, \mathbb{C})$$

with $\rho_2(G_2)$ non-elementary. This representation ρ_2 we may extend to an essential representation $\rho : G \to \mathrm{PSL}(2, \mathbb{C})$, and therefore G has a non-abelian free subgroup.

(c) Let

$$\begin{aligned} G &= \langle a, b, c \mid a^4 = b^2 = c^2 = (ab)^2 = (ac)^4 = (abc)^2 = 1 \rangle \\ &\cong \langle x, y, z \mid x^4 = y^2 = z^2 = (xy)^2 = (xz)^4 = (yz)^2 = 1 \rangle =: \overline{G}. \end{aligned}$$

The subgroup H of $\overline{G}$ generated by x and z has index 2 in $\overline{G}$ and a presentation

$$H = \langle x, z \mid x^4 = z^2 = (xz)^4 = 1 \rangle.$$

H is an euclidean planar discontinuous group and therefor solvable and infinite. Hence, altogether G is solvable and infinite.
The isomorphism between G and $\overline{G}$ follows as in Theorem 3.1. If we define $y = ab$, then $b = a^{-1}y$ and $(a^{-1}y)^{-1} = ya = a^{-1}aya$. Hence,

$$\begin{aligned} G &= \langle a, y, c \mid a^4 = y^2 = c^2 = (ay)^2 = (ac)^4 = (yc)^2 = 1 \rangle \\ &\cong \langle x, y, z \mid x^4 = y^2 = z^2 = (xy)^2 = (xz)^4 = (yz)^2 = 1 \rangle \\ &= \overline{G}. \end{aligned}$$

REMARK 4.2. Because of block length 1, the reduction to generalized tetrahedron groups as in example 7(c) can be used for many groups with a presentation (1.1) and block length 1 (see Theorem 3.1). It is worth mentioning that if a group G with a presentation (1.1) and block length 1 is solvable by finite and infinite then it is already solvable.

We think that we have presented a reasonable selection of examples where we could use straightforward the results and methods in the previous sections to get the desired result. More details for the case $1-7$ can be found in the dissertation by K. Rosenthal [**36**].

After all these considerations and applications we are left - up to equivalence or obvious isomorphism to generalized tetrahedron groups- with ten exceptional groups which we could not handle with the presented results and methods from the previous sections.

For these exceptional groups we used symbolic computations and were able to show that they all present finite groups. More concrete, we used Gröbner bases computations in non-commutative polynomial rings to calculate the dimensions of the respective group rings, considered as vector spaces.

Let $\mathbb{Q}\langle a, b, c \rangle$ be the non-commutative polynomial ring over the rational numbers. Choose a length compatible word ordering on the free monoid ring $\{a, b, c\}^\star$, for instance, the length lexicographic ordering $llex$ with $a >_{llex} b >_{llex} c$. For the group G with the presentation

$$\langle a, b, c \mid a^l = b^m = c^n = (a^\alpha b^\beta)^p = (a^\gamma c^\delta)^q = (abc)^r = 1 \rangle$$

with $l, m, n, p, q, r \geq 2$, $1 \leq \alpha, \gamma < l$, $1 \leq \beta < m$, $1 \leq \delta < n$ let $I \subseteq \mathbb{Q}\langle a, b, c\rangle$ be the ideal generated by the set

$$\{a^l - 1, b^m - 1, c^n - 1, (a^\alpha b^\beta)^p - 1, (a^\gamma c^\delta)^q - 1, (abc)^r - 1\}.$$

For the group ring $\mathbb{Q}\langle G\rangle$ we have $\mathbb{Q}\langle G\rangle \cong {}^{\mathbb{Q}\langle a,b,c\rangle}/_{I}$. Considered as a $\mathbb{Q}$-vector space we are able to compute the $\mathbb{Q}$-dimension of ${}^{\mathbb{Q}\langle a,b,c\rangle}/_{I}$ in the exceptional cases. For this, we compute a finite reduced $llex$ Gröbner basis for the ideal I. With this Gröbner basis we are able to compute the Hilbert-Dehn function of G to count the number of normal words modulo I with respect to $llex$, the $\mathbb{Q}$-dimension of ${}^{\mathbb{Q}\langle a,b,c\rangle}/_{I}$. So we get the number of words in G. For details see [**42**].

THEOREM 4.3. *Let GB_i be a reduced llex Gröbner basis of G_i, $1 \leq i \leq 10$. G_i is finite and we have:*

(1) $G_1 = \langle a, b, c \mid a^3 = b^3 = c^3 = (ab^2)^2 = (ac^2)^2 = (abc)^2 = 1\rangle$ *with* $|GB_1| = 32$ *and* $|G_1| = 96$.
(2) $G_2 = \langle a, b, c \mid a^3 = b^3 = c^5 = (ab^2)^2 = (ac^2)^2 = (abc)^2 = 1\rangle$ *with* $|GB_2| = 2806$ *and* $|G_2| = 7200$.
(3) $G_3 = \langle a, b, c \mid a^3 = b^3 = c^5 = (ab^2)^2 = (ac^3)^2 = (abc)^2 = 1\rangle$ *with* $|GB_3| = 1546$ *and* $|G_3| = 7200$.
(4) $G_4 = \langle a, b, c \mid a^4 = b^3 = c^3 = (ab^2)^2 = (ac^2)^2 = (abc)^2 = 1\rangle$ *with* $|GB_4| = 176$ *and* $|G_4| = 576$.
(5) $G_5 = \langle a, b, c \mid a^5 = b^2 = c^2 = (a^2b)^3 = (a^2c)^3 = (abc)^2 = 1\rangle$ *with* $|GB_5| = 2007$ *and* $|G_5| = 7200$.
(6) $G_6 = \langle a, b, c \mid a^5 = b^2 = c^3 = (a^2b)^3 = (a^2c)^2 = (abc)^2 = 1\rangle$ *with* $|GB_6| = 2246$ *and* $|G_6| = 7200$.
(7) $G_7 = \langle a, b, c \mid a^5 = b^2 = c^3 = (a^2b)^3 = (a^4c)^2 = (abc)^2 = 1\rangle$ *with* $|GB_7| = 2681$ *and* $|G_7| = 7200$.
(8) $G_8 = \langle a, b, c \mid a^5 = b^3 = c^3 = (a^2b)^2 = (a^3c)^2 = (abc)^2 = 1\rangle$ *with* $|GB_8| = 2398$ *and* $|G_8| = 7200$.
(9) $G_9 = \langle a, b, c \mid a^5 = b^3 = c^3 = (a^3b)^2 = (a^4c)^2 = (abc)^2 = 1\rangle$ *with* $|GB_9| = 1538$ *and* $|G_9| = 7200$.
(10) $G_{10} = \langle a, b, c \mid a^5 = b^3 = c^3 = (a^4b)^2 = (a^4c)^2 = (abc)^2 = 1\rangle$ *with* $|GB_{10}| = 1234$ *and* $|G_{10}| = 7200$.

PROOF. The computations were undertaken with the package *ncpoly* for ApCoCoA, version 1.9.1 (2013) [**1**]. First, for G_i, $1 \leq i \leq 10$, a reduced $llex$ Gröbner basis GB_i was computed. With this Gröbner basis the Hilbert-Dehn function of G_i was computed to eventually give the order of G_i. □

As we already mentioned, in all other cases the methods presented in the former sections give the result that the considered groups satisfy the Tits alternative. For the details see [**36**]. Hence, the proof of Theorem 4.1 is completed.

4.1. The final classification of the finite groups (1.1) with block length 1. As a direct consequence of the investigations of our groups with block length 1 concerning the Tits alternative in the last section, we obtain a complete classification of the finite ones among these groups. We do not have to take into account the groups which contain a non-abelian free subgroup and which are infinite solvable. In all other cases, our groups with block length 1 are isomorphic to a Tsaranov generalized tetrahedron group or are proved to be finite.

COROLLARY 4.4. *Let G be a group with a presentation of the form*

$$\langle a,b,c \mid a^l = b^m = c^n = \left(a^\alpha b^\beta\right)^p = \left(a^\gamma c^\delta\right)^q = (abc)^r = 1\rangle$$

with $l,m,n,p,q,r \geq 2$, $1 \leq \alpha,\gamma < l$, $1 \leq \beta < m$, $1 \leq \delta < n$. Moreover, assume that $m \leq n$ and that G is not isomorphic to a Tsaranov generalized tetrahedron group. If G is finite, then up to equivalence, G is one of the groups (1) to (10) in Theorem 4.3.

In addition, for the Tsaranov generalized tetrahedron groups which are not equivalent to an ordinary tetrahedron group there exists a complete classification, too. The finiteness of an ordinary tetrahedron group can be decided by computing the determinant of the so-called Coxeter matrix [**5**], [**6**].

THEOREM 4.5. [**40**], [**39**] *Let H be a Tsaranov generalized tetrahedron group which is not equivalent to an ordinary tetrahedron group. If G is finite, then up to equivalence, G is one of the following:*

(1) $\langle a,b,c \mid a^3 = b^3 = c^3 = (ab)^2 = (ac)^2 = (bc)^2 = 1\rangle$ *with* $|G| = 60$.
(2) $\langle a,b,c \mid a^3 = b^3 = c^4 = (ab)^2 = (ac)^2 = (bc)^2 = 1\rangle$ *with* $|G| = 192$.
(3) $\langle a,b,c \mid a^3 = b^3 = c^5 = (ab)^2 = (ac)^2 = (bc)^2 = 1\rangle$ *with* $|G| = 7200$.
(4) $\langle a,b,c \mid a^3 = b^3 = c^5 = (ab)^2 = (ac)^2 = \left(bc^2\right)^2 = 1\rangle$ *with* $|G| = 7200$.
(5) $\langle a,b,c \mid a^3 = b^3 = c^5 = (ab)^2 = (ac)^2 = \left(bc^3\right)^2 = 1\rangle$ *with* $|G| = 7200$.

As we saw in section 2 it is still an open problem if the Tits alternative holds for all generalized triangle and tetrahedron groups although we could use for these groups the theory of triangles of groups. This theory we can not use in general for groups with a presentation (1.1). The theory of triangles of groups was also very important for the classification of the finite generalized triangle and tetrahedron groups (see [**23**] [**9**], [**26**] and [**13**]). Also for this reason, the classification of all finite groups with a presentation (1.1) is at the moment out of reach.

5. The hyperbolicity of the groups with a presentation (1.1)

In this final short section we make a few remarks about the hyperbolicity of groups with a presentation (1.1).

A hyperbolic group is a finitely generated group equipped with a word metric satisfying certain properties characteristic of hyperbolic geometry (see [**20**]).

To prove a group is hyperbolic we can use Small Cancellation Theory, particularly the following theorem, which can be found also in [**3**].

THEOREM 5.1. *Let $G = \langle S \mid R\rangle$ be a finitely presented group and let R fulfill the condition $C'(\lambda)$ for $\lambda \leq \frac{1}{6}$. Then G is hyperbolic.*

Now, let G be a group with a presentation of the form

$$G = \langle a,b,c \mid a^l = b^m = c^n = R_1^p(a,b) = R_2^q(a,c) = (abc)^r = 1\rangle. \tag{1.1}$$

It is clear that these groups are hyperbolic if they are finite and if they are fundamental groups of hyperbolic orbifolds. Using the method in [**28**] we directly get the following.

PROPOSITION 5.2. *Let G be a group with a presentation* (1.1). *We assume that* $p, q > 6$, $r > 4$, $l > 6\max\{\alpha_1, \dots, \alpha_{k_1}, \gamma_1, \dots, \gamma_{k_2}\}$, $m > 6\max\{\beta_1, \dots, \beta_{k_1}\}$ *and* $n > 6\max\{\delta_1, \dots, \delta_{k_2}\}$. *Then G is hyperbolic.*

PROOF. The proof follows exactly the lines for the proof of the hyperbolicity of strict Pride groups as given in [**28**]. The general arguments there are easily applicable for the group G with these given period relations. We take R to be the smallest symmetric subset of the free group $F(a,b,c)$ on a,b,c which contains the relations $a^l, b^m, c^n, R_1^p(a,b), R_2^q(a,c)$ and $(abc)^r$. From [**28**] we see that under our assumptions the set R satisfies the small cancellation condition $C'(\frac{1}{6})$, and hence G is hyperbolic by Theorem 5.1. □

REMARK 5.3. The conditions in Theorem 5.2 are certainly not best possible. We just stated the result which we get directly from [**28**]. By a little thought we get for instance the following result for the special case of block length 1.

COROLLARY 5.4. *Let G be a group with a presentation* (1.1) *and block length* 1.

If $6 < l, m, n, p, q$ *and* $4 < r$ *then G is hyperbolic.* □

As a summary of our considerations we finish with two conjectures for a group G with a presentation (1.1).

Conjecture 1: The Tits alternative holds for G.

Conjecture 2: If $6 < l, m, n, p, q$ and $4 < r$ then G is hyperbolic in general.

References

[1] ApCoCoA Team, *ApCoCoA: Applied Computations in Commutative Algebra. Version 1.9.1*, Available at http://www.apcocoa.org (2013).

[2] G. Baumslag, J. W. Morgan, and P. B. Shalen, *Generalized triangle groups*, Math. Proc. Cambridge Philos. Soc. **102** (1987), no. 1, 25–31, DOI 10.1017/S0305004100067013. MR886432

[3] T. Camps, V. große Rebel, and G. Rosenberger, *Einführung in die kombinatorische und die geometrische Gruppentheorie* (German), Berliner Studienreihe zur Mathematik [Berlin Study Series on Mathematics], vol. 19, Heldermann Verlag, Lemgo, 2008. MR2378619

[4] D. Sjerve and M. Cherkassoff, *On groups generated by three involutions, two of which commute*, The Hilton Symposium 1993 (Montreal, PQ), CRM Proc. Lecture Notes, vol. 6, Amer. Math. Soc., Providence, RI, 1994, pp. 169–185. MR1290589

[5] H. S. M. Coxeter, *Discrete groups generated by reflections*, Ann. of Math. (2) **35** (1934), no. 3, 588–621, DOI 10.2307/1968753. MR1503182

[6] H. S. M. Coxeter, *The complete enumeration of finite groups of the form $R_i^2 = (R_iR_j)^{k_{ij}} = 1$*, J. London Math. Soc. **10** (1935), 21–25.

[7] M. Edjvet, J. Howie, G. Rosenberger, and R. M. Thomas, *Finite generalized tetrahedron groups with a high-power relator*, Proceedings of the Conference on Geometric and Combinatorial Group Theory, Part I (Haifa, 2000), Geom. Dedicata **94** (2002), 111–139, DOI 10.1023/A:1020912827484. MR1950876

[8] M. Edjvet, G. Rosenberger, M. Stille, and R. M. Thomas, *On certain finite generalized tetrahedron groups*, Computational and geometric aspects of modern algebra (Edinburgh, 1998), London Math. Soc. Lecture Note Ser., vol. 275, Cambridge Univ. Press, Cambridge, 2000, pp. 54–65, DOI 10.1017/CBO9780511600609.005. MR1776766

[9] B. Fine, M. Hahn, A. Hulpke, V. große Rebel, G. Rosenberger, and M. Scheer, *All finite generalized tetrahedron groups*, Algebra Colloq. **15** (2008), no. 4, 555–580, DOI 10.1142/S1005386708000540. MR2451990

[10] B. Fine, A. Hulpke, V. große Rebel, and G. Rosenberger, *The Tits alternative for spherical generalized tetrahedron groups*, Algebra Colloq. **15** (2008), no. 4, 541–554, DOI 10.1142/S1005386708000539. MR2451989

[11] B. Fine, J. Howie, and G. Rosenberger, *One-relator quotients and free products of cyclics*, Proc. Amer. Math. Soc. **102** (1988), no. 2, 249–254, DOI 10.2307/2045870. MR920981

[12] B. Fine, J. Howie, and G. Rosenberger, *Ree-Mendelsohn pairs in generalized triangle groups*, Comm. Algebra **17** (1989), no. 2, 251–258, DOI 10.1080/00927878908823726. MR978473

[13] B. Fine, A. Hulpke, and G. Rosenberger, *All finite generalized tetrahedron groups II*, Computational and combinatorial group theory and cryptography, Contemp. Math., vol. 582, Amer. Math. Soc., Providence, RI, 2012, pp. 105–106, DOI 10.1090/conm/582/11560. MR2987389

[14] B. Fine, F. Levin, F. Roehl, and G. Rosenberger, *The generalized tetrahedron groups*, Geometric group theory (Columbus, OH, 1992), Ohio State Univ. Math. Res. Inst. Publ., vol. 3, de Gruyter, Berlin, 1995, pp. 99–119. MR1355106

[15] B. Fine, F. Levin, and G. Rosenberger, *Free subgroups and decompositions of one-relator products of cyclics. I. The Tits alternative*, Arch. Math. (Basel) **50** (1988), no. 2, 97–109, DOI 10.1007/BF01194564. MR930108

[16] B. Fine and G. Rosenberger, *A note on generalized triangle groups*, Abh. Math. Sem. Univ. Hamburg **56** (1986), 233–244, DOI 10.1007/BF02941518. MR882417

[17] B. Fine and G. Rosenberger, *Algebraic generalizations of discrete groups*, Monographs and Textbooks in Pure and Applied Mathematics, vol. 223, Marcel Dekker, Inc., New York, 1999. A path to combinatorial group theory through one-relator products. MR1712997

[18] V. große Rebel, *Die Tits-Alternative für verallgemeinerte Tetraedergruppen*, Ph.D. thesis, Universität Dortmund, 2006.

[19] V. große Rebel, M. Hahn, and G. Rosenberger, *The Tits alternative for Tsaranov's generalized tetrahedron groups*, Groups Complex. Cryptol. **1** (2009), no. 2, 207–216, DOI 10.1515/GCC.2009.207. MR2598989

[20] M. Gromov, *Hyperbolic groups*, Essays in group theory, Math. Sci. Res. Inst. Publ., vol. 8, Springer, New York, 1987, pp. 75–263, DOI 10.1007/978-1-4613-9586-7_3. MR919829

[21] D. Hennig and G. Rosenberger, *Recent developments in the theory of Fuchsian and Kleinian groups*, Recent advances in group theory and low-dimensional topology (Pusan, 2000), Res. Exp. Math., vol. 27, Heldermann, Lemgo, 2003, pp. 51–64. MR2004631

[22] J. Howie and N. Kopteva, *The Tits alternative for generalized tetrahedron groups*, J. Group Theory **9** (2006), no. 2, 173–189, DOI 10.1515/JGT.2006.012. MR2220573

[23] J. Howie, V. Metaftsis, and R. M. Thomas, *Finite generalized triangle groups*, Trans. Amer. Math. Soc. **347** (1995), no. 9, 3613–3623, DOI 10.2307/2155028. MR1303121

[24] S. Katok, *Fuchsian groups*, Chicago Lectures in Mathematics, University of Chicago Press, Chicago, IL, 1992. MR1177168

[25] F. Levin and G. Rosenberger, *On free subgroups of generalized triangle groups. II*, Group theory (Granville, OH, 1992), World Sci. Publ., River Edge, NJ, 1993, pp. 206–228. MR1348902

[26] L. Lévai, G. Rosenberger, and B. Souvignier, *All finite generalized triangle groups*, Trans. Amer. Math. Soc. **347** (1995), no. 9, 3625–3627, DOI 10.2307/2155029. MR1303124

[27] C. L. May, *Large automorphism groups of compact Klein surfaces with boundary. I*, Glasgow Math. J. **18** (1977), no. 1, 1–10. MR0425113

[28] M. Neumann-Brosig, *A note on the hyperbolicity of strict Pride groups*, Computational and combinatorial group theory and cryptography, Contemp. Math., vol. 582, Amer. Math. Soc., Providence, RI, 2012, pp. 181–185, DOI 10.1090/conm/582/11566. MR2987395

[29] S. J. Pride, *The (co)homology of groups given by presentations in which each defining relator involves at most two types of generators*, J. Austral. Math. Soc. Ser. A **52** (1992), no. 2, 205–218. MR1143189

[30] G. Rosenberger, *Some remarks on a paper of A. F. Beardon and P . L. Waterman about: "Strongly discrete subgroups of* $\mathrm{SL}(2,\mathbf{C})$*" [J. London Math. Soc. (2) **24** (1981), no. 2, 325–328; MR0631943 (82i:30065)]*, J. London Math. Soc. (2) **27** (1983), no. 1, 39–42, DOI 10.1112/jlms/s2-27.1.39. MR686500

[31] G. Rosenberger, *Some remarks on the paper of A. Majeed on freeness of the group* $\langle a^n, b^n\rangle$ *for some integer* n*;* $a, b \in \mathrm{SL}(2,\mathbf{C})$, Results Math. **11** (1987), no. 3-4, 267–271, DOI 10.1007/BF03323274. MR897302

[32] G. Rosenberger, *Minimal generating systems of a subgroup of* $\mathrm{SL}(2,\mathbf{C})$, Proc. Edinburgh Math. Soc. (2) **31** (1988), no. 2, 261–265, DOI 10.1017/S0013091500003382. MR989759

[33] G. Rosenberger, *The SQ-universality of one-relator products of cyclics*, Results Math. **21** (1992), no. 3-4, 396–402, DOI 10.1007/BF03323096. MR1157342

[34] G. Rosenberger and C. Maclachlan, *Arithmetic Groups of Signature* $(0, m_1, m_2, m_3, m_4)$, Scienta, Series A Math. 24 (2013), 1–24.

[35] G. Rosenberger and M. Scheer, *Classification of the finite generalized tetrahedron groups*, Combinatorial and geometric group theory (New York, 2000/Hoboken, NJ, 2001), Contemp. Math., vol. 296, Amer. Math. Soc., Providence, RI, 2002, pp. 207–229, DOI 10.1090/conm/296/05076. MR1922275

[36] K. Rosenthal, *Die Tits-Alternative für eine relevante Klasse endlich präsentierter Gruppen unter besonderer Berücksichtigung computeralgebraischer Aspekte*, Ph.D. thesis, Universität Passau, 2014.

[37] A. Selberg, *On discontinuous groups in higher-dimensional symmetric spaces*, Contributions to function theory (internat. Colloq. Function Theory, Bombay, 1960), Tata Institute of Fundamental Research, Bombay, 1960, pp. 147–164. MR0130324

[38] J. Tits, *Free subgroups in linear groups*, J. Algebra **20** (1972), 250–270. MR0286898

[39] S. V. Tsaranov, *Finite generalized Coxeter groups*, Algebras Groups Geom. **6** (1989), no. 4, 421–452. MR1109560

[40] S. V. Tsaranov, *On a generalization of Coxeter groups*, Algebras Groups Geom. **6** (1989), no. 3, 281–318. MR1057647

[41] È. B. Vinberg, *Groups defined by periodic pairwise relations* (Russian, with Russian summary), Mat. Sb. **188** (1997), no. 9, 3–12, DOI 10.1070/SM1997v188n09ABEH000254; English transl., Sb. Math. **188** (1997), no. 9, 1269–1278. MR1481661

[42] X. Xiu, *Non-commutative Gröbner bases and applications*, Ph.D. thesis, Universität Passau, 2012.

Fachbereich Mathematik, Universität Hamburg, Bundesstrasse 55, 20146 Hamburg, Germany

E-mail address: anja.moldenhauer@uni-hamburg.de

Fachbereich Mathematik, Universität Hamburg, Bundesstrasse 55, 20146 Hamburg, Germany

E-mail address: gerhard.rosenberger@math.uni-hamburg.de

Fakultät für Wirtschaftswissenschaften, FernUniversität in Hagen, 58084 Hagen, Germany.

E-mail address: kristina.rosenthal@fernuni-hagen.de

Contemporary Mathematics
Volume **677**, 2016
http://dx.doi.org/10.1090/conm/677/13627

Geometry of the conjugacy problem in lamplighter groups

Andrew Sale

ABSTRACT. In this note we investigate the conjugacy problem in lamplighter groups with particular interest in the role of their geometry. In particular we show that the conjugacy length function is linear.

The conjugacy (search) problem of Max Dehn is a century-old algorithmic question in group theory that has received much attention of late due to potential applications to cryptography, see for example [**3**, **10**]. The conjugacy problem asks whether, given a group Γ with finite generating set A, there exists an algorithm which, on input words u, v on $A \cup A^{-1}$, determines whether u, v represent conjugate elements in Γ. The conjugacy search problem is similarly themed, but instead the input is two elements known to be conjugate, and the algorithm should produce a conjugating element.

We study the *conjugacy length function*, a quantitative version of the conjugacy problem. Given a group Γ and a length function $|\cdot| : \Gamma \to [0, \infty)$, (typically word length) the conjugacy length function is the minimal function

$$\mathrm{CLF}_\Gamma : [0, \infty) \to [0, \infty)$$

which satisfies the following: if u is conjugate to v in Γ and $|u| + |v| \leq n$ then there exists a conjugator $\gamma \in \Gamma$ such that $|\gamma| \leq \mathrm{CLF}_\Gamma(n)$. This function has been estimated for many classes of groups, a brief list includes hyperbolic groups [**11**] and mapping class groups [**7**, **12**, **18**] where in both it is known to be (at most) linear, it is bounded by an exponential function for CAT(0) groups [**8**], and in free solvable groups it is at most cubic [**16**].

In [**16**] the author investigated the behaviour of the conjugacy length function under wreath products. For the class of lamplighter groups $\mathbb{Z}_q \wr \mathbb{Z}$ or $\mathbb{Z} \wr \mathbb{Z}$, this result is not optimal, giving a cubic upper bound. We show here that it is linear. In fact, we consider the Diestel–Leader groups $\Gamma_2(R)$, which are defined in Section 1.2, that includes the lamplighter groups $\Gamma_2(\mathbb{Z}_q) = \mathbb{Z}_q \wr \mathbb{Z}$ and $\Gamma_2(\mathbb{Z}) = \mathbb{Z} \wr \mathbb{Z}$.

THEOREM 1. *Let R be a commutative ring with unity and let $\Gamma = \Gamma_2(R)$. Then there is a generating set S, which is finite when R is finite, such that with respect to the corresponding word length we have*

$$\mathrm{CLF}_\Gamma(n) \leq 3n.$$

2010 *Mathematics Subject Classification.* Primary 20F65, 20F16, 20F10.

©2016 American Mathematical Society

A key feature we use is the fact that the Cayley graphs of the groups $\mathrm{DL}_2(R)$ are the horocyclic product of two trees [**2**, **4**], a special case of a Diestel–Leader graph, introduced in [**9**]. The motivation for the geometric approach that we use is so that the method may generalise to a wider class of groups, the higher-rank Diestel–Leader groups $\Gamma_d(R)$ for $d \geq 2$. Indeed, C. Abbott has done precisely this, obtaining an exponential upper-bound for their conjugacy length functions when R is finite [**1**].

The tools developed for Theorem 1 are then applied to give an algorithm solving the conjugacy search problem. The algorithm will run in quadratic time with respect to the length of the input words when R is finitely generated as an abelian group.

Theorem 1 originally appeared with an algebraic proof in a preprint of the author [**14**], which was part of the inspiration for [**13**]. The latter looks at the behaviour of conjugacy length under group extensions, and the proof of Theorem 1 could be rewritten in the language used therein. This note has been modified from the geometric proof in the author's doctoral thesis [**15**].

The structure is as follows. We begin by defining the groups $\Gamma_d(R)$ in Section 1 and explain their geometry through Diestel–Leader graphs. A word length estimate, which follows from a formula of Stein and Taback [**17**], is given in Section 2, while Theorem 1 is proved in Section 3 and the algorithm is described in Section 4.

1. Lamplighter groups as Diestel–Leader groups

1.1. Horocyclic products, Diestel–Leader graphs and R–branching trees. We give here a brief introduction to horocyclic products and Diestel-Leader graphs. For a more complete description see for example [**4**].

Let T be a simplicial tree and $\omega \in \partial_\infty T$ a boundary point of T, that is ω is an equivalence class of asymptotic geodesic rays. Recall that in T, two geodesic rays $\rho_i : [0, \infty) \to T$, for $i = 1, 2$, are asymptotic if and only if they merge: there exists $x > 0$ and $s \in \mathbb{R}$ such that $\rho_1(t) = \rho_2(t + s)$, for all $t \geq x$.

For any vertex x in T there is a unique geodesic ray emerging from x that is in ω. Given a pair of vertices, x, y, the corresponding rays will coincide from some vertex $x \curlywedge y$ onwards. We call this the *greatest common ancestor* of x and y.

After fixing a basepoint o in the vertex set of T we can define a Busemann function $\mathfrak{h}$ on the vertices of T as

$$\mathfrak{h}(y) = d_T(y, o \curlywedge y) - d_T(o, o \curlywedge y).$$

The difference in the value of the Busemann function gives the distance from a vertex x to its common ancestor with some other vertex y. That is

$$d_T(x, x \curlywedge y) = \mathfrak{h}(x) - \mathfrak{h}(x \curlywedge y).$$

The k–th horocycle of T based at ω is $H_k = \{y \in T \mid \mathfrak{h}(y) = k\}$.

Given a collection $T_1, \ldots, T_n$ of simplicial trees together with a chosen collection of respective Busemann functions $\mathfrak{h}_1, \ldots, \mathfrak{h}_n$, we define the *horocyclic product* to be

$$\prod_{i=1}^{n} {}_{\mathfrak{h}} T_i = \left\{ (y_1, \ldots, y_n) \in T_1 \times \ldots \times T_n \mid \sum_{i=1}^{n} \mathfrak{h}_i(y_i) = 0 \right\}. \tag{1.1}$$

The *Diestel–Leader graph* $\mathrm{DL}(q_1, \ldots, q_d)$ is the 1–skeleton of the horocyclic product of trees $\mathbb{T}_{q_1}, \ldots, \mathbb{T}_{q_d}$, where $\mathbb{T}_q$ is the $q+1$ regular tree. When $q_1 = q_2 = \ldots = q_d = q$, the corresponding Diestel-Leader graph is also denoted by $\mathrm{DL}_d(q)$.

Following [**2**], we can extend this to the notion of an R–branching tree, where R is any commutative ring with unity. Such a tree, denoted $\mathbb{T}_R$, has the property that for every vertex v, the set $E(v)$ of edges touching v is in a bijection with $\{x_0\} \cup R$, where $x_0 \notin R$. We denote by $\mathrm{DL}_d(R)$ the horocyclic product of d copies of $\mathbb{T}_R$.

1.2. Diestel–Leader groups and generalised lamplighters. The fact that the Diestel–Leader graph $\mathrm{DL}_2(q)$ is a Cayley graph for the lamplighter group $\mathbb{Z}_q \wr \mathbb{Z}$ was explained in [**19**]. This is a special case of the following result of Bartholdi, Neuhauser and Woess:

THEOREM 2 (Bartholdi–Neuhauser–Woess [**4**, (3.14)]). *The Diestel–Leader graph $\mathrm{DL}_d(q)$ is a Cayley graph of a group, denoted $\Gamma_d(R)$, where R is a commutative ring with unity of order q.*

The groups $\Gamma_d(R)$ are called *Diestel–Leader groups.* When $d = 2$ they include the lamplighter groups $\mathbb{Z}_q \wr \mathbb{Z}$, and when $d = 3$ we obtain groups previously considered by Baumslag [**5**], [**6**]. The descriptions given below for the lamplighter groups can be extended to the Diestel–Leader groups of more than 2 trees. The word length in Diestel–Leader groups is studied by Stein and Taback [**17**] where they give a formula for its calculation.

As described by Amchislavska and Riley [**2**, Section 1.4] we can consider groups whose Cayley graph is the horocyclic product of R–branching trees $\mathbb{T}_R$. Mixing the notation of [**2**] and [**4**], we let $\Gamma_d(R)$ denote the group whose Cayley graph is the horocyclic product of d trees $\mathbb{T}_R$. In this note will focus on the case when $d = 2$. We can recognise these as groups of the form

$$\Gamma_2(R) \cong R[t, t^{-1}] \rtimes \mathbb{Z}, \tag{1.2}$$

where $\mathbb{Z}$ acts on $R[t, t^{-1}]$ by multiplication by t. We refer the reader to [**2**] for more details, particularly for when $d > 2$, where these groups are denoted by $\Gamma_{d-1}(R)$ — their subscript agrees with the number of copies of $\mathbb{Z}$ in the quotient.

The group $\Gamma_2(R)$ can be represented by matrices

$$\left\{ \begin{pmatrix} t^k & P \\ 0 & 1 \end{pmatrix} : k \in \mathbb{Z},\ P \in R[t^{-1}, t] \right\},$$

with generating set

$$S_0 = \left\{ \begin{pmatrix} t & b \\ 0 & 1 \end{pmatrix} : b \in R \right\}.$$

For the sake of notation, we use the identification from (1.2):

$$\begin{pmatrix} t^k & P \\ 0 & 1 \end{pmatrix} \mapsto (P, k).$$

Amchislavska and Riley generalised Theorem 2, giving

THEOREM 3 (Amchislavska and Riley [**2**]). *Let R be a ring with unity. There is a generating set S for $\Gamma_d(R)$, with respect to which the Cayley graph is $\mathrm{DL}_d(R)$. When $d = 2$, the generating set is $S = S_0$.*

When R is infinite, the generating set S is infinite too. This is to be expected, since the valency of the trees, and hence the Diestel–Leader graph, are infinite.

1.3. Realising $\mathrm{DL}_2(R)$ as the Cayley graph of a lamplighter group. We begin by labelling the edges of $\mathbb{T}_R$ with elements from R. We pick a bi-infinite geodesic $\rho : \mathbb{R} \to \mathbb{T}_R$ so that its restriction to $[0, \infty)$ is in $\omega \in \partial_\infty \mathbb{T}_R$. Let $o = \rho(0)$ and consider the corresponding Busemann function $\mathfrak{h}$. Recall that at each vertex v we have a bijection

$$E(v) \to \{x_0\} \cup R \tag{1.3}$$

where $E(v)$ is the set of edges touching v, and $x_0 \notin R$. At each vertex v there is a unique edge e_v in $E(v)$ which forms the first edge in a geodesic ray asymptotic to ρ. The labelling is an iterative method, as follows. First label each edge in ρ by 0. Then for vertices v in which e_v has been previously labelled, but some edges in $E(v)$ remain unlabelled, use the bijection (1.3) to label those remaining edges. Repeat this step.

We now take two copies of our decorated tree, T_1, T_2, with boundary points ω_1, ω_2 and geodesics ρ_1, ρ_2 respectively, and take their horocyclic product to get $\mathrm{DL}_2(R)$. Given a vertex $x = (x_1, x_2)$ of $\mathrm{DL}_2(R)$ we consider the two geodesic rays starting at x_1, x_2 respectively that are in ω. Similar to [**2**, Lemma 3.1] we get a finitely supported bi-infinite sequence (a_i) in R, obtained by reading the edge labels in these two rays as follows. For each i there is exactly one edge in the rays that travels between vertices in H_{i-1} and H_i if in T_1, and H_{-i+1} and H_{-i} if in T_2. The label of this edge determines a_i. Thus, $(a_{\mathfrak{h}(x_1)}, a_{\mathfrak{h}(x_1)-1}, \ldots)$ is the sequence of edge labels of the ray in T_1, and $(a_{\mathfrak{h}(x_1)+1}, a_{\mathfrak{h}(x_1)+2}, \ldots)$ is the sequence from the ray in T_2.

We can use the sequence (a_i) to determine an element of $R[t, t^{-1}]$:

$$f_x := \sum_{i=-\infty}^{\infty} a_i t^i. \tag{1.4}$$

An example is given in Figure 1. We use this to establish an identification

$$\mathrm{DL}_2(R) \ni x = (x_1, x_2) \longleftrightarrow (f_x, \mathfrak{h}(x_1)) \in \Gamma_2(R). \tag{1.5}$$

Note that comparing this identification with that in [**2**, Section 4], the role of our trees T_1, T_2 are exchanged.

Let $f \in R[t, t^{-1}]$. There exists some $n \in \mathbb{Z}$ such that the coefficient of t^k is zero for all $k < n$. The valuation $v_0(f)$ is defined to be the supremum of all such n (note that we define $v_0(0) = \infty$). Similarly we set $v_0^-(f)$ to be the largest n for which the coefficient of t^n in f is non-zero. Geometrically, $v_0(f) - 1$ gives the horocycle in which the geodesic ray emerging from x_1, asymptotic to ω_1, merges with ρ_1. Similarly, $-v_0^-(f)$ give the horocycle in T_2 where the geodesic ray in ω_2 that emerges from x_2 merges with ρ_2.

Before proceeding further, we make the following observation which will be useful when discussing word length.

LEMMA 4. *Let $x = (x_1, x_2), y = (y_1, y_2) \in \mathrm{DL}_2(R)$ and $\gamma \in \Gamma_2(R)$. Then*

$$\mathfrak{h}(\gamma x_i \curlywedge \gamma y_i) = \mathfrak{h}(x_i \curlywedge y_i) + \mathfrak{h}(\gamma o_i).$$

PROOF. We prove this for the first tree. For the second tree the proof is analogous.

Let $f_x = \sum a_i t^i$ and $f_y = \sum b_i t^i$ be as in (1.4). We claim that

$$\mathfrak{h}(x_1 \curlywedge y_1) = \min\{v_0(f_x - f_y) - 1, \mathfrak{h}(x_1), \mathfrak{h}(y_1)\}.$$

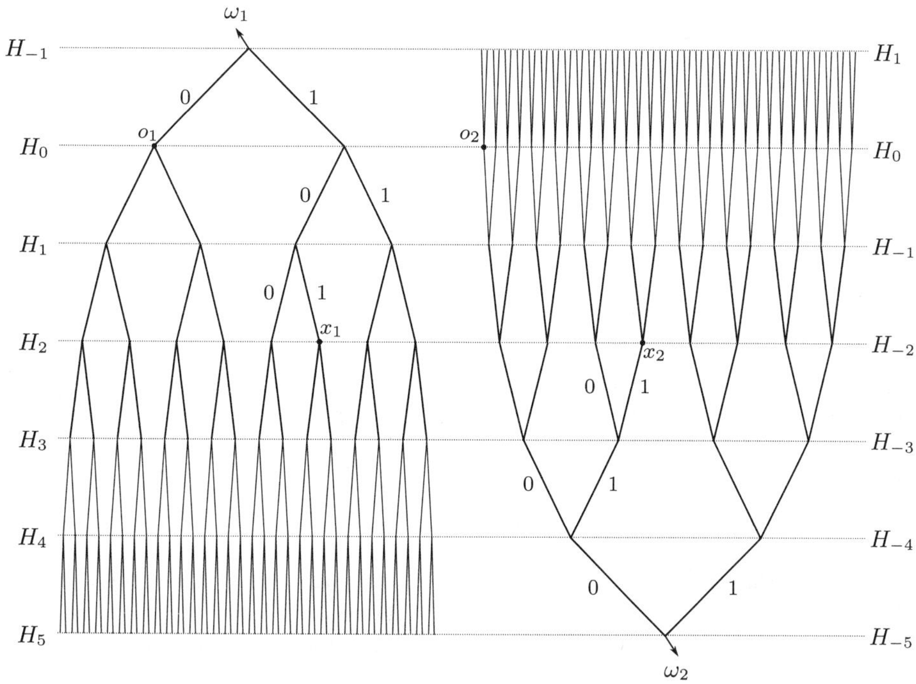

FIGURE 1. The tree T_1 is on the left, and T_2 on the right. From T_1 we get $a_2 = 1, a_1 = 0, a_0 = 1$ and $a_i = 0$ for $i < 0$. From T_2 we get $a_3 = 1, a_4 = 1$ and $a_i = 0$ for $i > 4$. Thus $f_x = 1 + t^2 + t^3 + t^4$.

Since both have finite support, there is some k such that for $i < k$ we have $a_i = b_i$. This k is the valuation $v_0(f_x - f_y)$. First suppose that $x_1 \curlywedge y_1$ is distinct from x_1 and y_1. Then $v_0(f_x - f_y)$ gives the last horocycle before the merging of the two geodesics in ω_1 emerging from x_1 and y_1 respectively. Hence $\mathfrak{h}(x_1 \curlywedge y_1) = v_0(f_x - f_y) - 1$.

On the other hand, if the common ancestor is one of the two given vertices, say x_1, then this valuation may not give the correct horocycle. Indeed it may be that f_x and f_y agree for coefficients of f_x that come from T_2 and have no consequence for the location of x_1. Hence, in this case we see that we have $\mathfrak{h}(x_1 \curlywedge y_1) = \mathfrak{h}(x_1)$. This proves the claim.

Suppose $\gamma = (P, s)$. Then $f_{\gamma x} = P + t^s f_x$ and $f_{\gamma y} = P + t^s f_y$. We thus obtain, using the above claim, the following for $\mathfrak{h}(\gamma x_1 \curlywedge \gamma y_1)$, noting that the terms in P will both cancel, leaving

$$\mathfrak{h}(\gamma x_1 \curlywedge \gamma y_1) = \min\{v_0(t^s f_x - t^s f_y) - 1, \mathfrak{h}(\gamma x_1), \mathfrak{h}(\gamma y_1)\}.$$

Since $v_0(t^s f_x - t^s f_y) = v_0(f_x - f_y) + s$, $\mathfrak{h}(\gamma x_1) = \mathfrak{h}(x_1) + s$, and similarly for y_1, the Lemma holds. □

We remark that Lemma 4 can be generalised to $\mathrm{DL}_d(q)$ for $d \geq 2$.

1.4. The (right) action of generators on $\mathrm{DL}_2(R)$**.** Recall we have two trees T_1, T_2, each isomorphic to $\mathbb{T}_R$ and coming with an equivalence class of rays ω_i in $\partial_\infty T_i$. Starting from a vertex $x_i \in T_i$, we will use the phrase *"go up in T_i"* to mean move along the (unique) edge in T_i which is in the geodesic ray emanating from x_i that is in ω_i. To *"go down in T_i"* means we move along any other edge. Recall this edge will be labelled by an element of R, as described in Section 1.3.

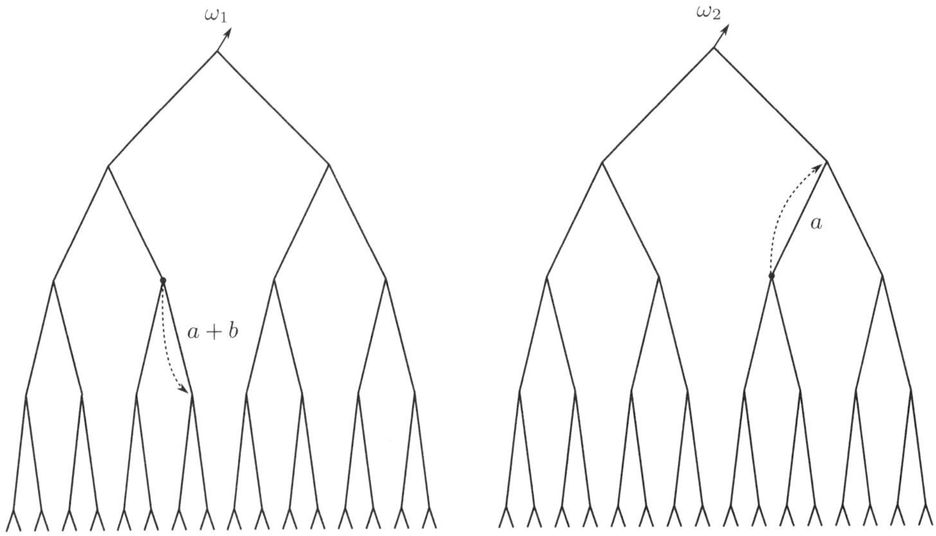

FIGURE 2. The action of a letter $(b, 1)$ within a word tells us to move down in T_1 and up in T_2. The label of the edge we move up in T_2 combines with b to tell us which edge to go down in T_1.

LEMMA 5. *A letter $(b, 1) \in S$ in a word is an instruction: move up one edge in tree T_2, suppose this has label a, and move down an edge labelled $a + b \in R$ in tree T_1.*

PROOF. We have an edge in the Cayley graph of $\Gamma_2(R)$ labelled by $(b, 1)$ starting at (f, k) and finishing at $(f, k)(b, 1) = (f + bt^k, k + 1)$. Thus, in T_1 we travel from a vertex in H_k down to H_{k+1}. Suppose a is the coefficient of t^k in f, then the edge we followed, by the identification (1.5), has label $a + b$, which is the coefficient of t^k in $f + bt^k$.

Meanwhile, in T_2 we have travelled up an edge, from H_{-k} to H_{-k-1}. The edge we followed was labelled by the coefficient of t^k in f, namely a. □

2. Word length

The word length of elements in a general Diestel–Leader group $\Gamma_d(R)$, when R is finite and $d \geq 2$, has been studied by Stein and Taback [**17**]. They give a formula for the word length of an element by looking at the climb and fall of a geodesic path in each tree from the basepoint to its image under the action of the element. Let o_i denote the basepoint of the i–th tree. For $g \in \Gamma_d(R)$, denote by $m_i(g)$ the length of the climb of the geodesic from o_i to go_i and $l_i(g)$ the length of the fall. More concretely we mean:

$$m_i(g) = d(o_i, o_i \curlywedge go_i), \quad l_i(g) = d(go_i, o_i \curlywedge go_i).$$

Stein and Taback's formula, see [**17**, Section 3], is precise, but can be used to give an estimate for the word length of g that is practical for our purpose. We include a proof of this estimate not only for completeness, but also because we wish to take R to be finite or infinite.

PROPOSITION 6 (Stein–Taback word length estimate [**17**]). *Let $g \in \Gamma_2(R)$. Then*

$$m_1(g) + m_2(g) \leq |g| \leq 2(m_1(g) + m_2(g))$$

where $|g|$ is the word length of g with respect to the generating set S of Theorem 3.

REMARK. We may replace m_i with l_i in the above inequality, since whenever we are in a horocyclic product we will have

$$\sum_{i=1}^{d} m_i(g) = \sum_{i=1}^{d} l_i(g).$$

PROOF OF PROPOSITION 6. With Lemma 5 in mind, to reach (go_1, go_2) from (o_1, o_2) in $\mathrm{DL}_2(R)$, we must climb at least $m_1(g)$ in T_1 and $m_2(g)$ in T_2, thus implying the lower bound.

Without loss of generality we may assume $\mathfrak{h}(go_1) \geq 0$. First apply $m_1(g)$ generators to get to $o_1 \curlywedge go_1$ in T_1. Second apply the appropriate sequence of downward movements in T_1 to reach go_1. This sequence will be a word of length $l_1(g)$. Meanwhile, in T_2 we have reached a vertex in the same horocycle as go_2. The third step is to climb to $o_2 \curlywedge go_2$ and drop down to go_2. During this third phase, we will drop down in T_1, but then climb, inevitably retracing our steps, back up to go_1. The third and final step requires $2l_2(g)$ elements from S. So in total, we have a word of length

$$m_1(g) + l_1(g) + 2l_2(g).$$

Since we are in a horocyclic product we have $m_1(g) + m_2(g) = l_1(g) + l_2(g)$. We also have $l_2(g) \leq m_2(g)$ since $\mathfrak{h}(go_1) \geq 0$. Hence we get

$$m_1(g) + l_1(g) + 2l_2(g) \leq 2(m_1(g) + m_2(g))$$

as required. □

The notion of climbing and falling can be extended to paths describing the concatenation of words. In particular, for $g, h \in \Gamma_d(R)$ let

$$m_{h,i}(g) = d(ho_i, ho_i \curlywedge go_i), \quad l_{h,i}(g) = d(go_i, ho_i \curlywedge go_i).$$

A consequence of Lemma 4 is that we can measure the length of the word by looking at its action on different points in each tree:

LEMMA 7. *For every $g, h \in \Gamma_2(R)$ and each $i = 1, 2$ we have the following:*

$$m_i(g) = m_{h,i}(hg), \quad l_i(g) = l_{h,i}(hg).$$

PROOF. By Lemma 4, in each tree, the height of the common ancestor $o_i \curlywedge go_i$ never travels too far from the height of $ho_i \curlywedge hgo_i$. To be precise:

$$\mathfrak{h}_i(o_i \curlywedge go_i) = \mathfrak{h}_i(ho_i \curlywedge hgo_i) - \mathfrak{h}_i(ho_i)$$

Hence

$$\begin{aligned} m_i(g) &= d(o_i, o_i \curlywedge go_i) \\ &= -\mathfrak{h}_i(o_i \curlywedge go_i) \\ &= -\mathfrak{h}_i(ho_i \curlywedge hgo_i) + \mathfrak{h}_i(ho_i). \end{aligned}$$

But $d(ho_i, ho_i \curlywedge hgo_i) = \mathfrak{h}_i(ho_i) - \mathfrak{h}_i(ho_i \curlywedge hgo_i)$, so we get $m_i(g) = m_{h,i}(hg)$. The result for $l_i(g)$ follows from the result for $m_i(g)$, the relationships:

$$m_i(g) - l_i(g) = \mathfrak{h}_i(go_i) \quad \text{and} \quad m_{h,i}(hg) - l_{h,i}(hg) = \mathfrak{h}_i(hgo_i) - \mathfrak{h}_i(ho_i)$$

and the fact that $\mathfrak{h}_i(go_i) = \mathfrak{h}_i(hgo_i) - \mathfrak{h}_i(ho_i)$. □

Using Lemma 7 we can deduce the following estimate from Proposition 6:

$$\sum_{i=1}^{d} m_{h,i}(g) \leq \left|g^{-1}h\right| \leq 2\sum_{i=1}^{d} m_{h,i}(g). \tag{2.1}$$

In these bounds for $\left|g^{-1}h\right|$ we may replace $m_{h,i}$ with $l_{h,i}$ since their sums are equal. To prove the estimate (2.1), we need

$$l_i(g^{-1}h) = l_{g,i}(h) = d(ho_i, go_i \curlywedge ho_i) = m_{h,i}(g)$$

where the first equality is an application of Lemma 7.

3. The conjugacy length function

We now prove that the conjugacy length function of $\Gamma_2(R)$ is linear with respect to the word length given by the generating set S defined in Section 1.2. The following tells us the structure of short conjugators, from which we deduce Theorem 1.

PROPOSITION 8. *Suppose $u = (P, s), v = (Q, r) \in \mathrm{DL}_2(R)$. If u is conjugate to v then $r = s$ and either*

(1) if $r = s = 0$ then there is a conjugator $\gamma = (0, k)$ with $|k| \leq |u| + |v|$;
(2) if $r = s \neq 0$ then there is a conjugator $\gamma = (f, k)$ with $0 \leq k < |r|$ and

$$m_i(\gamma) \leq \max\{m_i(u), m_i(v)\} + |r|\,. \tag{3.1}$$

PROOF. Suppose $\gamma = (f, k)$ is a conjugator. By direct calculation, $\gamma u = v\gamma$ if and only if the following equations hold:

$$s + k = k + r, \tag{3.2}$$

$$P + t^s f = f + t^k Q. \tag{3.3}$$

From the first of these equations we get $r = s$, and we thus split into the two cases.

Case 1: $r = 0$. This corresponds to the case when u and v map each basepoint to a vertex of the same height. Equation (3.3) becomes $P = t^k Q$, so we may set $f = 0$. This means that γ will act on each tree by either a sequence of consecutive up movements or a sequence of consecutive down movements, but never a mixture of both.

As long as v is non-trivial, we may assume that $vo_i \neq o_i$ for some i. Assume the action of γ on the i–th tree corresponds to a consecutive list of downward movements. This case is depicted in Figure 3 (a). Then $v\gamma o_i$ will be sitting below vo_i, to be precise we have $vo_i \curlywedge v\gamma o_i = vo_i$. Then the path from γo_i to γuo_i must pass through vo_i. Hence $m_{\gamma,i}(\gamma u) \geq |k|$. Using equation (2.1) we then obtain

$$|u| = \left|(\gamma u)^{-1}\gamma\right| \geq m_{\gamma,i}(\gamma u) \geq |k|\,.$$

If γ acts on the i–th tree in a purely upwards manner, then we will get a similar picture (see Figure 3 (b)), but γuo_i will be sitting directly above vo_i, and, provided $\gamma o_i \neq \gamma uo_i$, any path from o_i to vo_i must pass through γuo_i. Here we will get

$$|k| \leq m_i(v) \leq |v|\,.$$

The situation that remains is when, up to swapping the trees round, we have $\gamma o_1 = \gamma uo_1$, and $o_2 = vo_2$, and γ acts in a purely upward direction on T_1 and downward on T_2. When moving from $v(o_1, o_2)$ to $v\gamma(o_1, o_2)$ in $\mathrm{DL}_2(R)$ we get a pair

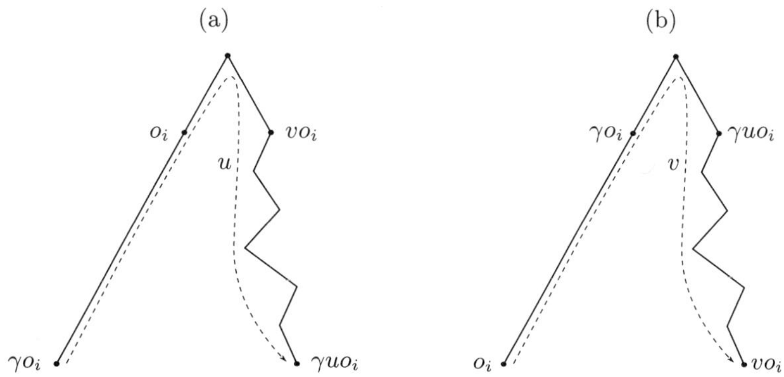

FIGURE 3. (a) The action of γ is purely downward; (b) the action of γ is purely upward.

of geodesics, one in each tree. The geodesic in T_1 will be from vo_1 to $v\gamma o_1 = \gamma uo_1$ and will merge with ρ_1 at $o_1 \curlywedge vo_1$, which is in the horocycle at level

$$\mathfrak{h}(o_1 \curlywedge vo_1) = -m_1(v).$$

Meanwhile, the geodesic in T_2 will separate from ρ_2, since $\gamma uo_2 \neq \gamma o_2$ as otherwise u would be the identity. Furthermore, this separation must occur before the geodesic in T_1 joins ρ_1, since once it has joined ρ_1, Lemma 5 tells us that in T_2 we must follow edges labelled 0, which would cause us to remain on ρ_2 and end up with $\gamma o_2 = \gamma uo_2$. In T_2 the geodesic separates from ρ_2 in the horocycle of level $|k| - m_2(u)$. Hence, by the above argument we must have

$$|k| - m_2(u) < m_1(v) \implies |k| < m_1(u) + m_2(v) \leq |u| + |v|.$$

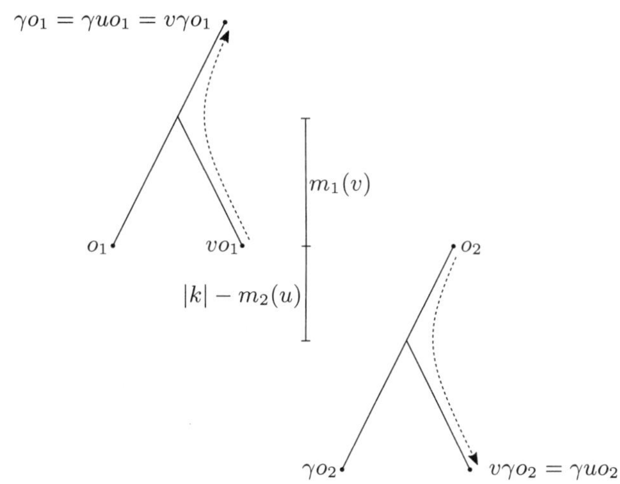

FIGURE 4. The path in T_1, on the left, merges with ρ after the path in T_2 separates from ρ.

Case 2: $r \neq 0$. By exchanging u and v with their inverses if necessary, we may assume that $r > 0$. The important step here is to pick the right conjugator. Take any conjugator γ', satisfying $\gamma' u = v\gamma'$, and premultiply it by a suitable power of u so that we obtain an element $\gamma = u^m\gamma'$, written as above, with $0 \leq k < r$. To prove the proposition, we just need to demonstrate the bound (3.1) on $m_i(\gamma)$. We will show that if this bound were not true then $\gamma u o_i$ and $v\gamma o_i$ would have to be on different branches of their respective trees, see Figure 5.

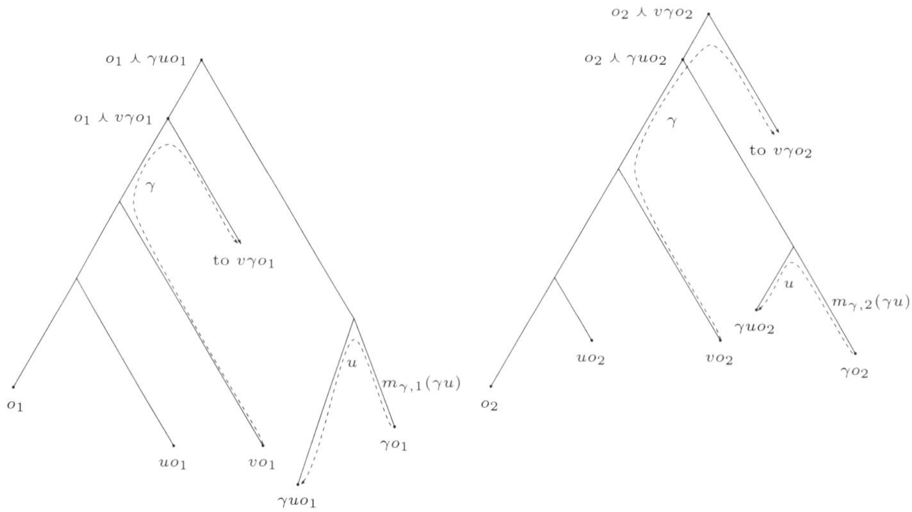

FIGURE 5. The common ancestor $o_i \curlywedge \gamma u o_i$ lies in a different horocycle to $o_i \curlywedge v\gamma o_i$.

Suppose $m_i(\gamma) > m_i(u) + r$. If this is the case, we will have

$$o_i \curlywedge \gamma o_i = o_i \curlywedge \gamma u o_i \tag{3.4}$$

since the climb $m_{\gamma,i}(\gamma u)$ from γo_i to $\gamma u o_i \curlywedge \gamma o_i$ will be less than the fall $l_i(\gamma)$ from $o_i \curlywedge \gamma o_i$ to γo_i. Indeed, the climb is given by $m_{\gamma,i}(\gamma u)$, which is equal to $m_i(u)$ by Lemma 7. By assumption

$$m_i(u) < m_i(\gamma) - r < m_i(\gamma) - k.$$

Since $\mathfrak{h}(\gamma o_i) = \pm k$, the fall $l_i(\gamma)$ from $o_i \curlywedge \gamma o_i$ to γo_i will be at least $m_i(\gamma) - k$. In summary, we get

$$m_{\gamma,i}(\gamma u) \leq l_i(\gamma)$$

which implies (3.4).

Suppose $m_i(\gamma) > m_i(v) + r$. In this case we claim

$$o_i \curlywedge v\gamma o_i = vo_i \curlywedge v\gamma o_i. \tag{3.5}$$

This follows from the fact that the climb $m_{v,i}(v\gamma)$ from vo_i to $vo_i \curlywedge v\gamma o_i$ is longer than the fall $l_i(v)$ from $o_i \curlywedge vo_i$ to vo_i. Indeed, by Lemma 7 we have $m_{v,i}(v\gamma) = m_i(\gamma)$, which is greater than $m_i(v) + r$. But the fall $l_i(v)$ is at most the climb $m_i(v) + \mathfrak{h}(vo_i) \leq m_i(v) + r$.

Thus, if we do not have (3.1) then we must have both (3.4) and (3.5). However this implies

$$o_i \curlywedge \gamma o_i = vo_i \curlywedge v\gamma o_i$$

which, by comparing their heights using Lemma 4, cannot occur when $r \neq 0$. □

PROOF OF THEOREM 1. With the structure of short conjugators understood from Proposition 8, the linear upper bound when $r = 0$ is clear since we may take $\gamma = (0, k)$ with $|k| \leq |u| + |v| = n$.

When $r \neq 0$, we use the conjguator $\gamma = (f, k)$ satisfying $0 \leq k < |r|$ and (3.1). Then Proposition 6 tells us that

$$|\gamma| \leq 2m_1(\gamma) + 2m_2(\gamma) \leq 2\max\{m_1(u), m_1(v)\} + 2\max\{m_2(u), m_2(v)\} + 4r.$$

As above, we may assume $r > 0$. Since $\mathfrak{h}(o_1 \curlywedge uo_1) \leq -r$, we get $m_1(u) \geq r$, and similarly for v. Thus we will get

$$\begin{aligned} |\gamma| &\leq 2\max\{m_1(u), m_1(v)\} + 2\max\{m_2(u), m_2(v)\} + 4\min\{m_1(u), m_1(v)\} \\ &\leq 3(m_1(u) + m_2(u) + m_1(v) + m_2(v)) \\ &\leq 3(|u| + |v|) = 3n. \end{aligned}$$

This proves the Theorem. □

4. A quadratic time algorithm

We finish by using the geometry developed in the preceding sections to describe an algorithm solving the conjugacy (search) problem in quadratic time.

Suppose R is generated by a finite set X as an abelian group. Then we can define a finite generating set for $\Gamma_2(R)$ as

$$Y = \left\{ \begin{pmatrix} t & x \\ 0 & 1 \end{pmatrix} : x \in X \right\}.$$

We write the corresponding word-lengths as $|\cdot|_Y$ or $|\cdot|_S$, where S is the generating set defined in Section 1.2. Note that when R is finite we may take $X = R$ and then $Y = S$. We naturally get the inequality

$$|u|_S \leq |u|_Y .$$

THEOREM 9. *Suppose R is finitely generated as an abelian group, with X, Y as above. Then there is an algorithm which determines whether two elements u, v in $\Gamma_2(R)$ are conjugate, and furthermore produces a conjugator, that runs in time $O(n^2)$, where $n = |u|_Y + |v|_Y$.*

PROOF. We describe the steps of the algorithm below.

Step 1. The input is given as words on the generating set Y. We may convert them into the form $u = (P, s)$, $v = (Q, r)$ in time linear in n. From this step we get a solution to the word problem in $\Gamma_2(R)$ which runs in linear time.

Step 2. Check if $r = s$. If not, then stop and conclude that u is not conjugate to v. If $r = s$, then continue to Step 3.

Step 3. If $r = s \neq 0$ then proceed now to Step 4. Otherwise, by Proposition 8 part (1), there will be a conjugator γ of the form $\gamma = (0, k)$ with $|k| \leq |u|_S + |v|_S \leq n$. Such k must satisfy $P = t^k Q$. Calculate $k = v_0(P) - v_0(Q)$. This can be done in linear time since we know $|v_0(P)|, |v_0(Q)| \leq |u|_S + |v|_S \leq n$, so we find the minimal non-zero coefficient in P or Q of t^i for $-n \leq i \leq n$. If $|k| > n$ then we stop and conclude u and v are not conjugate. Otherwise we check whether $\gamma u = v\gamma$ using the linear time solution to the word problem. If it is then we stop the algorithm with the output that u is conjugate to v and γ is the conjugator. If it is not then we stop and conclude that u and v are not conjugate.

Step 4. We have $r = s \neq 0$. Without loss of generality we assume $r > 0$, since otherwise we convert to u^{-1} and v^{-1}. From Proposition 8, we know that if u, v are conjugate then there is a conjugator γ of the form $\gamma = (f, k)$ where $0 \leq k < r$. Since the climbing of γ is limited by (3.1), we know we can write f in the form

$$f = \sum_{i=n_1}^{n_2} a_i t^i \quad \text{with} \quad |n_i| = \max\{m_i(u), m_i(v)\} + r \leq 2n.$$

Rearranging equation (3.3), we get $t^r f - f = t^k Q - P$. For each $k = 0, 1, 2, \ldots, r-1$, do the following: let

$$t^k Q - P = \sum_{i=-\infty}^{\infty} b_i t^i.$$

By comparing the coefficients of $t^{n_2+1}, \ldots, t^{n_2+r}$ in $t^r f - f = t^k Q - P$, we may set $a_i = b_{i+r}$ for $i = n_2 - r + 1, \ldots, n_2$. Now we can compare the coefficients of $t^{n_2-r+1}, \ldots, t^{n_2}$, and thus set $a_i = a_{i+r} + b_i$ for $i = n_2 - 2r + 1, \ldots, n_2 - r$. We repeat this process until all coefficients a_i have been assigned, and we check whether $a_i = b_i$ for $i = n_1, \ldots, n_2 - \alpha r$, where α is the maximal integer such that $n_1 \leq n_2 - \alpha r$.

Since $n_2 - n_1 \leq 4n$, the process of assigning elements of R to each a_i will run within $O(n)$ time. If $a_i = b_i$ for $i = n_1, \ldots, a_{n_2 - \alpha r}$ then we stop and γ is a conjugator for u, v. If for each k, there is some $i \in \{n_1, \ldots, n_2 - \alpha r\}$ for which $a_i \neq b_i$, then we stop and conclude that u, v are not conjugate. We have to do this at most r times, so in total the fourth step runs within quadratic time. □

Acknowledgements

Thanks are due to C. Abbott, Y. Antolín, C. Druţu, T. Riley and an anonymous referee for helpful discussions and comments.

References

[1] C. Abbott, *In preparation*, Ph.D. thesis, University of Wisconsin-Madison.

[2] M. Amchislavska and T. Riley, *Lamplighters, metabelian groups, and horocyclic products of trees*, to appear in L'Enseignement Mathématique.

[3] I. Anshel, M. Anshel, and D. Goldfeld, *An algebraic method for public-key cryptography*, Math. Res. Lett. **6** (1999), no. 3-4, 287–291, DOI 10.4310/MRL.1999.v6.n3.a3. MR1713130

[4] L. Bartholdi, M. Neuhauser, and W. Woess, *Horocyclic products of trees*, J. Eur. Math. Soc. (JEMS) **10** (2008), no. 3, 771–816, DOI 10.4171/JEMS/130. MR2421161

[5] G. Baumslag, *A finitely presented metabelian group with a free abelian derived group of infinite rank*, Proc. Amer. Math. Soc. **35** (1972), 61–62. MR0299662

[6] G. Baumslag, *Finitely presented metabelian groups*, Proceedings of the Second International Conference on the Theory of Groups (Australian Nat. Univ., Canberra, 1973), Springer, Berlin, 1974, pp. 65–74. Lecture Notes in Math., Vol. 372. MR0404462

[7] J. Behrstock and C. Druţu, *Divergence, thick groups, and short conjugators*, Illinois J. Math. **58** (2014), no. 4, 939–980. MR3421592

[8] M. R. Bridson and A. Haefliger, *Metric spaces of non-positive curvature*, Grundlehren der Mathematischen Wissenschaften [Fundamental Principles of Mathematical Sciences], vol. 319, Springer-Verlag, Berlin, 1999. MR1744486

[9] R. Diestel and I. Leader, *A conjecture concerning a limit of non-Cayley graphs*, J. Algebraic Combin. **14** (2001), no. 1, 17–25, DOI 10.1023/A:1011257718029. MR1856226

[10] K. H. Ko, S. J. Lee, J. H. Cheon, J. W. Han, J.-s. Kang, and C. Park, *New public-key cryptosystem using braid groups*, Advances in cryptology—CRYPTO 2000 (Santa Barbara, CA), Lecture Notes in Comput. Sci., vol. 1880, Springer, Berlin, 2000, pp. 166–183, DOI 10.1007/3-540-44598-6_10. MR1850042

[11] I. G. Lysënok, *Some algorithmic properties of hyperbolic groups* (Russian), Izv. Akad. Nauk SSSR Ser. Mat. **53** (1989), no. 4, 814–832, 912; English transl., Math. USSR-Izv. **35** (1990), no. 1, 145–163. MR1018749

[12] H. A. Masur and Y. N. Minsky, *Geometry of the complex of curves. II. Hierarchical structure*, Geom. Funct. Anal. **10** (2000), no. 4, 902–974, DOI 10.1007/PL00001643. MR1791145

[13] A. Sale, *Conjugacy length in group extensions*, Comm. Algebra **44** (2016), no. 2, 873–897, DOI 10.1080/00927872.2014.990021. MR3449958

[14] A. Sale, *Short conjugators in solvable groups*, arXiv:1112.2721v1 [math.GR] (2011).

[15] A. Sale, *The length of conjugators in solvable groups and lattices of semisimple lie groups*, DPhil thesis, University of Oxford, 2012.

[16] A. W. Sale, *The geometry of the conjugacy problem in wreath products and free solvable groups*, J. Group Theory **18** (2015), no. 4, 587–621, DOI 10.1515/jgth-2015-0009. MR3365819

[17] M. Stein and J. Taback, *Metric properties of Diestel-Leader groups*, Michigan Math. J. **62** (2013), no. 2, 365–386, DOI 10.1307/mmj/1370870377. MR3079268

[18] J. Tao, *Linearly bounded conjugator property for mapping class groups*, Geom. Funct. Anal. **23** (2013), no. 1, 415–466, DOI 10.1007/s00039-012-0206-3. MR3037904

[19] W. Woess, *Lamplighters, Diestel-Leader graphs, random walks, and harmonic functions*, Combin. Probab. Comput. **14** (2005), no. 3, 415–433, DOI 10.1017/S0963548304006443. MR2138121

Department of Mathematics, 1326 Stevenson Center, Vanderbilty University, Nashville, Tennessee 37240

E-mail address: andrew.sale@some.oxon.org

Contemporary Mathematics
Volume **677**, 2016
http://dx.doi.org/10.1090/conm/677/13628

A logspace solution to the word and conjugacy problem of generalized Baumslag-Solitar groups

Armin Weiß

ABSTRACT. Baumslag-Solitar groups were introduced in 1962 by Baumslag and Solitar as examples for finitely presented non-Hopfian two-generator groups. Since then, they served as examples for a wide range of purposes. As Baumslag-Solitar groups are HNN extensions, there is a natural generalization in terms of graph of groups.

Concerning algorithmic aspects of generalized Baumslag-Solitar groups, several decidability results are known. Indeed, a straightforward application of standard algorithms leads to a polynomial time solution of the word problem (the question whether some word over the generators represents the identity of the group). The conjugacy problem (the question whether two given words represent conjugate group elements) is more complicated; still decidability has been established by Anshel and Stebe for ordinary Baumslag-Solitar groups and for generalized Baumslag-Solitar groups independently by Lockhart and Beeker. However, up to now, no precise complexity estimates have been given.

In this work, we give a LOGSPACE algorithm for both problems. More precisely, we describe a uniform TC^0 many-one reduction of the word problem to the word problem of the free group. Then we refine the known techniques for the conjugacy problem and show it is AC^0-Turing-reducible to the word problem of the free group.

Finally, we consider uniform versions (where also the graph of groups is part of the input) of both word and conjugacy problem: while the word problem still is solvable in LOGSPACE the conjugacy problem becomes EXSPACE-complete.

1. Introduction

A *Baumslag-Solitar group* is a group of the form $\mathbf{BS}_{p,q} = \langle\, a, y \mid ya^py^{-1} = a^q \,\rangle$ for some $p, q \in \mathbb{Z} \setminus \{0\}$. These groups were introduced in 1962 by Baumslag and Solitar [**7**] as examples for finitely presented non-Hopfian two-generator groups. They showed that the class of Baumslag-Solitar groups comprises both Hopfian and non-Hopfian groups.

The usual presentation of a Baumslag-Solitar groups is as HNN extension of an infinite cyclic group with one stable letter. The different Baumslag-Solitar groups correspond to the different inclusions of the associated subgroup into the base group. HNN extensions are a special case of fundamental groups of a graph of

---

2010 *Mathematics Subject Classification.* Primary 20F10, 68Q25; Secondary 20E06, 03D15.
Key words and phrases. word problem, conjugacy problem, Baumslag-Solitar group, graph of groups, Logspace.

©2016 American Mathematical Society

groups – where the graph consists of exactly one vertex with one attached loop. Thus, there is a natural notion of generalized Baumslag-Solitar group (GBS group) as fundamental group of a graph of groups with infinite cyclic vertex and edge groups – see e. g. [**8**, **20**]. GBS groups were also studied in [**32**] and characterized as those finitely presented groups of cohomological dimension two which have an infinite cyclic subgroup whose commensurator is the whole group.

Algorithmic problems in group theory were introduced by Max Dehn more than 100 years ago. The two basic problems are the word problem and the conjugacy problem, which are defined as follows: Let G be a finitely generated group.
Word problem: On input of some word w written over the generators, decide whether $w = 1$ in G.
Conjugacy problem: On input of two words v and w written over the generators, decide whether v and w are conjugate, i. e., whether there exists $z \in G$ such that $zvz^{-1} = w$ in G.

In recent years, conjugacy played an increasingly important role in non-commutative cryptography, see e. g. [**15**, **22**, **47**]. These applications use that it is easy to create elements which are conjugated, but to check whether two given elements are conjugated might be difficult – even if the word problem is easy. In fact, there are groups where the word problem is easy but the conjugacy problem is undecidable [**41**].

It has been long known that both the word problem and the conjugacy problem in generalized Baumslag-Solitar groups are decidable. Actually, the standard application of Britton reductions leads to a polynomial time algorithm for the word problem (see e. g. [**34**]). Decidability of the conjugacy problem has been shown by Anshel and Stebe for ordinary Baumslag-Solitar groups [**5**] and for arbitrary GBS groups independently by Lockhart [**36**] and Beeker [**8**].

The probably first non-trivial complexity bounds for the word problem have been established by the general theorem by Lipton and Zalcstein [**35**] resp. Simon [**48**] that linear groups have word problem in $\mathsf{LOGSPACE}$ (although linear GBS groups form a small sub-class of all GBS groups). Later, Waack [**51**] examined the particular GBS group $\langle\, a, s, t \mid sas^{-1} = a, tat^{-1} = a^2 \,\rangle$ as an example of a non-linear group which has word problem in $\mathsf{LOGSPACE}$. In order to obtain the $\mathsf{LOGSPACE}$ bound for the word problem, he used the very special structure of this particular GBS group: the kernel under the canonical map onto the solvable Baumslag-Solitar group $\mathbf{BS}_{1,2}$ is a free group.

For solvable GBS groups – which are precisely the Baumslag-Solitar groups $\mathbf{BS}_{1,q}$ for $q \in \mathbb{Z}$ – the word problem was shown to be in (non-uniform) TC^0 by Robinson [**45**] (and also in $\mathsf{LOGSPACE}$). Moreover, in [**16**] it is shown that both the word and the conjugacy problem in $\mathbf{BS}_{1,2}$ is in uniform TC^0, indeed. It is straightforward to see that this proof also works for $\mathbf{BS}_{1,q}$ for arbitrary q, see [**52**]. The result for the conjugacy problem became possible because of the seminal theorem by Hesse [**23**, **24**] that integer division is in uniform TC^0 – a result which also plays a crucial role in this work.

Apart from these (and some other) special cases, no precise general complexity estimates have been given. In this work, we show that both the word problem and the conjugacy problem of every generalized Baumslag-Solitar group is in $\mathsf{LOGSPACE}$. More precisely, we establish the following results:

THEOREM A. *Let G be a GBS group. There is a uniform* TC^0 *many-one reduction from the word problem of G to the word problem of the free group F_2.*

Together with the well-known result that linear groups – in particular F_2 – have word problem in LOGSPACE [**35, 48**], this leads to a LOGSPACE algorithm of the word problem. Moreover, in view of [**12**], Theorem A shows that the word problem of GBS groups is in the complexity class $\mathsf{C_{=}NC^1}$ (for a definition see [**12**]).

THEOREM B. *Let G be a GBS group. The conjugacy problem of G is uniform-*AC^0*-Turing-reducible to the word problem of the free group.*

We also consider uniform versions of the word and conjugacy problem (where the GBS group is part of the input – for precise definitions see Section 3.2 and Section 4.3). This leads to the following contrasting theorem:

THEOREM C.

(i) *The uniform word problem for GBS groups is in* LOGSPACE. *Moreover, if the GBS groups are given as fundamental groups with respect to a spanning tree, the uniform word problem is* LOGSPACE*-complete.*
(ii) *The uniform conjugacy problem for GBS groups is* EXPSPACE*-complete.*

The paper is organized as follows: in Section 2, we fix our notation and recall some basic facts on complexity and graphs of groups – the reader who is familiar with these concepts might skip that section and only consult it for clarification. In Section 3, we give the proof of Theorem A, describe how to compute Britton-reduced words, and consider the uniform word problem. Finally, Section 4 deals with the non-uniform and uniform version of the conjugacy problem. Parts of this work are also part of the author's dissertation [**52**].

2. Preliminaries

Words. An *alphabet* is a (finite or infinite) set Σ; an element $a \in \Sigma$ is called a *letter.* The free monoid over Σ is denoted by Σ^*, its elements are called *words.* The multiplication of the monoid is concatenation of words. The identity element is the empty word 1. If w, p, x, q are words with $w = pxq$, then we call x a *factor* of w.

Rewriting systems. Let X be a set; a *rewriting system* over X is a binary relation $\Longrightarrow\, \subseteq X \times X$. If $(x, y) \in\, \Longrightarrow$, we write $x \Longrightarrow y$. The idea of the notation is that $x \Longrightarrow y$ indicates that x can be rewritten into y in one step. We denote the reflexive and transitive closure of $\Longrightarrow$ by $\overset{*}{\Longrightarrow}$; and by $\overset{*}{\Longleftrightarrow}$ its reflexive, transitive, and symmetric closure – it is the smallest equivalence relation such that x and y are in the same class for all $x \Longrightarrow y$.

Rewriting over words. Let Σ be an alphabet and $S \subseteq \Sigma^* \times \Sigma^*$ be a set of pairs. This defines a rewriting system $\underset{S}{\Longrightarrow}$ over Σ^* by $x \underset{S}{\Longrightarrow} y$ if $x = u\ell v$ and $y = urv$ for some $(\ell, r) \in S$. It is common to denote a rule $(\ell, r) \in S$ by $\ell \to r$ and we call S itself a rewriting system. Since $\underset{S}{\overset{*}{\Longleftrightarrow}}$ is an equivalence relation, we can form the set of equivalence classes $\Sigma^*/S = \{[x] \mid x \in \Sigma^*\}$, where $[x] = \{y \in \Sigma^* \mid x \underset{S}{\overset{*}{\Longleftrightarrow}} y\}$. Now, Σ^*/S becomes a monoid by $[x] \cdot [y] = [xy]$, and the mapping $x \to [x]$ yields a canonical homomorphism $\eta : \Sigma^* \to \Sigma^*/S$. The rewriting system S is called

- *confluent* if $x \underset{S}{\overset{*}{\Longrightarrow}} y$ and $x \underset{S}{\overset{*}{\Longrightarrow}} z$ implies $\exists\, w : y \underset{S}{\overset{*}{\Longrightarrow}} w$ and $z \underset{S}{\overset{*}{\Longrightarrow}} w$,

- *terminating* if there are no infinite chains $x_0 \underset{S}{\Longrightarrow} x_1 \underset{S}{\Longrightarrow} x_2 \underset{S}{\Longrightarrow} \cdots$,

A rewriting system S is confluent if and only if $x \overset{*}{\underset{S}{\Longleftrightarrow}} y$ implies $\exists\, w \,:\, x \overset{*}{\underset{S}{\Longrightarrow}} w$ and $y \overset{*}{\underset{S}{\Longrightarrow}} w$ (see [**9**, **29**]). Thus, if S is confluent and teminating, then in every class of Σ^*/S there is exactly one element to which no rule of S can be applied.

Groups. We consider a group G together with a surjective homomorphism $\eta : \Sigma^* \to G$ (a *monoid presentation*) for some (finite or infinite) alphabet Σ. In order to keep notation simple, we suppress the homomorphism η and consider words also as group elements. We write $w =_G w'$ as a shorthand of $\eta(w) = \eta(w')$ and $w \in_G A$ instead of $\eta(w) \in A$ for $A \subseteq G$ and $w \in \Sigma^*$.

For words (or group elements) v, w we write $v \sim_G w$ to denote conjugacy, i. e., $v \sim_G w$ if and only if there exists some $z \in G$ such that $zvz^{-1} =_G w$. If H is a subgroup of G, we write $v \sim_H w$ if there is some $z \in H$ such that $zvz^{-1} =_G w$.

Involutions. An involution on a set Σ is a mapping $x \mapsto \overline{x}$ such that $\overline{\overline{x}} = x$. We consider only fixed-point-free involutions, i. e., $x \neq \overline{x}$.

Free groups. Let Λ be some alphabet and set $\Sigma = \Lambda \cup \overline{\Lambda}$ where $\overline{\Lambda} = \{\overline{a} \mid a \in \Lambda\}$ is a disjoint copy of Λ. There is a fixed-point-free involution $\overline{\cdot} : \Sigma \to \Sigma$ defined by $a \mapsto \overline{a}$ and $\overline{a} \mapsto a$ (i. e., $\overline{\overline{a}} = a$). Consider the confluent and terminating rewriting system of *free reductions* $S = \{a\overline{a} \to 1 \mid a \in \Sigma\}$. Some word $w \in \Sigma^*$ is called *freely reduced* if there is no factor $a\overline{a}$ for any letter $a \in \Sigma$. The rewriting system S defines the *free group* $F_\Lambda = \Sigma^*/S$. We have $\overline{a} =_{F_\Lambda} a^{-1}$ for $a \in \Sigma$. We write F_2 as shorthand of $F_{\{a,b\}}$.

Graphs. For the notation of graphs we follow Serre's book [**46**]. A *graph* $Y = (V, E, \iota, \tau, \overline{\cdot})$ is given by the following data: a set of *vertices* $V = V(Y)$ and a set of *edges* $E = E(Y)$ together with two mappings $\iota, \tau : E \to V$ and an involution $e \mapsto \overline{e}$ without fixed points such that $\iota(e) = \tau(\overline{e})$.

An *orientation* of a graph Y is a subset $D \subseteq E$ such that E is the disjoint union $E = D \cup \overline{D}$. A *path* with start point u and end point v is a sequence of edges $e_1, \ldots, e_n$ such that $\tau(e_i) = \iota(e_{i+1})$ for all i and $\iota(e_1) = u$ and $\tau(e_n) = v$. A graph is *connected* if for every pair of vertices there is a path connecting them.

2.2. Complexity. Computation or decision problems are given by functions $f : \Delta^* \to \Sigma^*$ for some finite alphabets Δ and Σ. In case of a decision problem (or formal language) the range of f is the two element set $\{0, 1\}$.

$\mathsf{LOGSPACE}$ is the class of functions computable by a deterministic Turing machine with working tape bounded logarithmically in the length of the input.

Our result uses the following well-known theorem about linear groups (groups which can be embedded into a matrix group over some field). It was obtained by Lipton and Zalcstein [**35**] for fields of characteristic 0 and by Simon [**48**] for other fields.

THEOREM 1 ([**35**, **48**]). *Linear groups have word problem in* $\mathsf{LOGSPACE}$.

Circuit Complexity. The class AC^0 (resp. TC^0) is defined as the class of functions computed by families of circuits of constant depth and polynomial size with unbounded fan-in Boolean gates (and, or, not) (resp. unbounded fan-in Boolean and MAJORITY gates) – the alphabets Δ and Σ are encoded over the binary alphabet $\{0, 1\}$. In the following, we only consider $\mathsf{Dlogtime}$-uniform circuit families and we write uAC^0 (resp. uTC^0) as shorthand for Dlogtime-uniform AC^0 (resp. TC^0). Dlogtime-uniform means that there is a deterministic Turing machine which decides in time $\mathcal{O}(\log n)$ on input of two gate numbers (given in binary) and the string 1^n whether there is a wire between the two gates in the n-input circuit and also decides of which type some gates is. Note that the binary encoding of the gate numbers requires only $\mathcal{O}(\log n)$ bits – thus, the Turing machine is allowed to use time linear in the length of the encodings of the gates. For more details on these definitions we refer to [**50**].

Reductions. Let $K \subseteq \Delta^*$ and $L \subseteq \Sigma^*$ be languages and $\mathcal{C}$ a complexity class. Then K is called $\mathcal{C}$-many-one-reducible to L if there is a $\mathcal{C}$-computable function $f : \Delta^* \to \Sigma^*$ such that $w \in K$ if and only if $f(w) \in L$.

A function f is uAC^0*-reducible* (or uAC^0-Turing-reducible) to a function g if there is a Dlogtime-uniform family of AC^0 circuits computing f which, in addition to the Boolean gates, also may use oracle gates for g (i. e., gates which on input x output $g(x)$). We write $\mathsf{uAC}^0(F_2)$ for the family of problems which are uAC^0-reducible to the word problem of the free group F_2.

The Class uTC^0 *and Arithmetic.* Although uTC^0 is a very low parallel complexity class, it is still very powerful with respect to arithmetic. By the very definition of uAC^0 reducibility, MAJORITY is uTC^0-complete. As an immediate consequence, the word problem of $\mathbb{Z}$ with generators ± 1 is also uTC^0-complete (since a sequence over the alphabet $\{\pm 1\}$ sums up to 0 if and only if there is neither a majority of letters 1 nor of letters -1).

ITERATED ADDITION (resp. ITERATED MULTIPLICATION) are the following computation problems: On input of n binary integers $a_1, \ldots, a_n$ each having n bits (i. e., the input length is $N = n^2$), compute the binary representation of the sum $\sum_{i=0}^n a_i$ (resp. product $\prod_{i=0}^n a_i$). For INTEGER DIVISION, the input are two binary n-bit integers a, b; the binary representation of the integer $c = \lfloor a/b \rfloor$ has to be computed. The first statement of Theorem 2 is a standard fact, see [**50**]; the other statements are due to Hesse, [**23, 24**].

THEOREM 2 ([**23, 24, 50**]). *The problems* ITERATED ADDITION, ITERATED MULTIPLICATION, INTEGER DIVISION *are all in* uTC^0.

We have the following inclusions (note that even $\mathsf{uTC}^0 \subseteq \mathsf{P}$ is not known to be strict):

$$\mathsf{uTC}^0 \subseteq \mathsf{uAC}^0(F_2) \subseteq \mathsf{LOGSPACE} \subseteq \mathsf{P}.$$

The first inclusion is because there is a subgroup $\mathbb{Z}$ in F_2; the second inclusion is because of Theorem 1.

2.3. Graphs of Groups. Since generalized Baumslag-Solitar groups are defined as fundamental groups of graphs of groups, we give a brief introduction into this topic. Our presentation is a shortened version taken from [**17**], which in turn is based on Serre's book [**46**].

DEFINITION 3 (Graph of Groups). Let $Y = (V(Y), E(Y))$ be a connected graph. A *graph of groups* $\mathcal{G}$ over Y is given by the following data:

(i) For each vertex $a \in V(Y)$, there is a *vertex group* G_a.
(ii) For each edge $y \in E(Y)$, there is an *edge group* G_y such that $G_y = G_{\overline{y}}$.

(iii) For each edge $y \in E(Y)$, there is an injective homomorphism from G_y to $G_{\iota(y)}$, which is denoted by $c \mapsto c^y$. The image of G_y in $G_{\iota(y)}$ is denoted by G_y^y.

In the following, Y is always a finite graph. Since $G_y = G_{\overline{y}}$, there is also a homomorphism $G_y \to G_{\tau(y)}$. Thus, for $y \in E(Y)$ with $\iota(y) = a$ and $\tau(y) = b$, there are two isomorphisms and inclusions:

$$G_y \tilde{\to} G_y^y \leq G_a, \quad c \mapsto c^y, \qquad\qquad G_y \tilde{\to} G_{\overline{y}}^{\overline{y}} \leq G_b, \quad c \mapsto c^{\overline{y}}.$$

The fundamental group of $\mathcal{G}$ can be constructed as subgroup of the larger group $F(\mathcal{G})$: as an (possibly infinite) alphabet we choose a disjoint union

$$\Delta = E(Y) \cup \bigcup_{a \in V(Y)} (G_a \setminus \{1\}),$$

and we define the group

$$F(\mathcal{G}) = \Delta^* / \{gh = [gh],\, yc^{\overline{y}}\overline{y} = c^y \mid a \in V(Y),\, g, h \in G_a;\, y \in E(Y),\, c \in G_y\},$$

where $[gh]$ denotes the element obtained by multiplying g and h in G_a (where $1 \in G_a$ is identified with the empty word).

Let us define subsets of Δ^* as follows: for $a, b \in V(Y)$, we denote with $\Pi(\mathcal{G}, a, b)$ the set of words where the occurring edges form a path from a to b in Y and the elements of vertex groups between two edges are from the corresponding vertex in the path; more precisely,

$$\Pi(\mathcal{G}, a, b) = \{\, g_0 y_1 \cdots g_{n-1} y_n g_n \mid y_i \in E(Y),\, \iota(y_1) = a,\, \tau(y_n) = b, \\ \tau(y_i) = \iota(y_{i+1}),\, g_0 \in G_a,\, g_i \in G_{\tau(y_i)} \text{ for all } i \,\},$$

where again $1 \in G_a$ is identified with the empty word. Moreover, we set

$$\Pi(\mathcal{G}) = \bigcup_{a \in V(Y)} \Pi(\mathcal{G}, a, a).$$

In general, the image of $\Pi(\mathcal{G})$ in $F(\mathcal{G})$ is not a group but a so-called *groupoid.* If $w = g_0 y_1 \cdots g_{n-1} y_n g_n \in \Pi(\mathcal{G})$, then we call w a *$\mathcal{G}$-factorization* of the respective group element in $F(\mathcal{G})$; by saying this we implicitly require that $y_i \in E(Y)$, $\tau(y_i) = \iota(y_{i+1})$, $g_i \in G_{\tau(y_i)}$ for all i, $\tau(y_n) = \iota(y_1)$, and $g_0 \in G_{\iota(y_1)}$. We call $y_1 \cdots y_n$ the *underlying path* of w.

For all vertices $a \in V(Y)$, the image of $\Pi(\mathcal{G}, a, a)$ in $F(\mathcal{G})$ is a group.

DEFINITION 4.

(i) Let $a \in V(Y)$. The *fundamental group* $\pi_1(\mathcal{G}, a)$ of $\mathcal{G}$ with respect to the base point $a \in V(Y)$ is defined as the image of $\Pi(\mathcal{G}, a, a)$ in $F(\mathcal{G})$.

(ii) Let T be a spanning tree of Y (i.e., a subset of $E(Y)$ connecting all vertices and not containing any cycles). The *fundamental group* of $\mathcal{G}$ with respect to T is defined by

$$\pi_1(\mathcal{G}, T) = F(\mathcal{G}) / \{y = 1 \mid y \in T\}.$$

PROPOSITION 5 ([**46**]). *The canonical homomorphism from the subgroup $\pi_1(\mathcal{G}, a)$ of $F(\mathcal{G})$ to the quotient group $\pi_1(\mathcal{G}, T)$ is an isomorphism. In particular, the two definitions of the fundamental group are independent of the choice of the base point and the spanning tree.*

EXAMPLE 6. Let $\mathcal{G}$ be a graph of groups over the following graph:

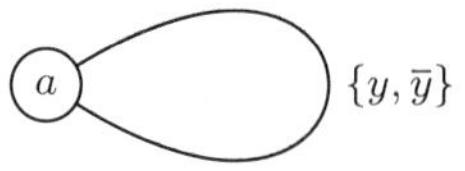

and let $G_a = \mathbb{Z} = \langle a \rangle$ and $G_y = G_{\overline{y}} = \mathbb{Z} = \langle c \rangle$ and the inclusions given by

$$c \mapsto a^p \qquad\qquad c \mapsto a^q$$

for some $p, q \in \mathbb{Z}\setminus\{0\}$. Then the fundamental group $\pi_1(\mathcal{G}, a)$ is the Baumslag-Solitar group

$$\pi_1(\mathcal{G}, a) = \mathbf{BS}_{p,q} = \left\langle\, a, y \;\middle|\; ya^py^{-1} = a^q \,\right\rangle.$$

Britton Reductions over Graphs of Groups. In [**10**], *Britton reductions* were originally defined for HNN extensions. They are given by the rewriting system $B_{\mathcal{G}} \subseteq \Delta^* \times \Delta^*$ with the following rules (see also [**37**, Sec. IV.2]):

$$\begin{aligned} gh &\longrightarrow [gh] && \text{for } a \in V(Y),\ g, h \in G_a \setminus \{1\}, \\ yc^{\overline{y}}\overline{y} &\longrightarrow c^y && \text{for } y \in E(Y),\ c \in G_y. \end{aligned}$$

As $B_{\mathcal{G}}$ is length-reducing, it is terminating. Furthermore, $F(\mathcal{G}) = \Delta^*/B_{\mathcal{G}}$. A word $w \in \Delta^*$ is called *Britton-reduced* if no rule from $\mathcal{B}_{\mathcal{G}}$ can be applied to it. As $\mathcal{B}_{\mathcal{G}}$ is terminating, there is a Britton-reduced $\hat{w}$ with $\hat{w} =_{F(\mathcal{G})} w$ for every w. However, this $\hat{w}$ might not be unique as $\mathcal{B}_{\mathcal{G}}$ is not confluent in general. Still, the following crucial facts hold:

LEMMA 7 (Britton's Lemma, [**10**]). *Let $w \in \Delta^*$ be Britton-reduced. If $w \in_{F(\mathcal{G})} G_a$, then w is the empty word or consists of a single letter of G_a. Moreover, if $w =_{F(\mathcal{G})} 1$, then $w = 1$ (i. e., w is the empty word).*

LEMMA 8. *If $v = h_0x_1 \cdots g_{n-1}x_nh_n, w = g_0y_1 \cdots g_{n-1}y_ng_n \in \Pi(\mathcal{G})$ with $v =_{F(\mathcal{G})} w$ are Britton-reduced, then $x_i = y_i$ for all i and there are $c_i \in G_{y_i}$ for $1 \leq i \leq n$ such that*

$$\begin{aligned} h_0 &=_{G_{\iota(y_1)}} g_0 (c_1^{-1})^{y_1}, \\ h_i &=_{G_{\tau(y_i)}} c_i^{\overline{y}_i} g_i (c_{i+1}^{-1})^{y_{i+1}}, && \textit{for } 1 \leq i \leq n-1, \textit{ and} \\ h_n &=_{G_{\tau(y_n)}} c_n^{\overline{y}_n} g_n. \end{aligned}$$

Using Lemma 7 one obtains a decision procedure for the word problem if the subgroup membership problem of G_y^y in $G_{\iota(y)}$ is decidable, the word problem of G_a is decidable for some $a \in V(Y)$, and the isomorphisms $G_y^y \to G_y^{\overline{y}}$ are effectively computable for all $y \in E(Y)$. However, this does not imply any bound on the complexity. The problem is that – even if all computations can be performed efficiently – the blow up due to the calculations of the isomorphisms $G_y^y \to G_y^{\overline{y}}$ might prevent an efficient solution of the word problem in the fundamental group. An example is the Baumslag group $\mathbf{G}_{1,2} = \left\langle\, a, t, b \;\middle|\; tat^{-1} = a^2, bab^{-1} = t \,\right\rangle$, which is an HNN extension of the Baumslag-Solitar group $\mathbf{BS}_{1,2}$. For $\mathbf{G}_{1,2}$, the straightforward algorithm of applying Britton reductions, leads to a non-elementary running time. However, in [**42**] it is shown that the word problem still can be solved in polynomial time.

For Baumslag-Solitar groups, the straighforward application of Britton reductions yields a polynomial time algorithm if the exponents are stored as binary integers.

Generalized Baumslag-Solitar Groups. A *generalized Baumslag-Solitar group* (*GBS group*) is a fundamental group of a finite graph of groups with only infinite cyclic vertex and edge groups. That means a GBS group is completely given by a finite graph Y and numbers $\alpha_y, \beta_y \in \mathbb{Z} \setminus \{0\}$ for $y \in E(Y)$ such that $\alpha_y = \beta_{\overline{y}}$. For $a \in V(Y)$ we write $G_a = \langle a \rangle$. Then we have

$$F(\mathcal{G}) = \left\langle\, V(Y), E(Y) \;\middle|\; \overline{y}y = 1,\ yb^{\beta_y}\overline{y} = a^{\alpha_y} \text{ for } y \in E(Y),\ a = \iota(y),\ b = \tau(y) \,\right\rangle$$

and $G = \pi_1(\mathcal{G}, a) \leq F(\mathcal{G})$ for any $a \in V(Y)$ as in Definition 4. (Note that $V(Y) \cup E(Y)$ generates $F(\mathcal{G})$ as a group, but in general, *not* as a monoid.)

As we have seen in Example 6, Baumslag-Solitar groups $\mathbf{BS}_{p,q}$ are the special case that Y consists of one vertex and one loop y with $\alpha_y = p$, $\beta_y = q$.

3. The Word Problem

In [**45**], Robinson showed that the word problem of non-cyclic free groups is NC^1-hard. Hence, for non-solvable GBS groups, we cannot expect the word or conjugacy problem to be in uTC^0 since they contain a free group of rank two. For ordinary Baumslag-Solitar groups, the word problem has recently been shown to be in NC^2 [**31**]. In the author's dissertation [**52**], this is improved to $\mathsf{LOGDCFL}$ – which means that it is $\mathsf{LOGSPACE}$-reducible to a deterministic context-free language. Here we aim for a $\mathsf{LOGSPACE}$ algorithm – or, more precisely, for a uTC^0 many-one reduction to the word problem of the free group F_2.

Let $G = \pi_1(\mathcal{G}, a)$ be a fixed GBS group given by a graph Y and numbers $\alpha_y, \beta_y \in \mathbb{Z} \setminus \{0\}$ for $y \in E(Y)$ and $a \in V(Y)$. Our alphabet is $\Delta = E(Y) \cup \left\{a^k \;\middle|\; a \in V(Y),\, k \in \mathbb{Z}\right\}$ – for simplicity we allow $k = 0$ and identify the letter a^0 with the empty word. We say that a word or $\mathcal{G}$-factorization w is represented in *binary* if the numbers k are written as binary integers (using a variable number of bits) – in the following we always assume this binary representation.

It turns out to be more convenient to work outside of G and to consider arbitrary $\mathcal{G}$-factorizations $w \in \Pi(\mathcal{G})$. Recall that a $\mathcal{G}$-factorization of some group element is a word

$$w = a_0^{k_0} y_1 a_1^{k_1} \cdots y_n a_n^{k_n}$$

with $a_i = \tau(y_i) = \iota(y_{i+1})$ for $0 < i < n$, $a_n = \tau(y_n) = a_0 = \iota(y_1)$, and $k_i \in \mathbb{Z}$. In the following, we always write a_i as shorthand of $\iota(y_{i+1})$.

LEMMA 9. *Let $w = a_0^{k_0} y_1 a_1^{k_1} \cdots y_n a_n^{k_n} \in \Pi(\mathcal{G})$. If $w \in_{F(\mathcal{G})} \langle a_0 \rangle$, then we have $w =_{F(\mathcal{G})} a_0^k$ for*

$$k = \sum_{\nu=0}^{n} k_\nu \cdot \prod_{\mu=1}^{\nu} \frac{\alpha_\mu}{\beta_\mu}$$

where $\alpha_\mu = \alpha_{y_\mu}$ and $\beta_\mu = \beta_{y_\mu}$ for $1 \leq \mu \leq n$.

PROOF. If $w = a_0^{k_0}$, then the formula is obviously correct. Hence, let $n > 0$. Then by Lemma 7, all the edges y_i can be cancelled by Britton reductions. In particular, we can find some $1 < i \leq n$ such that $w = a_0^{k_0}\, y_1 w' y_i\, w''$ with $y_1 = \overline{y}_i$ and $w' = a_1^{k_1} y_2 \cdots a_{i-1}^{k_{i-1}} \in_{F(\mathcal{G})} \left\langle a_1^{\beta_1} \right\rangle$ and $w'' = a_i^{k_i} y_{i+1} \cdots a_n^{k_n} \in_{F(\mathcal{G})} \langle a_i \rangle = \langle a_0 \rangle$.

By induction, we have $w' =_{F(\mathcal{G})} a_1^{k'}$ and $w'' =_{F(\mathcal{G})} a_0^{k''}$ where

$$k' = \sum_{\nu=1}^{i-1} k_\nu \cdot \prod_{\mu=2}^{\nu} \frac{\alpha_\mu}{\beta_\mu}, \qquad k'' = \sum_{\nu=i}^{n} k_\nu \cdot \prod_{\mu=i+1}^{\nu} \frac{\alpha_\mu}{\beta_\mu}.$$

Since $y_1 w' y_i \in_{F(\mathcal{G})} \langle a_0 \rangle$, we have $y_1 w' y_i =_{F(\mathcal{G})} a_0^{\frac{\alpha_1}{\beta_1} \cdot k'}$ and $\prod_{\mu=1}^{i} \frac{\alpha_\mu}{\beta_\mu} = 1$. Hence,

$$k = k_0 + \frac{\alpha_1}{\beta_1} k' + k'' = \sum_{\nu=0}^{n} k_\nu \cdot \prod_{\mu=1}^{\nu} \frac{\alpha_\mu}{\beta_\mu}.$$

□

For the rest of this section, we let $w = a_0^{k_0} y_1 a_1^{k_1} \cdots y_n a_n^{k_n}$ be a $\mathcal{G}$-factorization given in binary. For $0 \leq i \leq j \leq n$, we define

$$w_{i,j} = a_i^{k_i} y_{i+1} a_{i+1}^{k_{i+1}} \cdots y_j a_j^{k_j}$$

$$(1) \qquad k_{i,j} = \sum_{\nu=i}^{j} k_\nu \cdot \prod_{\mu=i+1}^{\nu} \frac{\alpha_\mu}{\beta_\mu} \qquad \in \mathbb{Q}$$

analogously to k in Lemma 9 where again $\alpha_\mu = \alpha_{y_\mu}$ and $\beta_\mu = \beta_{y_\mu}$ for $1 \leq \mu \leq n$. Note that we do not assume that $a_i^{k_i} y_{i+1} a_{i+1}^{k_{i+1}} \cdots y_j a_j^{k_j}$ lies in $\langle a_i \rangle$ – yet the numbers $k_{i,j}$ will play an important role in what follows. In particular, with the notation of Lemma 9, we have $k = k_{0,n}$. Moreover, by Lemma 9, we have

LEMMA 10. $\quad w_{i,j} \in_{F(\mathcal{G})} \langle a_i \rangle$ *if and only if* $w_{i,j} =_{F(\mathcal{G})} a_i^{k_{i,j}}$.

LEMMA 11. *The numbers $k_{i,j}$ (as fractions of binary integers) can be computed by a uniform family of TC^0 circuits – even if the numbers α_y, β_y are part of input.*

PROOF. ITERATED ADDITION and ITERATED MULTIPLICATION are in uTC^0, see Theorem 2; hence, the rational numbers $k_{i,j}$ can be computed in uTC^0 according to (1). Be aware that we do not require that the fractions are reduced. □

Now, pick some orientation $D \subseteq E(Y)$ of the edges (for every pair y, $\overline{y}$ choose exactly one of them to be in D). Consider the canonical map $\rho : G \to \mathbb{Z}^D$ onto the abelianization of the subgroup generated by the edges, which is defined by $a \mapsto 0$ for $a \in V(Y)$ and $y \mapsto e_y$, $\overline{y} \mapsto -e_y$ for $y \in D$ (where e_y is the unit vector having 1 at position y and 0 otherwise). With other words ρ counts the exponents of the edges. Consider the following observations:

- If $w =_{F(\mathcal{G})} 1$, then every edge y in w can be canceled with some $\overline{y}$ by Britton reductions.
- Consider a factor $yv\overline{y}$ for some word v. If y cancels with $\overline{y}$, then necessarily we have $\rho(v) = 0$, i. e., all edges occurring in between have exponent sum zero.

Now, the idea is to introduce colors and assign them to the letters y_i such that y_i and y_j get the same color only if they potentially might cancel. In order to do so, we start by defining a relation $\sim_c \subseteq \{1, \ldots, n\} \times \{1, \ldots, n\}$ and set

$$i \sim_c j \text{ if and only if } y_i = \overline{y}_j \text{ and } \begin{cases} \rho(w_{i,j-1}) = 0 \text{ and } k_{i,j-1} \in \beta_i \mathbb{Z} & \text{if } i < j, \\ \rho(w_{j,i-1}) = 0 \text{ and } k_{j,i-1} \in \beta_j \mathbb{Z} & \text{if } j < i. \end{cases}$$

Thus, $\sim_C$ is symmetric and we have $i \not\sim_C i$ for all i. Informally speaking, we have $i \sim_C j$ if and only if everything in between vanishes in the abelian quotient $\mathbb{Z}^D$ and y_i and y_j cancel given that everything in between cancels to something in $\langle a_i \rangle$ (the latter is a consequence of Lemma 10).

LEMMA 12. *If $i \sim_C \ell$, $\ell \sim_C m$, and $m \sim_C j$, then also $i \sim_C j$.*

PROOF. If two of the indices i, j, ℓ, m coincide, what can be the case only if $i = m$ or $\ell = j$, we are done. Otherwise, we have to show that $y_i = \overline{y}_j$, $\rho(w_{i,j-1}) = 0$, and $k_{i,j-1} \in \beta_i\mathbb{Z}$ (resp. $\rho(w_{j,i-1}) = 0$ and $k_{j,i-1} \in \beta_j\mathbb{Z}$ for $j < i$). We have $y_i = \overline{y}_\ell = y_m = \overline{y}_j$.

In order to see the other two conditions, we put the indices i, j, ℓ, m in ascending order. That means we fix $\lambda_1 < \lambda_2 < \lambda_3 < \lambda_4$ such that $\{\lambda_1, \lambda_2, \lambda_3, \lambda_4\} = \{i, j, \ell, m\}$. There are three situations to consider, as depicted in Figure 1:

(i) $y_{\lambda_1} = y_{\lambda_2}$ and $y_{\lambda_3} = y_{\lambda_4} = \overline{y}_{\lambda_1}$,
(ii) $y_{\lambda_1} = y_{\lambda_3}$ and $y_{\lambda_2} = y_{\lambda_4} = \overline{y}_{\lambda_1}$,
(iii) $y_{\lambda_1} = y_{\lambda_4}$ and $y_{\lambda_2} = y_{\lambda_3} = \overline{y}_{\lambda_1}$.

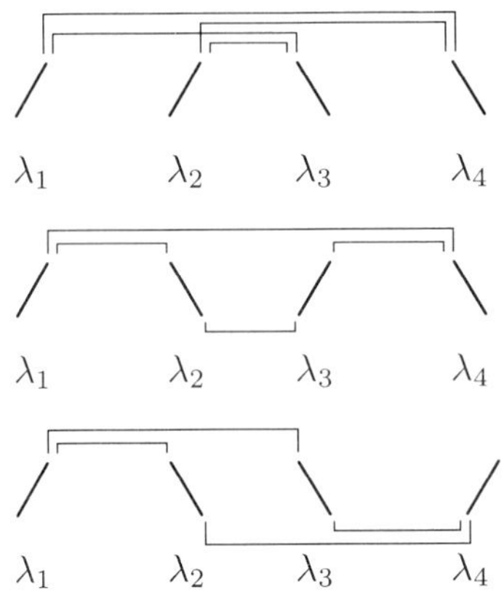

FIGURE 1. Three different situations. The four pairings of each situation are depicted as brackets.

All these cases have in common that there are exactly four pairings $\{\lambda_r, \lambda_s\}$ with $y_{\lambda_r} = \overline{y}_{\lambda_s}$, and these four pairings correspond to the four pairings $\{i, \ell\}$, $\{\ell, m\}$, $\{m, j\}$, and $\{i, j\}$. In each case, the conditions $\rho(w_{\lambda_r,\lambda_s-1}) = 0$ and $k_{\lambda_r,\lambda_s-1} \in \beta_{\lambda_r}\mathbb{Z}$ hold for three of the $\{\lambda_r, \lambda_s\}$, and we have to show it for the fourth.

In case (i), we have

$$\rho(w_{\lambda_1,\lambda_4-1}) = \rho(w_{\lambda_1,\lambda_3-1}) + \rho(w_{\lambda_2,\lambda_4-1}) - \rho(w_{\lambda_2,\lambda_3-1}).$$

Thus, since three of these vectors are zero, so is the fourth (i.e., we have shown that $\rho(w_{i,j-1}) = 0$ resp. $\rho(w_{j,i-1}) = 0$). In particular, we have $\rho(w_{\lambda_1,\lambda_2}) = \rho(w_{\lambda_1,\lambda_4-1}) - \rho(w_{\lambda_2,\lambda_4-1}) = 0$. Hence,

$$\prod_{\mu=\lambda_1+1}^{\lambda_2} \frac{\alpha_\mu}{\beta_\mu} = \prod_{y \in D} \left(\frac{\alpha_y}{\beta_y}\right)^{\rho(w_{\lambda_1,\lambda_2})_y} = 1,$$

where $\rho(w_{\lambda_1,\lambda_2})_y$ denotes the component of the vector belonging to y (recall $D \subseteq E(Y)$ is the orientation) – the first equality is because $\frac{\alpha_{\overline{y}}}{\beta_{\overline{y}}} = \left(\frac{\alpha_y}{\beta_y}\right)^{-1}$ and $\rho(w_{\lambda_1,\lambda_2})_y$ simply counts the number of occurrences of y (positive) and $\overline{y}$ (negative) in w_{λ_1,λ_2}. It follows that

$$\begin{aligned}
k_{\lambda_1,\lambda_4-1} &= \sum_{\nu=\lambda_1}^{\lambda_4-1} k_\nu \cdot \prod_{\mu=\lambda_1+1}^{\nu} \frac{\alpha_\mu}{\beta_\mu} \\
&= \sum_{\nu=\lambda_1}^{\lambda_3-1} k_\nu \cdot \prod_{\mu=\lambda_1+1}^{\nu} \frac{\alpha_\mu}{\beta_\mu} + \sum_{\nu=\lambda_2}^{\lambda_4-1} k_\nu \cdot \prod_{\mu=\lambda_1+1}^{\nu} \frac{\alpha_\mu}{\beta_\mu} - \sum_{\nu=\lambda_2}^{\lambda_3-1} k_\nu \cdot \prod_{\mu=\lambda_1+1}^{\nu} \frac{\alpha_\mu}{\beta_\mu} \\
&= \sum_{\nu=\lambda_1}^{\lambda_3-1} k_\nu \cdot \prod_{\mu=\lambda_1+1}^{\nu} \frac{\alpha_\mu}{\beta_\mu} + \sum_{\nu=\lambda_2}^{\lambda_4-1} k_\nu \cdot \prod_{\mu=\lambda_2+1}^{\nu} \frac{\alpha_\mu}{\beta_\mu} - \sum_{\nu=\lambda_2}^{\lambda_3-1} k_\nu \cdot \prod_{\mu=\lambda_2+1}^{\nu} \frac{\alpha_\mu}{\beta_\mu} \\
&= k_{\lambda_1,\lambda_3-1} + k_{\lambda_2,\lambda_4-1} - k_{\lambda_2,\lambda_3-1}.
\end{aligned}$$

Hence, since three of them are in $\beta_{\lambda_1}\mathbb{Z} = \beta_{\lambda_2}\mathbb{Z}$, so is the fourth.

The other cases follow with the same arguments: in case (ii) we have

$$\begin{aligned}
\rho(w_{\lambda_1,\lambda_4-1}) &= \rho(w_{\lambda_1,\lambda_2-1}) + \rho(y_{\lambda_2}) + \rho(w_{\lambda_2,\lambda_3-1}) + \rho(y_{\lambda_3}) + \rho(w_{\lambda_3,\lambda_4-1}) \\
&= \rho(w_{\lambda_1,\lambda_2-1}) + \rho(w_{\lambda_2,\lambda_3-1}) + \rho(w_{\lambda_3,\lambda_4-1})
\end{aligned}$$

because $y_{\lambda_2} = \overline{y}_{\lambda_3}$, what again implies that all of them are zero. Like in the first case, we have $\prod_{\mu=\lambda_1+1}^{\lambda_2} \frac{\alpha_\mu}{\beta_\mu} = \frac{\alpha_{\lambda_2}}{\beta_{\lambda_2}}$ (because $\rho(w_{\lambda_1,\lambda_2-1}) = 0$) and $\prod_{\mu=\lambda_1+1}^{\lambda_3} \frac{\alpha_\mu}{\beta_\mu} = 1$ (because $\rho(w_{\lambda_1,\lambda_3}) = \rho(w_{\lambda_1,\lambda_4-1}) - \rho(w_{\lambda_3,\lambda_4-1}) = 0$). It follows that

$$\begin{aligned}
k_{\lambda_1,\lambda_4-1} &= \sum_{\nu=\lambda_1}^{\lambda_4-1} k_\nu \cdot \prod_{\mu=\lambda_1+1}^{\nu} \frac{\alpha_\mu}{\beta_\mu} \\
&= \sum_{\nu=\lambda_1}^{\lambda_2-1} k_\nu \cdot \prod_{\mu=\lambda_1+1}^{\nu} \frac{\alpha_\mu}{\beta_\mu} + \frac{\alpha_{\lambda_2}}{\beta_{\lambda_2}} \cdot \sum_{\nu=\lambda_2}^{\lambda_3-1} k_\nu \cdot \prod_{\mu=\lambda_2+1}^{\nu} \frac{\alpha_\mu}{\beta_\mu} + \sum_{\nu=\lambda_3}^{\lambda_4-1} k_\nu \cdot \prod_{\mu=\lambda_3+1}^{\nu} \frac{\alpha_\mu}{\beta_\mu} \\
&= k_{\lambda_1,\lambda_2-1} + \frac{\alpha_{\lambda_2}}{\beta_{\lambda_2}} \cdot k_{\lambda_2,\lambda_3-1} + k_{\lambda_3,\lambda_4-1}.
\end{aligned}$$

Since $y_{\lambda_1} = y_{\lambda_3} = \overline{y}_{\lambda_2}$, we have $\alpha_{\lambda_2} = \beta_{\lambda_1}$ and $\beta_{\lambda_3} = \beta_{\lambda_1}$. That means we have $\frac{\alpha_{\lambda_2}}{\beta_{\lambda_2}} \cdot k_{\lambda_2,\lambda_3-1} \in \beta_{\lambda_1}\mathbb{Z}$ if and only if $k_{\lambda_2,\lambda_3-1} \in \beta_{\lambda_2}\mathbb{Z}$, and $k_{\lambda_3,\lambda_4-1} \in \beta_{\lambda_1}\mathbb{Z}$ if and only if $k_{\lambda_3,\lambda_4-1} \in \beta_{\lambda_3}\mathbb{Z}$. Thus, since for three of the $k_{\lambda,\lambda'}$ we have $k_{\lambda,\lambda'} \in \beta_\lambda\mathbb{Z}$, this is true also for the fourth.

Finally, in case (iii), because of $y_{\lambda_2} = y_{\lambda_3}$, we have

$$\begin{aligned}
\rho(w_{\lambda_1,\lambda_3-1}) - \rho(w_{\lambda_1,\lambda_2-1}) &= \rho(w_{\lambda_2-1,\lambda_3-1}) \\
&= \rho(w_{\lambda_2,\lambda_3}) \\
&= \rho(w_{\lambda_2,\lambda_4-1}) - \rho(w_{\lambda_3,\lambda_4-1}).
\end{aligned}$$

Therefore, they are all 0. As before, $\rho(w_{\lambda_1,\lambda_2-1}) = 0$ implies that $\prod_{\mu=\lambda_1+1}^{\lambda_2-1} \frac{\alpha_\mu}{\beta_\mu} = 1$ and $\rho(w_{\lambda_2,\lambda_3}) = 0$ implies that $\prod_{\mu=\lambda_2+1}^{\lambda_3} \frac{\alpha_\mu}{\beta_\mu} = 1$. Thus, we have

$$\begin{aligned}
k_{\lambda_1,\lambda_3-1} - k_{\lambda_1,\lambda_2-1} &= \sum_{\nu=\lambda_1}^{\lambda_3-1} k_\nu \cdot \prod_{\mu=\lambda_1+1}^{\nu} \frac{\alpha_\mu}{\beta_\mu} - \sum_{\nu=\lambda_1}^{\lambda_2-1} k_\nu \cdot \prod_{\mu=\lambda_1+1}^{\nu} \frac{\alpha_\mu}{\beta_\mu} \\
&= \frac{\alpha_{\lambda_2}}{\beta_{\lambda_2}} \cdot \sum_{\nu=\lambda_2}^{\lambda_3-1} k_\nu \cdot \prod_{\mu=\lambda_2+1}^{\nu} \frac{\alpha_\mu}{\beta_\mu} \\
&= \frac{\alpha_{\lambda_2}}{\beta_{\lambda_2}} \cdot \left(\sum_{\nu=\lambda_2}^{\lambda_4-1} k_\nu \cdot \prod_{\mu=\lambda_2+1}^{\nu} \frac{\alpha_\mu}{\beta_\mu} - \sum_{\nu=\lambda_3}^{\lambda_4-1} k_\nu \cdot \prod_{\mu=\lambda_3+1}^{\nu} \frac{\alpha_\mu}{\beta_\mu} \right) \\
&= \frac{\alpha_{\lambda_2}}{\beta_{\lambda_2}} \cdot (k_{\lambda_2,\lambda_4-1} - k_{\lambda_3,\lambda_4-1})
\end{aligned}$$

with $\alpha_{\lambda_2} = \beta_{\lambda_1}$ and $\beta_{\lambda_3} = \beta_{\lambda_2}$. So, again since for three of the $k_{\lambda,\lambda'}$ we have $k_{\lambda,\lambda'} \in \beta_\lambda \mathbb{Z}$, this is true also for the fourth. □

Now, we define a new relation $\approx \subseteq \{1, \dots, n\} \times \{1, \dots, n\}$ as $i \approx j$ if and only if there is some ℓ with $i \sim_C \ell$ and $\ell \sim_C j$. Moreover, we set $i \approx i$ for all i.

LEMMA 13. *$\approx$ is an equivalence relation.*

PROOF. By definition, $\approx$ is reflexive. Because $\sim_C$ is symmetric, $\approx$ is also symmetric. Transitivity follows from Lemma 12. □

Denote by $\Sigma_w = \{[i] \mid i \in \{1, \dots, n\}\}$ the set of equivalence classes of $\approx$. For $[i] \in \Sigma_w$ define $\overline{[i]} = [j]$ if $i \sim_C j$ – if no such j exists, we add a new element $\overline{[i]}$ to Σ_w. From the definition of $\approx$ it follows that $\bar{\cdot}$ is well-defined. Moreover, we have $\overline{\overline{[i]}} = [i]$ and $\overline{[i]} \neq [i]$ for all $[i] \in \Sigma_w$. In particular, Σ_w is an alphabet with fixed-point-free involution. We can think of each class $[i] \cup \overline{[i]}$ as a color assigned to the edges y_i. From the definition of $\sim_C$ and Lemma 10 it is clear that only edges with the same color can cancel. Let $\Lambda_w \subseteq \Sigma_w$ such that $\Sigma_w = \Lambda_w \cup \overline{\Lambda}_w$ as a disjoint union, i. e., for every pair $[i], \overline{[i]}$ exactly one of them is in Λ_w. Then we have $\Sigma_w^* / \left\{ [i]\overline{[i]} = \overline{[i]}[i] = 1 \,\middle|\, i \in \{1, \dots, n\} \right\} = F_{\Lambda_w}$. Now, we define

$$\mathcal{C}(w) = [1] \cdots [n], \qquad \mathcal{C}(w_{i,j}) = [i+1] \cdots [j].$$

LEMMA 14. *$w_{i,j} \in_{F(\mathcal{G})} \langle a_i \rangle$ if and only if $\mathcal{C}(w_{i,j}) =_{F(\Lambda_w)} 1$.*

Before we prove Lemma 14, we present an example and some consequences.

EXAMPLE 15. Consider the group $\mathbf{BS}_{2,3}$ and the word

$$w = yaya\bar{y}a^3ya\bar{y}a\bar{y}\,ya^2\bar{y}.$$

Then we have

$$\begin{aligned}
\mathcal{C}(w) &= [1][2][3][4][5][6][7][8] \\
&= [1][2][3]\overline{[3]}\,\overline{[2]}\,\overline{[1]}[1]\overline{[1]} =_{F_{\Lambda_w}} 1.
\end{aligned}$$

Indeed, consider for example the factor $\bar{y}a^3y$. As $k_{3,3} = 3 \in 3\mathbb{Z}$, it follows that $3 \sim_C 4$ and thus $[4] = \overline{[3]}$; however, $2 \not\sim_C 3$ since $k_{2,2} = 1 \notin 2\mathbb{Z}$, see Figure 2. By Lemma 14, we know that $w \in \langle a \rangle$.

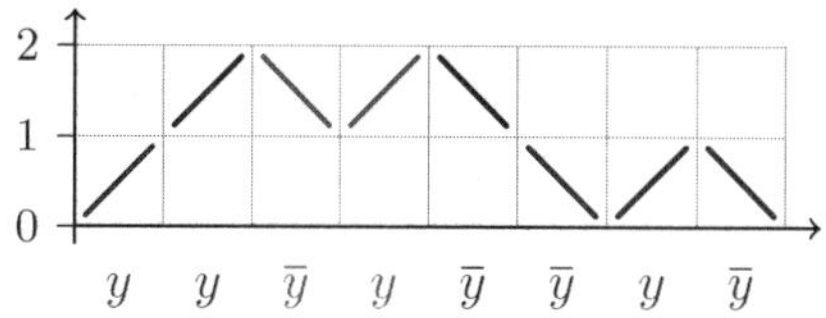

FIGURE 2. $\rho(w)$ and $\mathcal{C}(w)$ depicted graphically – each color represents one $[i] \cup \overline{[i]}$. **(The color version of this figure appears in the online version of the paper.)**

As immediate consequences of Britton's Lemma, Lemma 9, and Lemma 14, we obtain:

COROLLARY 16. *$w =_{F(\mathcal{G})} 1$ if and only if $\mathcal{C}(w) =_{F(\Lambda_w)} 1$ and $k_{0,n} = 0$.*

COROLLARY 17. *For $w = a_0^{k_0} y_1 a_1^{k_1} \cdots y_n a_n^{k_n}$, let $[i_1] \cdots [i_j] \in \Sigma_w^*$ be freely reduced with $\mathcal{C}(w) = [1] \cdots [n] =_{F_{\Lambda_w}} [i_1] \cdots [i_j]$. Then the $\mathcal{G}$-factorization*

$$\hat{w} = a_0^{k_{0,i_1-1}} y_{i_1} a_{i_1}^{k_{i_1,i_2-1}} \cdots y_{i_j} a_{i_j}^{k_{i_j,n}}$$

is Britton-reduced and $w =_{F(\mathcal{G})} \hat{w}$.

Note that Corollary 17 is independent of the choice of the representatives of $[i_1], \ldots, [i_j]$.

PROOF OF LEMMA 14. Let $w_{i,j} \in_{F(\mathcal{G})} \langle a_i \rangle$. By Britton's Lemma, we can write

$$w_{i,j} = a_i^{k_i}\, y_{i+1} w_{i+1,\ell-1} y_\ell\, w_{\ell,j}$$

with $y_\ell = \overline{y}_{i+1}$, $w_{i+1,\ell-1} \in_{F(\mathcal{G})} \left\langle a_{i+1}^{\beta_{i+1}} \right\rangle$, and $w_{\ell,j} \in_{F(\mathcal{G})} \langle a_i \rangle$. By Lemma 10, we have $k_{i+1,\ell-1} \in \beta_{i+1}\mathbb{Z}$. As also $y_\ell = \overline{y}_{i+1}$ and $\rho(w_{i+1,\ell-1}) = 0$, this implies $i+1 \sim_c \ell$. By induction, we know that $\mathcal{C}(w_{i+1,\ell-1}) =_{F(\Lambda_w)} \mathcal{C}(w_{\ell,j}) =_{F(\Lambda_w)} 1$. Thus, we obtain

$$\mathcal{C}(w_{i,j}) = [i+1]\, \mathcal{C}(w_{i+1,\ell-1})\, \overline{[i+1]}\, \mathcal{C}(w_{\ell,j}) =_{F(\Lambda_w)} 1.$$

For the other direction let $\mathcal{C}(w_{i,j}) =_{F(\Lambda_w)} 1$. Then $\mathcal{C}(w_{i,j})$ is not freely reduced and we can write it in the form

$$\mathcal{C}(w_{i,j}) = [i+1]\, \mathcal{C}(w_{i+1,\ell-1})\, [\ell]\, \mathcal{C}(w_{\ell,j})$$

for some ℓ with $[i+1] = \overline{[\ell]}$ and $\mathcal{C}(w_{i+1,\ell-1}) =_{F(\Lambda_w)} \mathcal{C}(w_{\ell,j}) =_{F(\Lambda_w)} 1$.

By induction, we know that $w_{i+1,\ell-1} \in_{F(\mathcal{G})} \langle a_{i+1} \rangle$ and $w_{\ell,j} \in_{F(\mathcal{G})} \langle a_\ell \rangle$; thus, by Lemma 10, $w_{i+1,\ell-1} =_{F(\mathcal{G})} a_{i+1}^{k_{i+1,\ell-1}}$ and $w_{\ell,j} =_{F(\mathcal{G})} a_\ell^{k_{\ell,j}}$. Since $[i+1] \sim_c [\ell]$, we have $y_{i+1} = \overline{y}_\ell$ and $k_{i+1,\ell-1} \in \beta_{i+1}\mathbb{Z}$. As, in particular, $a_i = a_\ell$, we obtain

$$\begin{aligned} w_{i,j} &= a_i^{k_i}\, y_{i+1}\, w_{i+1,\ell-1}\, y_\ell\, w_{\ell,j} \\ &=_{F(\mathcal{G})} a_i^{k_i}\, y_{i+1} a_{i+1}^{k_{i+1,\ell-1}} y_\ell\, a_\ell^{k_{\ell,j}} \\ &= a_i^{k_i}\, a_i^{\frac{\alpha_{i+1}}{\beta_{i+1}} k_{i+1,\ell-1}}\, a_i^{k_{\ell,j}} \in \langle a_i \rangle . \end{aligned}$$

□

Now, we are ready to describe a uTC^0-many-one reduction of the word problem for $\mathcal{G}$-factorizations to the free group $F_2 = \langle a, b \rangle$. The input is a $\mathcal{G}$-factorization w, the output some word in $\tilde{w} \in \{a, \bar{a}, b, \bar{b}\}^*$ such that $w =_{F(\mathcal{G})} 1$ if and only if $\tilde{w} =_{F_2} 1$. The circuit computes the following steps:

ALGORITHM 18.

(i) Compute $k_{0,n}$. If $k_{0,n} \neq 0$, then output a (or some arbitrary other non-identity element of F_2).

(ii) Otherwise, compute and output an encoding of $\mathcal{C}(w)$ in F_2 as follows:

  (a) For all pairs $i < j$ check independently in parallel whether $i \sim_{\mathcal{C}} j$ in uTC^0:

    1. check whether $y_i = \bar{y}_j$,
    2. compute $\rho(w_{i,j-1})$ and check whether $\rho(w_{i,j-1}) = 0$,
    3. compute $k_{i,j-1}$, check whether $k_{i,j-1} \in \mathbb{Z}$ and, if yes, whether $\beta_i \mid k_{i,j-1}$.

    If all points hold, then $i \sim_{\mathcal{C}} j$, otherwise not.

  (b) For every index i compute in parallel the smallest j with $j \in [i] \cup \overline{[i]}$ as representative of $[i]$ – depending on whether $j \in [i]$ or $j \in \overline{[i]}$ the corresponding output is $b^j a \bar{b}^j$ or $b^j \bar{a} \bar{b}^j$.

  (c) Concatenate all output words of the previous step.

By Lemma 11 and Hesse's result Theorem 2, step (i) and (ii) (a) can be computed in uTC^0. Steps (ii) (b) and (ii) (c) are straightforward in uTC^0. Indeed, the smallest $j \in [i] \cup \overline{[i]}$ satisfies the first order formula

$$\left(i = j \vee i \sim_{\mathcal{C}} j \vee \bigvee_{k} (i \sim_{\mathcal{C}} k \wedge k \sim_{\mathcal{C}} j) \right) \wedge \bigwedge_{k<j} \neg \left(i \sim_{\mathcal{C}} k \vee \bigvee_{\ell} (i \sim_{\mathcal{C}} \ell \wedge \ell \sim_{\mathcal{C}} k) \right),$$

which describes an uAC^0 circuit in the obvious way (see [**6**] for the general correspondence between circuits and formulas). Step (ii) (c) can be seen as the application of a homomorphism of free monoids, what can be done in uTC^0 (see [**33**]). Thus, we have established a uTC^0 many-one reduction to the word problem of F_2.

Note that in none of the above steps the actual graph played a role – only the numbers α_y, β_y were used. This is because, up to now, we assumed that the input is already given as $\mathcal{G}$-factorization. But also the transformation of elements of $\pi_1(\mathcal{G}, T)$ into $\mathcal{G}$-factorizations can be done in uTC^0 as we see in the next theorem, which proves Theorem A.

THEOREM 19. *Let $G = \pi_1(\mathcal{G}, a) \cong \pi_1(\mathcal{G}, T)$ be a GBS group with graph Y and $\Delta = E(Y) \cup \{a^k \mid a \in V(Y),\, k \in \mathbb{Z}\}$. There is a many-one reduction computed by a uniform family of TC^0-circuits from each of the problems*

(i) *given a word $w \in \Delta^*$, decide whether w is a $\mathcal{G}$-factorization and, if so, decide whether $w =_{F(\mathcal{G})} 1$.*

(ii) *given a word $w \in \Delta^*$, decide whether $w =_{\pi_1(\mathcal{G},T)} 1$,*

to the word problem of the free group F_2. In particular, the word problem of G is in $\mathsf{LOGSPACE}$.

PROOF. In order to decide whether w is a $\mathcal{G}$-factorization, one simply needs to verify whether w is of the form $a_0^{k_0} y_1 a_1^{k_1} \cdots y_n a_n^{k_n}$ and then check whether $a_0 = a_n$

and $a_{i-1} = \iota(y_i)$ and $\tau(y_i) = a_i$ for all $1 \le i \le n$. This can be done in uAC^0. Then it remains to apply Algorithm 18 – which we already have seen to be in uTC^0.

For (ii), one needs to compute the isomorphism $\pi_1(\mathcal{G}, T) \to \pi_1(\mathcal{G}, a)$. For $a, b \in V(Y)$ let $T[a, b]$ denote the unique path from a to b in the spanning tree T. We read $T[a, b]$ as a group element. To compute the isomorphism, every letter $y \in E(Y)$ has to be replaced by the word $T[a, \iota(y)]\, y\, T[\tau(y), a]$ and every letter b^k with $b \in V(Y)$, $k \in \mathbb{Z}$ by $T[a, b]\, b^k\, T[b, a]$. This means we apply a homomorphism of free monoids, what can be done in uTC^0 (see [**33**]). Moreover, this replacement produces a $\mathcal{G}$-factorization as output. □

3.1. Computing Britton-reduced words. Before we consider the problem of computing Britton-reduced words, we focus on the analog problem in free groups, the computation of freely reduced words. Since already in the solution of the word problem free groups of arbitrary rank were appearing, we consider the alphabet Λ as part of the input and assume that it is properly encoded over the binary alphabet $\{0, 1\}$. In particular, we assume that the involution $\Lambda \cup \overline{\Lambda} \to \Lambda \cup \overline{\Lambda}$ can be computed in uAC^0 – e. g. by a bit-flip.

PROPOSITION 20. *The following problem is uAC^0-reducible to the word problem of F_2: given a finite alphabet Λ and a word $w \in (\Lambda \cup \overline{\Lambda})^*$, compute a freely reduced word $\hat{w} \in (\Lambda \cup \overline{\Lambda})^*$ with $\hat{w} =_{F(\Lambda)} w$.*

PROOF. We follow a similar approach as for the solution of the word problem of GBS groups. For $w = w_1 \cdots w_n$ with $w_i \in \Lambda \cup \overline{\Lambda}$, we set $w_{i,j} = w_{i+1} \cdots w_j$. We define an equivalence relation $\approx_{\mathcal{F}} \subseteq \{1, \dots, n\} \times \{1, \dots, n\}$ by

$$i \approx_{\mathcal{F}} j \text{ if and only if } w_i = w_j \text{ and } \begin{cases} w_{i,j} =_{F(\Lambda)} 1 & \text{if } i < j, \\ w_{j,i} =_{F(\Lambda)} 1 & \text{if } j < i. \end{cases}$$

By using the embedding of F_Λ into F_2, it can be checked in $\mathsf{uAC}^0(F_2)$ for all pairs i, j whether $i \approx_{\mathcal{F}} j$. Furthermore, let us define a partial map

$$\begin{aligned} \overline{\cdot} : \{1, \dots, n\}/{\approx_{\mathcal{F}}} &\to \{1, \dots, n\}/{\approx_{\mathcal{F}}} \\ [i] &\mapsto \overline{[i]} = [j] \quad \text{if there is some } j \text{ with } w_i = \overline{w}_j \text{ and} \\ &\qquad\qquad\quad w_{i,j-1} =_{F_\Lambda} 1 \text{ (resp. } w_{j,i-1} =_{F_\Lambda} 1). \end{aligned}$$

To see that this map is well defined, we have to verify two points:

(i) that the map $i \mapsto \overline{[i]}$ is well-defined;
(ii) that $\overline{[i]} = \overline{[j]}$ if $i \approx_{\mathcal{F}} j$.

For the first point, consider $i < j < k$ with $w_i = \overline{w}_j$, $w_{i,j-1} =_{F_\Lambda} 1$ and $w_i = \overline{w}_k$, $w_{i,k-1} =_{F_\Lambda} 1$. Then we have $w_j = \overline{w}_i = w_k$ and $w_{j,k} =_{F_\Lambda} (w_{i,j-1} w_j)^{-1} w_{i,k-1} w_k =_{F_\Lambda} 1$ – hence, $j \approx_{\mathcal{F}} k$. Likewise all other orderings of i, j, k can be dealt with; hence, the image $\overline{[i]}$ is uniquely defined for each i.

For the second point, let $i \approx_{\mathcal{F}} j$ and $k \in \overline{[j]}$ with $i < j < k$. Then we have $w_i = w_j = \overline{w}_k$ and $w_{i,k-1} = w_{i,j} w_{j,k-1} \overline{[i]}$ and, thus, $\overline{[i]} = \overline{[j]}$. Again all other orderings of i, j, k follow the same way.

Since $\overline{\overline{[i]}} = [i]$ for all i, we have a well-defined partial involution $\overline{\cdot}$. In the following, if $\overline{[i]}$ is not defined, we consider it to be the empty set.

When looking at the indices in $[i] \cup \overline{[i]}$ in ascending order, indices from $[i]$ and indices from $\overline{[i]}$ always alternate. This is because if $w_{i,j} = 1$, then there must be some $k \in \{i+1, \ldots, j-1\}$ such that w_k cancels with w_j by free reductions. In particular, $w_k = \overline{w}_j$ and $w_{k,j-1} = 1$. Thus, $k = \overline{[j]}$.

Therefore, we have $\left| |[i]| - \left|\overline{[i]}\right| \right| \leq 1$ for all i. Moreover, if two equivalence classes $[i]$ and $\overline{[i]}$ have the same number of members, then all corresponding letters w_j for $j \in [i] \cup \overline{[i]}$ can be canceled by free reductions. On the other hand, if $|[i]| - \left|\overline{[i]}\right| = 1$, then after any sequence of free reductions, there remains still one letter w_j for some $j \in [i] \cup \overline{[i]}$ which cannot be canceled. This is because a letter w_i can only cancel with a letter w_j if $w_i = \overline{w}_j$ and $w_{i,j-1} =_{F_\Lambda} 1$ (resp. $w_{j,i-1} =_{F_\Lambda} 1$) – with other words, w_i can only cancel with letters w_j for $j \in \overline{[i]}$.

Thus, for each i with $|[i]| - \left|\overline{[i]}\right| = 1$, denote by $j_{[i]}$ the maximal index in $[i]$. Now, the freely reduced word $\hat{w}$ consists of exactly those w_j with $j = j_{[i]}$ for some i. All other letters are deleted. Apart from the computation of $\approx_{\mathcal{F}}$ and $\bar{\cdot}$, everything can be done in uTC^0 (with the same arguments as steps (ii) (b) and (ii) (c) of Algorithm 18); hence, the whole procedure is in $\mathsf{uAC}^0(F_2)$. □

COROLLARY 21. *The following problems are* uAC^0*-reducible to the word problem of the free group* F_2*:*

(i) *given a word* $w \in \Delta^*$*, decide whether* w *is a* $\mathcal{G}$*-factorization and, if so, compute a Britton-reduced* $\mathcal{G}$*-factorization* $\hat{w}$ *with* $\hat{w} =_{F(\mathcal{G})} w$*.*

(ii) *given a word* $w \in \Delta^*$*, compute a Britton-reduced* $\mathcal{G}$*-factorization* $\hat{w}$ *with* $\hat{w} =_{\pi_1(\mathcal{G},T)} w$*.*

Moreover, the number of bits required for $\hat{w}$ *is linear in the number of bits of* w*.*

PROOF. As in the proof of Theorem 19 it can be checked in uAC^0 whether w is a $\mathcal{G}$-factorization (resp. a $\mathcal{G}$-factorization can be computed from w via the isomorphism $\pi_1(\mathcal{G},T) \to \pi_1(\mathcal{G},a)$ in uTC^0). Thus, we can assume that w is a $\mathcal{G}$-factorization.

We can compute $\mathcal{C}(w) \in \Lambda_w^*$ by step (ii) of Algorithm 18 in uTC^0 (or more precisely, a proper encoding of $\mathcal{C}(w)$ over an alphabet of fixed size). By Proposition 20, a Britton-reduced word $\widehat{\mathcal{C}(w)} = [i_1] \cdots [i_j] \in \Lambda_w^*$ can be computed in $\mathsf{uAC}^0(F_2)$. By Corollary 17, the desired output is $\hat{w} = a_0^{k_{0,i_1-1}} y_{i_1} a_{i_1}^{k_{i_1,i_2-1}} \cdots y_{i_j} a_{i_j}^{k_{i_j,n}}$; as before, it can be computed from $[i_1] \cdots [i_j]$ in uTC^0. According to (1), the number of bits of $k_{i,j}$ is linear in $j - i + \max\{\log |k_\nu| \mid \nu \in \{i, \ldots, j\}\}$. Thus, the number of bits of $\hat{w}$ is linear in the number of bits of w. □

3.2. Uniform versions of the word problem. It is not obvious what the uniform version of the word problem of GBS groups is (i. e., a version of the word problem where the group is part of the input). Indeed, there are different ways how to define a uniform version of the word problem – and they lead to slightly different complexity bounds. We consider a uniform version of Theorem 19 (i) and a uniform version of Theorem 19 (ii).

In the uniform versions we assume that the graph of groups is given in a proper encoding. For instance we assume that the encoding consists of the numbers $|V(Y)|$ and $|E(Y)|$ and a list of tuples $(y, \iota(y), \tau(y), \alpha_y, \beta_y, \overline{y})$ for the edges. Here $y, \overline{y} \in \{0, \ldots, |E(Y)| - 1\}$ and $\iota(y), \tau(y) \in \{0, \ldots, |V(Y)| - 1\}$ and all numbers (also the α_y, β_y) are encoded as binary integers using the same number of bits for all y. The graph of groups also defines the alphabet $\Delta = E(Y) \cup \{a^k \mid a \in V(Y),\, k \in \mathbb{Z}\}$. Recall that the integer exponents k are represented in binary using a variable number of bits.

We say an encoding is *valid*, if all tuples are properly formed, for every edge y, there is an inverse edge $\overline{y}$ satisfying $\iota(y) = \tau(\overline{y})$ and $\alpha_y = \beta_{\overline{y}}$, and the graph is connected.

COROLLARY 22. *The following problem is* uTC^0*-many-one-reducible to the word problem of* F_2. *Input: a valid encoding of a graph of groups* $\mathcal{G}$ *and a word* $w \in \Delta^*$. *Decide whether* w *is a* $\mathcal{G}$*-factorization and, if so, decide whether* $w =_{F(\mathcal{G})} 1$.

Note that we need the promise in Corollary 22 that the input is a valid encoding of a graph of groups. Indeed, it cannot be checked whether the graph is connected in uTC^0 unless $\mathsf{uTC}^0 = \mathsf{LOGSPACE}$ (by [**14**], already connectivity for forests is LOGSPACE-complete with respect to NC^1-reductions – the reduction is actually a uAC^0 reduction[1], see also [**30**]).

On the other hand, by the seminal paper by Reingold [**44**], connectivity of undirected graphs can be checked in LOGSPACE. Hence, as the other points can be easily verified in uAC^0, it can be checked in LOGSPACE whether an encoding of a graph of groups is valid.

PROOF OF COROLLARY 22. We need to verify two things, namely, that for some word $w \in \Delta^*$ it can be checked in uTC^0 whether it is a $\mathcal{G}$-factorization and, second, that Algorithm 18 is still in uTC^0 also if the graph is part of the input.

For the first point, it only needs to be checked whether w is of the form $a_0^{k_0} y_1 a_1^{k_1} \cdots y_n a_n^{k_n}$ and, if so, whether $a_0 = a_n$ and $\iota(y_i) = a_i$ and $\tau(y_i) = a_{i+1}$ for all i. This can be done in uAC^0.

Algorithm 18 is almost independent of the graph. Indeed, there is only the lookup of the numbers α_y, β_y, the check whether $y = \overline{x}$ for edges x, y, and the choice of the orientation D in order to compute the homomorphism ρ. The first two points are straightforward and also the last point is no difficulty: for a pair $y, \overline{y}$, simply choose the one with smaller index in the list coding the graph to be in D. □

The uniform version of Theorem 19 (ii) is not so immediate. The difficulty lies in the computation of the paths $T[a, b]$. This problem is complete for LOGSPACE under NC^1 reductions [**14**] (and indeed under uAC^0 reductions as remarked above). Thus, together with the computation of the isomorphism $\pi_1(\mathcal{G}, T) \to \pi_1(\mathcal{G}, a)$, the algorithm of Theorem 19 (ii) is no longer a uTC^0 many-one reduction (or at least it is not known whether it is). Still, we can prove the following result, which together with Corollary 22 yields the proof for the first part of Theorem C.

COROLLARY 23. *The following problem is complete for* LOGSPACE *under* uAC^0 *reductions: Given a (valid) encoding of graph of groups* $\mathcal{G}$ *with a spanning tree* T *(given as list of edges) and a word* $w \in \Delta^*$. *Decide whether* $w =_{\pi_1(\mathcal{G},T)} 1$.

[1]For a problem $\mathcal{P}$ in LOGSPACE, the reduction computes the full configuration graph of the LOGSPACE Turing machine for $\mathcal{P}$. By using a time-stamp and an additional sink vertex, every vertex except the sink vertex and the accepting configuration can be made to have out-degree exactly one. Now, for every edge an inverse edge can be introduced without changing connectivity properties – the resulting graph has precisely two connected components and in each component one vertex is known.

Note that the question whether $w =_{\pi_1(\mathcal{G},T)} 1$ depends on the spanning tree T. For instance assume that $w = y_1$ consists of a single edge. Then we have $w =_{\pi_1(\mathcal{G},T)} 1$ if and only if y_1 is part of the spanning tree. Therefore, we require T to be part of the input, although by [**43**] (together with [**44**]), for a given graph a spanning tree can be computed in LOGSPACE.

PROOF. Since apart from the computation of the isomorphism $\pi_1(\mathcal{G},T) \to \pi_1(\mathcal{G},a)$ (which is in LOGSPACE by [**14**]), we are in the same situation as in Corollary 22, it remains to prove the hardness part.

We reduce the following special version of *Undirected Forest Accessibility* (see [**14**]) to our problem. The problem receives an undirected forest (i. e., an acyclic graph) Γ with precisely two connected components and three vertices $s, t, u \in V(\Gamma)$ as input such that t and u are in two different connected components – we may assume that the graph is given as list of tuples $(y, \iota(y), \tau(y), \overline{y})$ representing edges where each tuple has the same bit-length. The question is whether s and t are connected by a path.

In order to obtain an instance for the uniform word problem of GBS groups $(\mathcal{G}, T, w)$, we take the input forest Γ and assign to every edge y the numbers $\alpha_y = \beta_y = 1$. That means each tuple $(y, \iota(y), \tau(y), \overline{y})$ has to be replaced by $(y, \iota(y), \tau(y), 1, 1, \overline{y})$. As we assumed all tuples to have the same bit-length, this can be done hard-wired in the circuit. Finally, we create a new edge y_{tu} connecting t and u with $\alpha_{y_{tu}} = \beta_{y_{tu}} = 2$. The spanning tree T consists of all edges. The input word is $w = st^{-1}$.

Now, $\pi_1(\mathcal{G},T)$ is isomorphic to the amalgamated product $\langle t\rangle *_{t^2=u^2} \langle u\rangle$ and we have either $s =_{\pi_1(\mathcal{G},T)} t$ or $s =_{\pi_1(\mathcal{G},T)} u$ – depending on the connected component of Γ in which s lies. In particular, $s =_{\pi_1(\mathcal{G},T)} t$ if and only if s and t are connected by some path in Γ. □

4. The Conjugacy Problem

Decidability of conjugacy in Baumslag-Solitar groups was established by Anshel and Stebe [**5**]. In [**1**] this was generalized to the special case of GBS groups where the graph Y consists of only one vertex (i. e., an HNN extension with several stable letters). Later in [**26**], Horadam showed that the conjugacy problem is decidable in GBS groups if there is some constant $c \in \mathbb{Z}$ with $\alpha_y = c$ for all $y \in E(Y)$. In [**27**], this was further generalized to some other class of GBS groups which contains the linear GBS groups (with the generalization that they also considered infinite graphs); in [**36**], Lockhart gave a solution for all GBS groups. Finally, in [**8**], Beeker independently gave a solution of the conjugacy problem in all GBS groups.

Before we start with the solution of the conjugacy problem in GBS groups, we recall some general facts about conjugacy in fundamental groups of graphs of groups.

4.1. Conjugacy and Graphs of Groups.

Let $\mathcal{G}$ again be an arbitrary graph of groups with graph Y and $a \in V(Y)$.

LEMMA 24. *Let $g, h \in \pi_1(\mathcal{G},a) \leq F(\mathcal{G})$. If $g \sim_{F(\mathcal{G})} h$, then already $g \sim_{\pi_1(\mathcal{G},a)} h$.*

PROOF. Let $\varphi : F(\mathcal{G}) \to \pi_1(\mathcal{G},T)$ be the projection and $\psi : \pi_1(\mathcal{G},T) \to \pi_1(\mathcal{G},a)$ be the canonical isomorphism. If $z \in F(\mathcal{G})$ is a conjugator, then $\psi(\varphi(z)) \in \pi_1(\mathcal{G},a)$ is also a conjugator. □

By Lemma 24, instead of testing conjugacy in the fundamental group $\pi_1(\mathcal{G}, a)$, we can test it in the larger group $F(\mathcal{G})$. This simplifies the algorithms substantially because for $\mathcal{G}$-factorizations in $F(\mathcal{G})$ there is good notion of cyclically Britton-reduced elements.

Let $w = g_0 y_1 g_1 \cdots y_n g_n \in \Pi(\mathcal{G})$. We say that v is a *cyclic permutation* of w if there are $u, u' \in \Delta^*$ such that $w = uu'$ and $v = u'u$. A word $w \in \Delta^*$ is called *cyclically Britton-reduced* if every cyclic permutation of w is Britton-reduced. That means w is cyclically Britton-reduced if and only if ww is Britton-reduced or $w \in G_a$ for some $a \in V(Y)$. The following lemma provides a tool to compute cyclically Britton-reduced $\mathcal{G}$-factorizations.

LEMMA 25. *Let $w = g_0 y_1 g_1 \cdots y_n g_n \in \Pi(\mathcal{G})$ with $n \geq 1$ be Britton-reduced. Then for*

$$y_{\lfloor n/2+1 \rfloor} g_{\lfloor n/2+1 \rfloor} \cdots y_n g_n g_0 y_1 g_1 \cdots y_{\lfloor n/2 \rfloor} g_{\lfloor n/2 \rfloor} \underset{B_\mathcal{G}}{\overset{*}{\Longrightarrow}} \hat{w},$$

if $\hat{w}$ is Britton-reduced, then $\hat{w}$ is cyclically Britton-reduced and $w \sim \hat{w}$.

PROOF. It is clear that $w \sim \hat{w}$. If $\hat{w}$ does not contain any $y \in E(Y)$, we are done. In the other case, we have to show that $\hat{w}\hat{w}$ is Britton-reduced. When computing $\hat{w}$, Britton reductions may only occur in the middle; thus, we know that $y_{\lfloor n/2+1 \rfloor}$ is still present in $\hat{w}$. If $\hat{w}\hat{w}$ is not Britton-reduced, then the occurrence of $y_{\lfloor n/2+1 \rfloor}$ in the second factor $\hat{w}$ must cancel with something in the first factor. This can be either $y_{\lfloor n/2+1 \rfloor}$ or $y_{\lfloor n/2 \rfloor}$ depending on whether $y_{\lfloor n/2 \rfloor}$ has been canceled when computing $\hat{w}$. However, the first case would mean that $y_{\lfloor n/2+1 \rfloor}$ is self-inverse; the second case is a contradiction to the assumption that w was Britton-reduced. □

Let $\mathcal{C}$ denote the union of all G_y^y. The following result is due to Horadam [**25**]; it is the main tool for deciding the conjugacy problem. For amalgamated products, it first appeared in [**38**]; the special case for HNN extensions is known as *Collins' Lemma* [**13**] – see also [**37**, Thm. IV.2.5].

THEOREM 26 (Conjugacy Criterion, [**25**]). *Let $w \in \Pi(\mathcal{G})$ be cyclically Britton-reduced. Then one of the following cases holds:*

- (i) *There is some $a \in V(Y)$ with $w \in G_a$ (w is called* elliptic*).*
  - (a) *If $w \sim_{F(\mathcal{G})} c$ for some $c \in \mathcal{C}$, then there exists a sequence of elements $c = c_0, c_1, \ldots, c_m \in \mathcal{C}$ such that $c_m \sim_{G_a} w$ and for every i there is some $b_i \in \Delta$ with $c_i = b_i c_{i-1} \bar{b}_i$.*
  - (b) *If w is not conjugate to any $c \in \mathcal{C}$ and $w \sim_{F_\mathcal{G}} v$ for some cyclically Britton-reduced v, then $v \in G_a$ and $v \sim_{G_a} w$.*
- (ii) *We have $w \notin G_a$ for any $a \in V(Y)$ (w is called* hyperbolic*), i. e., w has the form $w = y_1 g_1 \cdots y_n g_n h'_{s-1} t^{\eta_s}$ with $n \geq 1$. If w is conjugate to a cyclically Britton-reduced $\mathcal{G}$-factorization $v = x_1 h_1 \cdots x_m h_m$, then $m = n$ and there are $i \in \{1, \ldots, n\}$ and $c \in G_{y_i}^{y_i} \subseteq \mathcal{C}$ such that*

$$v =_{F(\mathcal{G})} c\, y_i g_i \cdots y_n g_n\, y_1 g_1 \cdots y_{i-1} g_{i-1}\, c^{-1},$$

  i. e., w can be transformed into v by a cyclic permutation followed by a conjugation with an element of $\mathcal{C}$.

4.2. Conjugacy in GBS groups. The input for the conjugacy problem are two words $v, w \in \Delta^*$. As we have seen in the proof of Theorem 19, we may assume that v and w are either words representing group elements of the fundamental groups with respect to some spanning tree $\pi_1(\mathcal{G}, T)$ or $\mathcal{G}$-factorizations of elements of $\pi_1(\mathcal{G}, a)$. In view of Theorem 26, a first step towards the solution of the conjugacy problem is the computation of cyclically Britton-reduced $\mathcal{G}$-factorizations. By Corollary 21 we can compute Britton-reduced $\mathcal{G}$-factorizations in $\mathsf{uAC}^0(F_2)$. Thus, by Lemma 25, also cyclically Britton-reduced $\mathcal{G}$-factorizations can be computed in $\mathsf{uAC}^0(F_2)$.

Before we start to examine conjugacy, we need a technical lemma:

LEMMA 27. *Let $\mathcal{P}$ be some fixed finite set of prime numbers. The following problem is solvable in uTC^0: Given $c_i, d_i \in \mathbb{Z}$ (in binary) for $i = 0, \dots, n$ such that d_i has only prime factors in $\mathcal{P}$. Decide whether the system of congruences*

$$x \equiv c_i \mod d_i \qquad \text{for } i = 0, \dots, n$$

has a solution.

PROOF. Since $\mathcal{P}$ is finite, the following can be done for all $p \in \mathcal{P}$ in parallel. Considering only powers of p, the system of congruences transforms into

$$x \equiv c_i \mod p^{e_i} \qquad \text{for } i = 0, \dots, n \tag{2}$$

where e_i is maximal such that p^{e_i} divides d_i. Such e_i can be determined in uTC^0 by checking whether p^e divides d_i for all $0 \le e \le \log |d_i|$ in parallel using Theorem 2 for INTEGER DIVISION. If there is some $i \neq j$ with $e_i \le e_j$ and $c_i \not\equiv c_j \mod p^{e_i}$, then (2) obviously does not have a solution. Again this can be checked in parallel for all pairs i, j. If there is no such pair $i \neq j$, (2) is equivalent to a single congruence $x \equiv c \mod p^e$ where $e = \max_{i \in \{0,\dots,n\}} e_i$ and $c = c_i$ for the respective i.

If (2) has a solution for all $p \in \mathcal{P}$, then there is a solution for the original congruence by the Chinese Remainder Theorem. □

PROPOSITION 28. *The following problem is in uTC^0: Given two cyclically Britton-reduced hyperbolic $\mathcal{G}$-factorizations $v, w \in \Pi(\mathcal{G})$ in binary representation, decide whether $v \sim_{F(\mathcal{G})} w$.*

PROOF. By assumption, we are in case (ii) of the Conjugacy Criterion, Theorem 26. Let

$$v = y_1 a_1^{k_1} \cdots y_n a_n^{k_n}$$

be a $\mathcal{G}$-factorization. By Theorem 26 (ii) we know that if v and w are conjugate, then the underlying path $y_1 \cdots y_n$ of v is a cyclic permutation of the underlying path of w. Since in uTC^0 all these cyclic permutation can be checked in parallel, we may assume that w is of the form

$$w = y_1 a_1^{\ell_1} \cdots y_n a_n^{\ell_n}.$$

and (also by Theorem 26 (ii))

$$v \sim_{F(\mathcal{G})} w \iff \exists x \in \mathbb{Z} \text{ such that } a^x v a^{-x} =_{F(\mathcal{G})} w.$$

By Lemma 8, we have $a^x v a^{-x} =_{F(\mathcal{G})} w$ if and only if there are $x_1, \ldots, x_n \in \mathbb{Z}$ such that

$$\begin{aligned} x - \alpha_1 x_1 &= 0, \\ \beta_i x_i + k_i - \alpha_{i+1} x_{i+1} &= \ell_i && \text{for } i = 1, \ldots, n-1, \\ \beta_n x_n + k_n - x &= \ell_n. \end{aligned}$$

Like in [**26**], these equations imply that it is decidable whether v and w are conjugate. As we aim for a good complexity bound, we have to take a closer look. By solving these equations for x_{i+1}, we obtain

$$\begin{aligned} x_1 &= \frac{x}{\alpha_1}, \\ x_{i+1} &= \frac{k_i - \ell_i + \beta_i x_i}{\alpha_{i+1}} && \text{for } i = 1, \ldots, n-1, \\ x &= k_n - \ell_n + \beta_n x_n. \end{aligned}$$

By induction follows

$$x_i = \frac{1}{\alpha_i} \left(x \cdot \prod_{\mu=1}^{i-1} \frac{\beta_\mu}{\alpha_\mu} + \sum_{\nu=1}^{i-1} (k_\nu - \ell_\nu) \cdot \prod_{\mu=\nu+1}^{i-1} \frac{\beta_\mu}{\alpha_\mu} \right) \qquad \text{for } i = 1, \ldots, n, \tag{3}$$

and the last equation becomes

$$x = k_n - \ell_n + x \cdot \prod_{\mu=1}^{n} \frac{\beta_\mu}{\alpha_\mu} + \sum_{\nu=1}^{n-1} (k_\nu - \ell_\nu) \cdot \prod_{\mu=\nu+1}^{n} \frac{\beta_\mu}{\alpha_\mu}. \tag{4}$$

We distinguish two cases:

First, assume that (4) has a unique solution. Then, the rational values x_i are also determined uniquely and we have $v \sim_{F(\mathcal{G})} w$ if and only if x and the x_i are all integers. In this case, we have $\prod_{\mu=1}^{n} \frac{\beta_\mu}{\alpha_\mu} \neq 1$ and

$$x = \frac{\sum_{\nu=1}^{n} (k_\nu - \ell_\nu) \cdot \prod_{\mu=\nu+1}^{n} \frac{\beta_\mu}{\alpha_\mu}}{1 - \prod_{\mu=1}^{n} \frac{\beta_\mu}{\alpha_\mu}}.$$

All occurring numbers are rationals; hence, they can be represented as fractions of binary integers. Since Iterated Multiplication is in uTC^0 (Theorem 2), the products can be computed. A common denominator for the sums can be computed by Iterated Multiplication, again. Thus, calculating the sum is just Iterated Addition (Theorem 2). Let $c, d, e, f \in \mathbb{Z}$ be such that $\frac{c}{d} = \sum_{\nu=1}^{n} (k_\nu - \ell_\nu) \cdot \prod_{\mu=\nu+1}^{n} \frac{\beta_\mu}{\alpha_\mu}$ and $\frac{e}{f} = 1 - \prod_{\mu=1}^{n} \frac{\beta_\mu}{\alpha_\mu}$. In case $x = \frac{cf}{de}$ is an integer, we can determine this by applying Hesse's circuit for Integer Division (Theorem 2) to cf and de. If x is not an integer, we can notice that by multiplying the result of the division with de; if the result is not cf, there is no x with $a^x v a^{-x} =_{F(\mathcal{G})} w$.

If x is an integer, the numbers x_i can be computed in uTC^0 with the same technique, and it can be checked whether $x_i \in \mathbb{Z}$ for all i. Thus, we are done with the case that (4) has a unique solution.

In the second case, we have $\prod_{\mu=1}^{n} \frac{\beta_\mu}{\alpha_\mu} = 1$. Then (4) is equivalent to

$$k_n - \ell_n + \sum_{\nu=1}^{n-1} (k_\nu - \ell_\nu) \cdot \prod_{\mu=\nu+1}^{n} \frac{\beta_\mu}{\alpha_\mu} = 0.$$

Again, this equality can be checked in uTC^0 as before. If the equality does not hold, then there is no x with $a^x v a^{-x} =_{F(\mathcal{G})} w$. Otherwise, by (3), we have $a^x v a^{-x} =_{F(\mathcal{G})} w$ for $x \in \mathbb{Z}$ if and only if

$$x_i = \frac{1}{\alpha_i}\left(x \cdot \prod_{\mu=1}^{i-1} \frac{\beta_\mu}{\alpha_\mu} + \sum_{\nu=1}^{i-1}(k_\nu - \ell_\nu) \cdot \prod_{\mu=\nu+1}^{i-1} \frac{\beta_\mu}{\alpha_\mu}\right) \in \mathbb{Z} \quad \text{for all } i \in \{1, \ldots, n\}.$$

By solving for x, we obtain

$$x \in \mathbb{Z} \cap \bigcap_{i=1}^{n} \left(\alpha_i \cdot \prod_{\mu=1}^{i-1} \frac{\alpha_\mu}{\beta_\mu}\right) \cdot \left(-\frac{1}{\alpha_i} \sum_{\nu=1}^{i-1}(k_\nu - \ell_\nu) \cdot \prod_{\mu=\nu+1}^{i-1} \frac{\beta_\mu}{\alpha_\mu} + \mathbb{Z}\right). \tag{5}$$

Let $M \in \mathbb{Z}$ be the product of the denominators of all terms in this intersection and $c_i, d_i \in \mathbb{Z}$ such that

$$\frac{c_i}{M} = -\left(\prod_{\mu=1}^{i-1} \frac{\alpha_\mu}{\beta_\mu}\right) \cdot \sum_{\nu=1}^{i-1}(k_\nu - \ell_\nu) \cdot \prod_{\mu=\nu+1}^{i-1} \frac{\beta_\mu}{\alpha_\mu} \qquad \text{and}$$

$$\frac{d_i}{M} = \alpha_i \cdot \prod_{\mu=1}^{i-1} \frac{\alpha_\mu}{\beta_\mu} \qquad \text{for } i = 1, \ldots, n.$$

In addition, we set $c_0 = 0$ and $d_0 = M$. Now, (5) is equivalent to

$$x \in \frac{1}{M}(c_i + d_i\mathbb{Z}) \qquad \text{for } i = 0, \ldots, n.$$

We substitute Mx by z. Because of the choice of c_0 and d_0, the existence of an integer solution x is equivalent to the system of congruences

$$z \equiv c_i \mod d_i \qquad \text{for } i = 0, \ldots, n \tag{6}$$

having a solution. Let $\mathcal{P}$ be the finite set of prime divisors of the α_y and β_y for $y \in E(Y)$. As M as well as the d_is are products of the α_y and β_y, they have only prime factors in $\mathcal{P}$. Furthermore, as before, the numbers c_i, d_i and M can be computed in uTC^0. By Lemma 27, it can be checked in uTC^0 whether (6) has a solution. □

Before we examine the conjugacy problem for elliptic elements in GBS group, we consider the special case of Baumslag-Solitar groups $\mathbf{BS}_{p,q}$, where the solution is straightforward.

PROPOSITION 29. *The following problem is in* uTC^0*: Given* $v = a^k$ *and* $w = a^\ell$ *with* $k, \ell \in \mathbb{Z}$ *given in binary, decide whether* $v \sim_{\mathbf{BS}_{p,q}} w$.

PROOF. We are in case (i)a or (i)b of the Conjugacy Criterion, Theorem 26. Since a conjugation with a has no effect and a conjugation with $y^{\pm 1}$ multiplies the exponent by $\frac{p}{q}$ resp. $\frac{q}{p}$, we have

$$a^k \sim a^\ell \iff \exists j \in \mathbb{Z} \text{ such that } k \cdot \left(\frac{q}{p}\right)^j = \ell \text{ and } \begin{cases} k \in p\mathbb{Z},\ \ell \in q\mathbb{Z}, & \text{if } j > 0, \\ k \in q\mathbb{Z},\ \ell \in p\mathbb{Z}, & \text{if } j < 0. \end{cases}$$

Since we have $|j| \leq \log_{|q/p|} \max\{|k|, |\ell|\}$ if such j exists, only polynomially many (in the input size) values for j need to be tested, what can be done in parallel. As ITERATED MULTIPLICATION and INTEGER DIVISION are in uTC^0 ([**23, 24**], see Theorem 2), we have concluded the proof of Proposition 29. □

Thus, for ordinary Baumslag-Solitar groups, we have solved the conjugacy problem completely by combining Corollary 21 with Proposition 28 and Proposition 29.

For arbitrary GBS groups, it remains to examine elliptic elements (cases (i)a and (i)b of Theorem 26). We follow the ideas of Anshel [**1–3**] in order to describe a $\mathsf{uAC}^0(F_2)$ solution to the conjugacy problem in this case.

Let $v = a^k$, $w = b^\ell$ for some $a, b \in V(Y)$, $k, \ell \in \mathbb{Z}$. By Theorem 26, we know that $v \sim_{F(\mathcal{G})} w$ if and only if there is some $z = a_0^{k_0} y_1 a_1^{k_1} \cdots y_n a_n^{k_n} \in \Pi(\mathcal{G}, b, a)$ such that $z a^k z^{-1} =_{F(\mathcal{G})} w$ and

$$y_i a_i^{k_i} \cdots y_n\, a_n^{k_n} \cdot a^k \cdot a_n^{-k_n} \overline{y}_n \cdots a_i^{-k_i} \overline{y}_i \in G_{y_i}^{y_i} \qquad \text{for all } i.$$

Since a conjugation with a_i has no effect on elements of $G_{a_i} = \langle a_i \rangle$, we may assume that $z = y_1 \cdots y_n$ if v and w are conjugate.

Let $\mathcal{P} = \{p_1, \ldots, p_m\}$ as before be the set of prime divisors occurring in the α_y for $y \in E(Y)$. Here and in what follows, we treat -1 as a prime number. Let

$$(7) \qquad k = r_k \cdot \prod_{i=1}^m p_i^{e_i(k)}, \qquad\qquad \ell = r_\ell \cdot \prod_{i=1}^m p_i^{e_i(\ell)},$$

such that $r_k, r_\ell > 0$ are not divisible by any $p \in \mathcal{P} \setminus \{-1\}$. The numbers r_k, r_ℓ and the exponents $e_i(k), e_i(\ell)$ can be computed in uTC^0 as before by checking for all $p \in \mathcal{P}$ and $e \leq \log|k|$ (or, more precisely, for all e at most the number of bits used to represent k) in parallel whether p^e divides k using Hesse's uTC^0 circuit for Integer Division, Theorem 2 (for $p = -1$, it has to be checked whether $k > 0$) – and likewise for ℓ. If $v \sim_{F(\mathcal{G})} w$, then $r_k = r_\ell$. Hence, all the information it remains to consider is given by the the vectors $(e_1(k), \ldots, e_m(k)), (e_1(\ell), \ldots, e_m(\ell)) \in \mathbb{N}^m$ and the vertices $a, b \in V(Y)$. In order to code also the vertices as vectors, we consider vectors in $\mathbb{N}^m \times \mathbb{N}^{V(Y)}$ where a vertex a is encoded by the unit vector $\vec{u}_a \in \mathbb{N}^{V(Y)}$ (which has a 1 at position a and 0 otherwise).

Let us define an equivalence relation on $\mathbb{N}^m \times \mathbb{N}^{V(Y)}$ which reflects conjugacy in $F(\mathcal{G})$. For $\vec{e} = (e_1, \ldots, e_m, \vec{u}_a)$, $\vec{f} = (f_1, \ldots, f_m, \vec{u}_b) \in \mathbb{N}^m \times \mathbb{N}^{V(Y)}$ with arbitrary $(e_1, \ldots, e_m)$, $(f_1, \ldots, f_m) \in \mathbb{N}^m$ and $a, b \in V(Y)$, we define $\vec{e} \sim \vec{f}$ if

$$a^{\prod_{i=1}^m p_i^{e_i}} \sim_{F(\mathcal{G})} b^{\prod_{i=1}^m p_i^{f_i}};$$

for $\vec{e} = (e_1, \ldots, e_m, \vec{e}\,'), \vec{f} = (f_1, \ldots, f_m, \vec{f}\,') \in \mathbb{N}^m \times \mathbb{N}^{V(Y)}$ with $\vec{e}\,'$ and $\vec{f}\,'$ not being zero nor a unit vector, we define $\vec{e} \sim \vec{f}$ regardless what the e_i, f_i are. As an immediate consequence of this definition, we have

Lemma 30. *Let $a, b \in V(Y)$, $k, \ell \in \mathbb{Z}$. Then $a^k \sim_{F(\mathcal{G})} b^\ell$ if and only if*

$$r_k = r_\ell \text{ and } (e_1(k), \ldots, e_m(k), \vec{u}_a) \sim (e_1(\ell), \ldots, e_m(\ell), \vec{u}_b).$$

The numbers $e_i(k), e_i(\ell)$ of (7) are bounded by a linear function in the input size. In particular, we have a uTC^0-many-one reduction from the question whether $a^k \sim_{F(\mathcal{G})} b^\ell$ to the question whether $(e_1(k), \ldots, e_m(k), \vec{u}_a) \sim (e_1(\ell), \ldots, e_m(\ell), \vec{u}_b)$ where the numbers $e_i(k), e_i(\ell)$ are represented in unary. Thus, we aim for a $\mathsf{uAC}^0(F_2)$ circuit to decide whether $\vec{e} \sim \vec{f}$ for vectors $\vec{e}, \vec{f} \in \mathbb{N}^m \times \mathbb{N}^{V(Y)}$. This can be achieved by using the following crucial observation, which is another immediate consequence of the definition of $\sim$.

Lemma 31. *If $\vec{e} \sim \vec{f}$, then also $\vec{e} + \vec{g} \sim \vec{f} + \vec{g}$ for all $g \in \mathbb{N}^m \times \mathbb{N}^{V(Y)}$. In particular, $\sim$ defines a congruence on $\mathbb{N}^m \times \mathbb{N}^{V(Y)}$.*

Thus, $(\mathbb{N}^m \times \mathbb{N}^{V(Y)})/{\sim}$ is a commutative monoid and it remains to solve the word problem of this monoid. Malcev [**39**] and Emelichev [**19**] showed that the word problem for finitely generated commutative monoids is decidable – even if the congruence is part of the input.

In [**18**, Thm. II], Eilenberg and Schützenberger showed that every congruence on $\mathbb{N}^M$ is a semilinear subset of $\mathbb{N}^M \times \mathbb{N}^M$ (this follows also from the results [**49**], that congruences are definable by Presburger formulas, and [**21**], that Presburger definable sets are semilinear – for definition of all these notions we refer to the respective papers). In [**28**, Thm. 1], Ibarra, Jiang, Chang, and Ravikumar showed that membership in a fixed semilinear set can be decided in uniform NC^1. As the word problem of F_2 is hard for uniform NC^1 under uAC^0 reductions [**45**], this means that for every fixed congruence $\sim\, \subseteq \mathbb{N}^M \times \mathbb{N}^M$, on input of $u, v \in \mathbb{N}^M$, it can be decided in $\mathsf{uAC}^0(F_2)$ whether $u \sim v$.

Thus, by Lemma 30, it can be decided in $\mathsf{uAC}^0(F_2)$ whether $a^k \sim_{F(\mathcal{G})} b^\ell$ for $a, b \in V(Y)$, $k, \ell \in \mathbb{Z}$. Now, we can combine this result with Corollary 21 (calculation of Britton-reduced $\mathcal{G}$-factorizations) and Proposition 28 (solution to conjugacy in the hyperbolic case) and we obtain a proof of the main result on conjugacy, Theorem B.

THEOREM 32. *Let G be a generalized Baumslag-Solitar group. Then the conjugacy problem of G is in $\mathsf{uAC}^0(F_2)$.*

4.3. The Uniform Conjugacy Problem. In Section 3.2, we have seen that the uniform version of the word problem for GBS groups was essentially as difficult as the word problem for a fixed GBS group. For conjugacy this picture changes dramatically. Like for the word problem in Section 3.2, the *uniform conjugacy problem* for GBS groups receives as input a graph of groups $\mathcal{G}$ consisting of a finite graph Y and numbers $\alpha_y, \beta_y \in \mathbb{Z}\setminus\{0\}$ for $y \in E(Y)$ and two $\mathcal{G}$-factorizations $v, w \in \Delta^*$, where as before $\Delta = E(Y) \cup \{a^k \mid a \in V(Y), k \in \mathbb{Z}\}$. The question is whether $v \sim_{F(\mathcal{G})} w$ (what by Lemma 24 is equivalent to conjugacy in the fundamental group with respect to a base point).

In [**4**], Anshel and McAloon considered a special (more difficult) variant of the uniform conjugacy problem; they showed that the so-called finite special equality problem for some GBS groups is decidable but not primitive recursive. However, they did not consider the uniform conjugacy problem. By following the ideas for the non-uniform case (which themselves are based on Anshel's work [**1–3**]), we obtain a precise complexity estimate for the uniform conjugacy problem.

THEOREM 33. *The uniform conjugacy problem for GBS groups is* $\mathsf{EXPSPACE}$*-complete – even if the numbers α_y, β_y are given in unary.*

This concludes the proof of Theorem C. The proof of Theorem 33 is an application of the next theorem by Cardoza, Lipton and Meyer [**11**] resp. Mayr and Meyer [**40**].

THEOREM 34 ([**11, 40**]). *The uniform word problem for finitely presented commutative semigroups is* $\mathsf{EXPSPACE}$*-complete.*

PROOF OF THEOREM 33. For the hardness part, we give a $\mathsf{LOGSPACE}$ reduction from the uniform word problem of f. g. commutative semigroups to the uniform conjugacy problem for GBS groups. W. l. o. g. we only consider commutative monoids. Let $m \in \mathbb{N}$, $e, f \in \mathbb{N}^m$, $(r_i, s_i)_{i \in \{1 \dots n\}}$ with $r_i, s_i \in \mathbb{N}^m$ be some instance for the uniform word problem of commutative monoids (i. e., the question is whether $e \sim f$ for the smallest congruence $\sim$ satisfying $r_i \sim s_i$ for all i).

We construct an instance for the uniform conjugacy problem as follows: The graph Y consists of a single vertex a; for all $i \in \{1, \ldots, n\}$ there is a pair of edges $y_i, \overline{y}_i \in E(Y)$. Let $\mathcal{P} = \{p_1, \ldots, p_m\}$ be the set of the first m prime numbers. The numbers p_j can be computed in LOGSPACE since each of them requires a logarithmic (in m) number of bits, only (by the prime number theorem there are enough primes). Now, for every relator (r_i, s_i), we define $\alpha_{y_i} = \prod_{j=1}^{m} p_j^{(r_i)_j}$ and $\beta_{y_i} = \prod_{j=1}^{m} p_j^{(s_i)_j}$, where $(r_i)_j$ denotes the jth component of the vector r_i, and $k = \prod_{j=1}^{m} p_j^{e_j}$ and $\ell = \prod_{j=1}^{m} p_j^{f_j}$. According to the proof in [**40**], we may assume that all the vectors e, f, r_i and s_i (for all i) have at most four non-zero entries and these non-zero entries are at most 2. Thus, the results k, ℓ, α_{y_i}, and β_{y_i} are bounded polynomially in the input length and they can be written down in unary on the output tape. In particular, the products can be computed in LOGSPACE. Now we have $a^k \sim_{F(\mathcal{G})} a^\ell$ if and only if $e \sim f$.

It remains to show that the uniform conjugacy problem is in EXPSPACE. The two input words for an instance of the uniform conjugacy problem for GBS groups can be cyclically Britton-reduced as in Corollary 21. Note, however, that the linear bound on the size of the cyclically Britton-reduced words does not hold anymore. Still the size remains bounded polynomially.

The algorithm of Proposition 28 can be executed in polynomial time even if the graph of groups is part of the input. This gives a polynomial time bound for hyperbolic elements. However, we do not know a better bound as the proof of Proposition 28 involves a computation of greatest common divisors (or prime factorizations) of the numbers α_y, β_y. For elliptic elements, by Lemma 30, we obtain an instance of the uniform word problem of commutative semigroups, which is in EXPSPACE. □

Acknowledgments. Many thanks goes to my thesis advisor, Volker Diekert, as well as to Jonathan Kausch with whom I had so many inspiring conversations about complexity and Baumslag-Solitar groups. Special thanks to the anonymous referee for many helpful comments and, in particular, for pointing out that the word problem for a fixed commutative monoid actually can be solved in NC^1 – thus, providing the last piece for the solution of the conjugacy problem in $\mathsf{uAC}^0(F_2)$ and not only in LOGSPACE.

References

[1] Michael Anshel, *The conjugacy problem for HNN groups and the word problem for commutative semigroups*, Proc. Amer. Math. Soc. **61** (1976), no. 2, 223–224. MR0422457

[2] Michael Anshel, *Conjugate powers in HNN groups*, Proc. Amer. Math. Soc. **54** (1976), 19–23. MR0393249

[3] Michael Anshel, *Decision problems for HNN groups and vector addition systems*, Math. Comput. **30** (1976), no. 133, 154–156. MR0396766

[4] Michael Anshel and Kenneth McAloon, *Reducibilities among decision problems for HNN groups, vector addition systems and subsystems of Peano arithmetic*, Proc. Amer. Math. Soc. **89** (1983), no. 3, 425–429, DOI 10.2307/2045489. MR715859

[5] Michael Anshel and Peter Stebe, *The solvability of the conjugacy problem for certain HNN groups*, Bull. Amer. Math. Soc. **80** (1974), 266–270. MR0419615

[6] David A. Mix Barrington, Neil Immerman, and Howard Straubing, *On uniformity within* NC^1, J. Comput. System Sci. **41** (1990), no. 3, 274–306, DOI 10.1016/0022-0000(90)90022-D. MR1079468

[7] Gilbert Baumslag and Donald Solitar, *Some two-generator one-relator non-Hopfian groups*, Bull. Amer. Math. Soc. **68** (1962), 199–201. MR0142635

[8] Benjamin Beeker, *Problèmes géométriques et algorithmiques dans des graphes de groupes*, Ph.D. thesis, Université de Caen Basse-Normandie, 2011.

[9] Ronald V. Book and Friedrich Otto, *String-rewriting systems*, Texts and Monographs in Computer Science, Springer-Verlag, New York, 1993. MR1215932

[10] John L. Britton, *The word problem*, Ann. of Math. (2) **77** (1963), 16–32. MR0168633

[11] E. Cardoza, R. Lipton, and A. R. Meyer, *Exponential space complete problems for Petri nets and commutative semigroups: preliminary report*, Eighth Annual ACM Symposium on Theory of Computing (Hershey, Pa., 1976), Assoc. Comput. Mach., New York, 1976, pp. 50–54. MR0445912

[12] Hervé Caussinus, Pierre McKenzie, Denis Thérien, and Heribert Vollmer, *Nondeterministic* NC^1 *computation*, J. Comput. System Sci. **57** (1998), no. 2, 200–212, DOI 10.1006/jcss.1998.1588. Complexity 96—The Eleventh Annual IEEE Conference on Computational Complexity (Philadelphia, PA). MR1661791

[13] Donald J. Collins, *On embedding groups and the conjugacy problem*, J. London Math. Soc. (2) **1** (1969), 674–682. MR0252489

[14] Stephen A. Cook and Pierre McKenzie, *Problems complete for deterministic logarithmic space*, J. Algorithms **8** (1987), no. 3, 385–394, DOI 10.1016/0196-6774(87)90018-6. MR905994

[15] Matthew J. Craven and Henri C. Jimbo, *Evolutionary algorithm solution of the multiple conjugacy search problem in groups, and its applications to cryptography*, Groups Complex. Cryptol. **4** (2012), no. 1, 135–165, DOI 10.1515/gcc-2012-0002. MR2921158

[16] Volker Diekert, Alexei G. Myasnikov, and Armin Weiß, *Conjugacy in Baumslag's group, generic case complexity, and division in power circuits*, LATIN 2014: theoretical informatics, Lecture Notes in Comput. Sci., vol. 8392, Springer, Heidelberg, 2014, pp. 1–12, DOI 10.1007/978-3-642-54423-1_1. MR3188091

[17] Volker Diekert and Armin Weiß, *Context-Free Groups and Bass-Serre Theory*, ArXiv e-prints (2013).

[18] Samuel Eilenberg and M. P. Schützenberger, *Rational sets in commutative monoids*, J. Algebra **13** (1969), 173–191. MR0246985

[19] Vladimir A. Emelichev, *Commutative semigroups with one defining relation*, Shuya Gosudarstvennyi Pedagogicheskii Institut Uchenye Zapiski **6** (1958), 227–242.

[20] Max Forester, *On uniqueness of JSJ decompositions of finitely generated groups*, Comment. Math. Helv. **78** (2003), no. 4, 740–751, DOI 10.1007/s00014-003-0780-y. MR2016693

[21] Seymour Ginsburg and Edwin H. Spanier, *Semigroups, Presburger formulas, and languages*, Pacific J. Math. **16** (1966), 285–296. MR0191770

[22] Dima Grigoriev and Vladimir Shpilrain, *Authentication from matrix conjugation*, Groups Complex. Cryptol. **1** (2009), no. 2, 199–205, DOI 10.1515/GCC.2009.199. MR2598988

[23] William Hesse, *Division is in uniform* TC^0, Automata, languages and programming, Lecture Notes in Comput. Sci., vol. 2076, Springer, Berlin, 2001, pp. 104–114, DOI 10.1007/3-540-48224-5_9. MR2065855

[24] William Hesse, Eric Allender, and David A. Mix Barrington, *Uniform constant-depth threshold circuits for division and iterated multiplication*, J. Comput. System Sci. **65** (2002), no. 4, 695–716, DOI 10.1016/S0022-0000(02)00025-9. Special issue on complexity, 2001 (Chicago, IL). MR1964650

[25] K. J. Horadam, *The word problem and related results for graph product groups*, Proc. Amer. Math. Soc. **82** (1981), no. 2, 157–164, DOI 10.2307/2043301. MR609643

[26] K. J. Horadam, *The conjugacy problem for graph products with central cyclic edge groups*, Proc. Amer. Math. Soc. **91** (1984), no. 3, 345–350, DOI 10.2307/2045299. MR744626

[27] K. J. Horadam and G. E. Farr, *The conjugacy problem for HNN extensions with infinite cyclic associated groups*, Proc. Amer. Math. Soc. **120** (1994), no. 4, 1009–1015, DOI 10.2307/2160210. MR1185267

[28] Oscar H. Ibarra, Tao Jiang, Jik H. Chang, and Bala Ravikumar, *Some classes of languages in* NC^1, Inform. and Comput. **90** (1991), no. 1, 86–106, DOI 10.1016/0890-5401(91)90061-6. MR1088807

[29] Matthias Jantzen, *Confluent string rewriting*, EATCS Monographs on Theoretical Computer Science, vol. 14, Springer-Verlag, Berlin, 1988. MR972260

[30] Birgit Jenner, Klaus-Jörn Lange, and Pierre McKenzie, *Tree isomorphism and some other complete problems for deterministic logspace*, publication #1059, DIRO, Université de Montréal, 1997.

[31] Jonathan Kausch, *Private conversation*, 2013.

[32] P. H. Kropholler, *Baumslag-Solitar groups and some other groups of cohomological dimension two*, Comment. Math. Helv. **65** (1990), no. 4, 547–558, DOI 10.1007/BF02566625. MR1078097

[33] Klaus-Jörn Lange and Pierre McKenzie, *On the complexity of free monoid morphisms*, Algorithms and computation (Taejon, 1998), Lecture Notes in Comput. Sci., vol. 1533, Springer, Berlin, 1998, pp. 247–256, DOI 10.1007/3-540-49381-6_27. MR1733965

[34] Jürn Laun, *Solving algorithmic problems in Baumslag-Solitar groups and their extensions using data compression*, Dissertation, Institut für Formale Methoden der Informatik, Universität Stuttgart, 2012.

[35] Richard J. Lipton and Yechezkel Zalcstein, *Word problems solvable in logspace*, J. Assoc. Comput. Mach. **24** (1977), no. 3, 522–526. MR0445901

[36] Jody Meyer Lockhart, *The conjugacy problem for graph products with infinite cyclic edge groups*, Proc. Amer. Math. Soc. **114** (1992), no. 3, 603–606, DOI 10.2307/2159376. MR1072088

[37] Roger C. Lyndon and Paul E. Schupp, *Combinatorial group theory*, Classics in Mathematics, Springer-Verlag, Berlin, 2001. Reprint of the 1977 edition. MR1812024

[38] Wilhelm Magnus, Abraham Karrass, and Donald Solitar, *Combinatorial group theory*, 2nd ed., Dover Publications, Inc., Mineola, NY, 2004. Presentations of groups in terms of generators and relations. MR2109550

[39] Anatolij I. Malcev, *On homomorphisms of finite groups*, Ivano Gosudarstvennyi Pedagogicheskii Institut Uchenye Zapiski **18** (1958), 49–60.

[40] Ernst W. Mayr and Albert R. Meyer, *The complexity of the word problems for commutative semigroups and polynomial ideals*, Adv. in Math. **46** (1982), no. 3, 305–329, DOI 10.1016/0001-8708(82)90048-2. MR683204

[41] Charles F. Miller III, *On group-theoretic decision problems and their classification*, Princeton University Press, Princeton, N.J.; University of Tokyo Press, Tokyo, 1971. Annals of Mathematics Studies, No. 68. MR0310044

[42] Alexei Myasnikov, Alexander Ushakov, and Dong Wook Won, *The word problem in the Baumslag group with a non-elementary Dehn function is polynomial time decidable*, J. Algebra **345** (2011), 324–342, DOI 10.1016/j.jalgebra.2011.07.024. MR2842068

[43] Noam Nisan and Amnon Ta-Shma, *Symmetric Logspace is closed under complement*, Chicago J. Theoret. Comput. Sci., posted on 1995, Article 1, approx. 11pp. (electronic), DOI 10.4086/cjtcs.1995.001. MR1345937

[44] Omer Reingold, *Undirected connectivity in log-space*, J. ACM **55** (2008), no. 4, Art. 17, 24, DOI 10.1145/1391289.1391291. MR2445014

[45] David Hill Robinson, *Parallel algorithms for group word problems*, ProQuest LLC, Ann Arbor, MI, 1993. Thesis (Ph.D.)–University of California, San Diego. MR2688969

[46] Jean-Pierre Serre, *Arbres, amalgames*, SL_2 (French), Société Mathématique de France, Paris, 1977. Avec un sommaire anglais; Rédigé avec la collaboration de Hyman Bass; Astérisque, No. 46. MR0476875

[47] Vladimir Shpilrain and Gabriel Zapata, *Combinatorial group theory and public key cryptography*, Appl. Algebra Engrg. Comm. Comput. **17** (2006), no. 3-4, 291–302, DOI 10.1007/s00200-006-0006-9. MR2233788

[48] Hans-Ulrich Simon, *Word problems for groups and contextfree recognition*, Fundamentals of computation theory (Proc. Conf. Algebraic, Arith. and Categorical Methods in Comput. Theory, Berlin/Wendisch-Rietz, 1979), Math. Res., vol. 2, Akademie-Verlag, Berlin, 1979, pp. 417–422. MR563704

[49] M. A. Taĭclin, *Algorithmic problems for commutative semigroups* (Russian), Dokl. Akad. Nauk SSSR **178** (1968), 786–789. MR0227293

[50] Heribert Vollmer, *Introduction to circuit complexity*, Texts in Theoretical Computer Science. An EATCS Series, Springer-Verlag, Berlin, 1999. A uniform approach. MR1704235

[51] S. Waack, *Tape complexity of word problems*, Fundamentals of computation theory (Szeged, 1981), Lecture Notes in Comput. Sci., vol. 117, Springer, Berlin-New York, 1981, pp. 467–471. MR653014

[52] Armin Weiß, *On the complexity of conjugacy in amalgamated products and HNN extensions*, Dissertation, Institut für Formale Methoden der Informatik, Universität Stuttgart, 2015.

INSTITUT FÜR FORMALE METHODEN DER INFORMATIK (FMI), UNIVERSITÄT STUTTGART, GERMANY

E-mail address: `armin.weiss@fmi.uni-stuttgart.de`

Contemporary Mathematics
Volume **677**, 2016
http://dx.doi.org/10.1090/conm/677/13629

Cryptographic hash functions from sequences of lifted Paley graphs

Serena Yuan

ABSTRACT. The use of expander graphs to generate pseudo-random behavior is widely known to complexity theorists and cryptographers. Many have proposed constructing collision resistant hash functions from exapnder graphs. We introduce the cryptographic hash function schemes that utilize graph-theoretic tools introduced by Bilu-Linial [**2**] to improve expander properties and optimize the hash functions. As examples, we investigate an elementary as well as efficiently constructable family of optimal expander graphs: the families of Ramanujan graphs named Paley graphs, with vertex set in $GF(q)$ for q a prime or a prime power. Especially for q prime, the number of operations is minimized in constructing the graph and computing hash values.

The first proposal of expander hashes was by Zemor in the early 1990s, a scheme that was a Cayley hash on a directed Cayley graph built from the set of 2×2 matrices over F_p with determinant 1. Three years after the first proposal of an expander hash, Jean-Pierre Tillich and Gilles Zemor proposed a new family of hash functions based on computations over a finite field of characteristic 2, constructing the expander hash using $SL_2(F_{2^n})$ instead of $SL_2(F_p)$. After 15 years, Charles, Goren and Lauter [**3**] have proposed expander hash functions based on the families of Ramanujan graphs constructed by Lubotzky-Phillips-Sarnak and Pizer at the 14th Annual International Cryptography Conference. Charles et al. proposed two constructions of expander hashes based on undirected graphs, contrasting from the directed graphs in Zemor's first proposal as well as the Zemor-Tillich hash function.

There are many applications that use d-regular sparse expanders, from the design of efficient communication networks, explicit error-correcting codes with efficient encoding and decoding schemes, and derandomization of randomized algorithms. The survey of Hoory, Linial, and Wigderson illustrates many applications of expanders [**8**]. Expansion is related to the ratio between d and λ, the second largest eigenvalue in absolute value of the adjacency matrix. Graphs with smaller λ have better graph expansion. From the Alon-Boppana bound [**14**] $\lambda \geq 2\sqrt{d-1} - o(1)$ where $o(1) \to 0$ as the vertices of the graph tend to infinity. Optimal expanders have property $\lambda \leq 2\sqrt{d-1}$ and are called *Ramanujan graphs.*

2010 *Mathematics Subject Classification.* Primary 05Cxx, 05C50, 05C76, 05C81, 05C22, 05E18; Secondary 15A18, 15A42, 20C07, 20F10, 03B25.

Key words and phrases. Hash functions, Expander graphs, Graph theory, Graphs and linear algebra, Graph operations, Random walks on graphs, Signed and weighted graphs, Nearly optimal spectral gap.

©2016 American Mathematical Society

We wish to answer the question of finding ways to efficiently improve the expander properties of graphs while increasing their size to increase randomness. A major step in the direction of finding constructions of good expanders is a construction that involves iterative use of the zig-zag product [**2**]. This tool is able to increase the size of the graph while decreasing the degrees of nodes. Constructions of the zig-zag product generate graphs with second eigenvalue $O(d^{3/4})$ which can be improved with additional analysis to $O(d^{2/3})$. Bilu-Linial's proposed 2-lift tool also increases the graph size while decreasing degrees and will be able to improve the bounds to a close-to-optimal $\lambda = O(\sqrt{dlog^3 d})$ [**2**].

Our Contributions. In this paper we apply a graph-theoretic tool that increases the size of the d-regular graph while keeping its eigenvalues within the original bound to the practice of expander hashes. We propose a cryptographic hash scheme that combines the graph operator called 2-lifts with expander graphs to optimize the randomness of the hash. We illustrate the efficiency of our scheme and the usefulness of 2-lifts for other expander hash functions. We combine ideas from block cipher hash functions, expander hash functions. In summary, we provide an efficient application of 2-lifts to expander graph based hash functions.

Paper Outline: Our paper is organized as follows. In section 2, we describe basic definitions, 2-lift models, and results in spectral theory required to understand the rest of the paper. In section 3, we briefly describe how to construct a hash function from an expander graph and hard problems. In section 4, we describe a family of Ramanujan graphs called Paley graphs and their properties. In section 5, we describe our scheme for lifted Paley graphs. In section 6, we outline the expander properties that are kept by 2-lifts. In section 7, we prove that our scheme is efficient, and in section 8, we provide a short cryptanalysis of our scheme.

1. Preliminaries

1.1. Hash Functions. A hash function maps bit strings of an arbitrary finite length to bit strings of some fixed finite length, and they must be easy to compute. We are concerned with keyed hash functions in the sense that an initialization vector is applied to the expander graph adjacency matrix to transform it into a 2-lifted graph.

1.2. Graph Theory. We refer the reader to [**9**] for some basic definitions in graph theory.

1.3. Cayley Graphs. We define Cayley graphs [**9**].

DEFINITION 1.1. Let G be a group and generating set $S \subseteq G$. The *Cayley graph* that is associated to (G, S) is the directed graph that has one vertex for each group element. We assign to each generator $s \in S$ a color c_s and draw a directed edge of color c_s between any vertices g and gs for $g \in G, s \in S$. The edge set is given by (g, gs).

1.4. Expander Graphs. We will look at undirected graphs and we say a graph is d-regular if each vertex has d edges incident on it. [**20**]

DEFINITION 1.2. A d-regular graph is called a λ-expander, if all its eigenvalues but the largest are in $[-\lambda, \lambda]$.

We may express the graph's expansion property in an algebraic way. The adjacency matrix of an undirected graph is symmetric and this means that all its eigenvalues are real.

An expander graph is a k-regular graph, a graph such that any vertex is incident to k edges. For this graph, the elements of any column of the adjacency matrix of a k-regular graph sum up to k, and any k-regular graph has an eigenvalue equal to k, its largest eigenvalue, which corresponds to the left eigenvector $(1, 1, ..., 1)$.

For a connected graph G we order its eigenvalues starting from its largest k:

$$k > \mu_1 \geq \mu_2 \geq \cdots \geq \mu_{N-1} \tag{1.1}$$

So we have the expansion constant c that is able to be expressed in terms of the eigenvalues:

$$c \geq \frac{2(k - \mu_1)}{3k - 2\mu_1}. \tag{1.2}$$

We may observe that the expansion constant is improved as the eigenvalue μ_1 is smaller.

A family of graphs $\{G_n\}$ is defined to be a (d, α) expander family if the following three conditions hold for all n: [**10**]

(1) $\lim_{n\to\infty}|V(G_n)| = \infty$.
(2) G_n is d-regular.
(3) The bottleneck ratio of a simple random walk on G_n satisfies $\Phi_*(G_n) \geq \alpha$.

Therefore graphs with smaller 2nd eigenvalues μ_1 have better expansion constants. For an infinite family X_n of connected, k-regular graphs with the number of vertices tending to infinity, a theorem of Alon-Boppana states that lim inf $\mu_1(X_n) \geq 2\sqrt{k-1}$. This motivates us to define and make use of Ramanujan graphs, since a family of k-regular Ramanujan graphs is optimal with respect to the size of μ_1. [**3**]

DEFINITION 1.3. A *Ramanujan graph* is a k-regular connected graph such that

$$\mu_1 \leq 2\sqrt{k-1}. \tag{1.3}$$

1.5. 2-lifts. We present the tools that are explicit constructions for expander graphs with a nearly optimal spectral graph [**2**].

DEFINITION 1.4. A *signing* of the edges of a graph $G = (V, E)$ is a function given by $s : E(G) \to \{-1, 1\}$. This gives us a *signed adjacency matrix* where the (x, y) entry is $s(x, y)$ for $(x, y) \in E$ and 0 otherwise.

In our construction scheme, the signing plays the role of an initialization vector/key. A signing is also a real symmetric matrix with entries $0, -1, 1$.

DEFINITION 1.5. The *spectral radius of a signing* is the spectral radius of the corresponding signed adjacency matrix.

DEFINITION 1.6. When the spectral radius of a signing of a d-regular graph is $\tilde{O}(\sqrt{d})$ then we say the signing (or lift) is *Quasi-Ramanujan*.

DEFINITION 1.7. (2-lift construction) The *2-lift* of G associated with a signing s is a graph $\hat{G}$ as follows: we associate with every vertex $x \in V$ two vertices x_0, x_1 called the *fiber* of x. If $(x, y) \in E$ and $s(x, y) = 1$ then the corresponding edges in $\hat{G}$ are given by (x_0, y_0) and (x_1, y_1). On the other hand, if $s(x, y) = -1$ then the corresponding edges in $\hat{G}$ are given by (x_0, y_1) and (x_1, y_0) and the edges are 'crossed' instead of 'parallel' as in the original graph.

We can describe the process as a random choice of edges of the original graph to be left with their original edge connections or to be "crossed," or flipped with another edge, except both these edges connect to the pairs of vertices in the copy of the graph instead of vertices in the original graph. The signing vector determines the random choices, where $+1$ corresponds to keeping the edge the same way and -1 corresponds to flipping the edge.

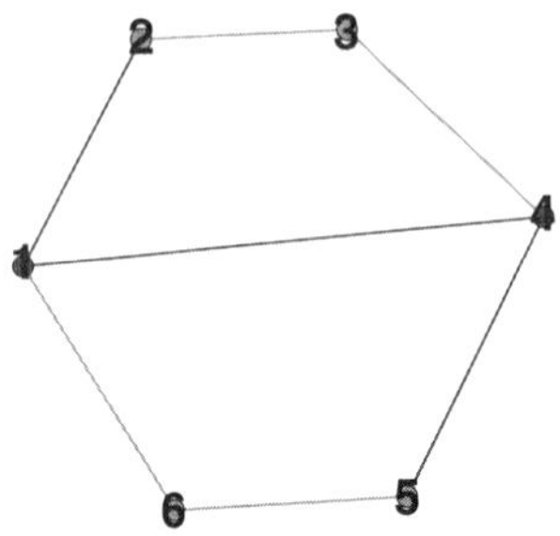

FIGURE 1. Base graph

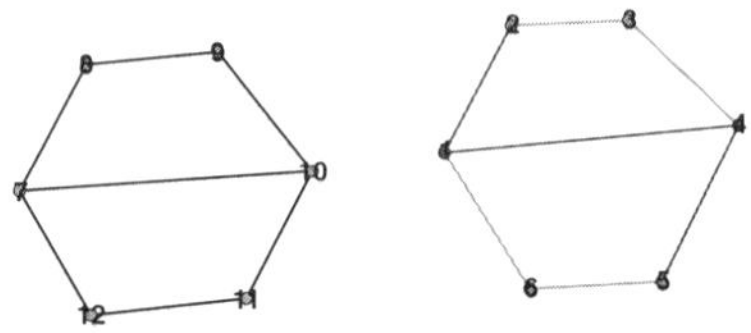

FIGURE 2. Graphs with edges in parallel

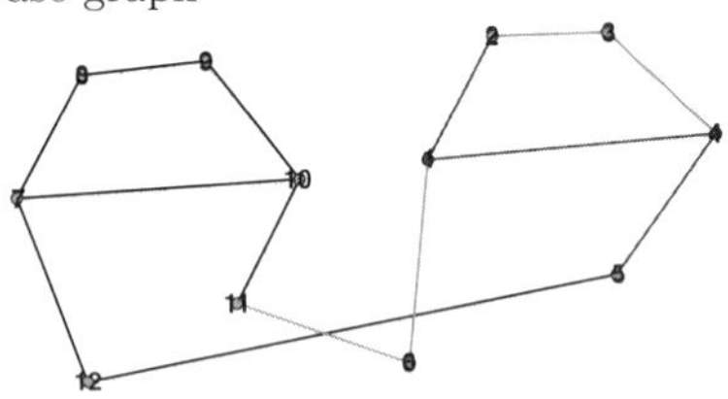

FIGURE 3. Graphs with one pair of edges crossed

In Figure 1, 2, and 3, we see the action of the 2-lift by seeing how it leaves edges in parallel or crossed [**19**]. Figure 2 and Figure 3 are two versions of 2-lifts of the initial graph in Figure 1, with signings $[1, 1, 1, 1, 1, 1, 1]$ and $[1, 1, 1, -1, 1, 1, 1]$, respectively.

We may give a definition based on [**2**] for evaluating how good a signing is.

DEFINITION 1.8. A signing s of a d-regular graph G is (n, d)-good if the spectral radius of A_s is $O(\sqrt{d \log^3 d})$ and $\hat{G}$ is $(\gamma(d), 1 + \log_2 n)$-sparse. A d-regular graph G is an (n, d)-good expander if it is an $(n, d, O(\sqrt{d \log^3 d}))$-expander, and is $(\gamma(d), 1 + \log_2 n)$-sparse.

A proof that finding a good signing is equivalent to finding a signing for which X does not become greater than its expected value is given by [**2**]. It is possible to show that the conclusion holds when we choose the signing from an (ϵ, k)-wise independent sample space instead of choosing the sign of each edge uniformly and independently. Here, we must have $k = d \log_2 n$ and $\epsilon = d^{-2d \log_2 n}$.

DEFINITION 1.9. Let Ω_m denote a sample space of m-bit strings, and let $S = s_1 \cdots s_m$ be chosen uniformly at random from Ω_m. We say that Ω_m is an (ϵ, k)-wise independent sample space if for any $k' \leq k$ and positions $i_1 < i_2 < \cdots < i_{k'}$,

$$\sum_{\alpha \in \{-1,1\}^{k'}} |\Pr s_{i_1}, \cdots s_{i_{k'}} = \alpha - 2^{-k'}| < \epsilon.$$

1.6. Spectral Expansion. We present some of the theory associated to spectral expansion. The Expander Mixing Lemma states that the number of edges between two subsets of vertices in an expander graph is equivalent to the edges of a random graph, up to an additive error depending on the second largest eigenvalue.

LEMMA 1.10. *(Expander Mixing Lemma) Let G be a D-regular, N-vertex digraph with second largest eigenvalue λ and with spectral expansion $1 - \lambda$. Then for all sets of vertices S, T of densities $\alpha = |S|/N$ and $\beta = |T|/N$, we have*

$$| \tfrac{e(S,T)}{ND} - \alpha\beta| \leq \lambda\sqrt{\alpha(1-\alpha)\beta(1-\beta)} \leq \lambda\sqrt{\alpha\beta} \leq \lambda.$$

We may see that the denominator ND counts all edges of the graph, as ordered pairs. The lemma states that the difference between the fraction of edges from S to T and the the expected value from choosing G randomly is "small;" approximately λ times the square root of this fraction. Note that Theorem 1 (1) follows from the Expander Mixing Lemma by setting $T = S^c$ so that $\beta = 1 - \alpha$ and $e(S,T)/ND \geq (1-\lambda)\alpha(1-\alpha) \geq \gamma\alpha/2$.

Given a digraph $G = (V, E)$ with the property that $|e(S,T)/|E| - \alpha\beta| = O(1)$ for all sets S, T with densities α, β, then we call this graph *quasirandom*. The Expander Mixing Lemma implies that a regular digraph with $\lambda(G) = O(1)$ is quasirandom.

THEOREM 1.11. *(Converse to Expander Mixing Lemma)*
Let G be a D-regular, N-vertex undirected graph. Suppose that for all pairs of disjoint vertex sets S, T, we have $|e(S,T)/(ND) - \mu(S)\mu(T)| \leq \theta\sqrt{\mu(S)\mu(T)}$ for some $\theta \in [0,1]$ where $\mu(R) = |R|/N$ for any set R of vertices. Then $\lambda(G) = O(\theta \log(1/\theta))$.

With an adjacency matrix of an n-vertex graph G that we denote by $A = A(G)$, we have n real eigenvalues given by $\lambda_1 \geq \lambda_2 \geq \cdots \geq \lambda_n$. We give some facts relating the spectrum of $A(G)$ to properties of the graph and expansion. [**1**]

LEMMA 1.12. *For a regular directed graph G on N vertices with the transition matrix of the random walk M, then G has spectral expansion $\gamma = 1 - \lambda$ if and only if $M = \gamma J + \lambda E$ where J is the $N \times N$ matrix where each entry is $\frac{1}{N}$ (the matrix of the random walk on the complete graph with self-loops) and $||E|| \leq 1$.*

- If G is d-regular, then $\lambda_1 = d$
- The graph is connected if and only if $\lambda_1 > \lambda_2$. The spectral gap is given by $\lambda_1 - \lambda_2$. Having a larger spectral gap implies that there is larger expansion.
- A substantial spectral gap implies logarithmic diameter.
- For d-regular graphs with the second eigenvalue $\lambda_2 \leq 2\sqrt{d-1}$, they are Ramanujan.
- An ideal expander example is the infinite d-regular tree. With some spectral theory, it is possible to show that the spectrum of the d-regular infinite tree spans the interval $(-2\sqrt{d-1}, 2\sqrt{d-1})$.

2. Construction of a hash function from an expander graph

Based on proposals by [**3**] we describe how to construct the hash function based on expander graphs.

The input to our hash function is used as directions for walking around a graph, without backtracking. To avoid backtracking, it is preferable to walk on a directed graph. The output of the hash function is the ending vertex of the walk. For a certain fixed hash function, the walk starts at a fixed vertex in the given graph. A family of hash functions is defined by choosing varying starting vertices for the walk. We can perform a walk on a k-regular expander graph by converting the input to the hash function to a base $(k-1)$ number whose digits then will dictate which edge to take at each step. The input to the hash function must be broken into chunks of size c such that $2^c = k-1$ for undirected k-regular graphs and $2^c = k$ for directed k-regular graphs. This is to prevent backtracking for undirected graphs. Starting at the first vertex, each step of the walk will choose an edge from the vertex to reach the next vertex. The choice of the edge to follow from the previous is determined by the next c bits of the input. Since there is no backtracking in the walk, only $k-1$ choices for the next edge are allowed at each step for an undirected graph and k choices for a directed graph.

According to Lemma 1.10 random walk on an expander graph mixes very fast, so this tells us that the output of the hash function is uniform given a uniform input.

DEFINITION 2.1. A *Cayley hash* is constructed from a directed Cayley graph $C_{G,S}$. We let $\{1,...,k\}^*$ be a set of sequences $(m_1,...,m_\ell)$ of arbitrary length with elements in $\{1,...,k\}$. If we fix an initial value $g_0 \in G$ and an ordering $\sigma : \{1,...,k\} \to S$, we may determine a Cayley hash function $H : \{1,...,k\}^* \to G$. This is given by $H() = g_0, H(m_1) = g_0\sigma(m_1), ..., H(m_1,...,m_\ell) = H(m_1,...,m_{\ell-1})\sigma(m_\ell)$. These computations, $g_0\sigma(m_1), g_0\sigma(m_1)\sigma(m_2)$, are the same as the walk from v_{g_0} to $v_{g_0\sigma(m_1)}$, etc.

2.1. Required Properties. Graphs used in expander hashes should at least satisfy the following requirements [**15**]:

- **Large expansion** This guarantees that hash values of relatively short messages are well-distributed in the output set. We see that the magnitude of the second largest eigenvalue will let us determine the spectral expansion of the graph.
- **Short diameter** A short diameter guarantees that all vertices correspond to short messages of the output.
- **Large girth** This guarantees that no short collision exists because it will bound the "distance" between any two colliding messages. For Cayley hashes, a large girth is always required.
- **Efficiency** Computation of the neighbors of any given vertex should be very efficient.
- **Collision, preimage, second preimage resistance** Problems in collision, preimage, and second preimage resistances must be hard.

3. Paley Graphs

We now introduce Paley Graphs as a family of nested Ramanujan expander graphs, as discussed by [**3**].

A d-regular graph is called a λ-expander if all its eigenvalues but the largest are in interval $[-\lambda, \lambda]$. We know that random d-regular graphs are good expanders. As stated by [**13**], Paley graphs are a family of graphs that exhibit properties that greatly resemble random graph properties; they are quasirandom [**6**].

In fact, they show that for every n and p there exists a specific efficiently computable graph of size $N' = 2^{\theta(n)}$ and edge density $p' = p \pm \epsilon$ which is $D(n)$-equivalent to $G(N', p')$. In their lemma 2 [**13**], they show that a sequence of Paley graphs satisfy $D(n) - 0/1$ laws with $D(n)$ that is related to the size of nonzero-certifying sets, based on Graham and Spencer's proof of connections between similar Paley graphs and tournament problems [**7**].

A $D(n)-0/1$ law is a natural generalization of the basic 0/1 law: for a number of choices of $p(n)$, any fixed first-order property ϕ holds for random graph $G(2^n, p(n))$ with probability that tends either to 0 or 1 as n grows to infinity [**13**]. A first-order property is a property of a graph expressed in first-order language, where variables stand for vertices and the only relations are equality and adjacency.

Therefore we have that Paley graphs are random-looking graphs that preserve the properties of random graphs, relating to chromatic number, clique number and independence number, maximal and minimal degree, connectivity number, etc. In particular, satisfying the D(0/1) law as much as possible is a property of random graphs and, if one-way functions exist, by Theorem 7 [**13**] this implies the construction is computationally indistinguishable from a random graph $G(2^n, p(n))$. Naor and Naor use sequences of Paley graphs in their optimal construction of a single huge graph that is deterministically constructible and behaves randomly.

Sequences of Ramanujan maps may be constructed as a consequence of the existence of natural maps between the graphs. For every inclusion of totally real fields $L \subset M$, a rational prime p and prime ideals $m_1 m_2 .... m_\alpha = \ell$ where ℓ is a prime of L and m_i is a prime of M, there is a natural "map"

$$G(L; p, \ell) \to G(M; p, m_1, m_2, ..., m_\alpha). \tag{3.1}$$

The map of superspecial graphs $G(L; p, \ell) \to G(M; p, \ell)$ is injective on vertices [**3**].

DEFINITION 3.1. Let p be a prime number and n be a positive integer such that $p^n \equiv 1 \mod 4$. The graph $P = (V, E)$ with

$$V(P) = \mathbb{F}_{p^n} \text{ and } E(P) = \{(x, y) : x, y \in \mathbb{F}_{p^n}, x - y \in (\mathbb{F}_{p^n}*)^2\}$$

is the *Paley graph* of order p^n.

So for $q = p^n$, the vertex set of this graph are elements of this field, and the edge set consist of edges connecting two vertices $x \neq y$ for the case that $x - y$ is a square in $\mathbb{F}_q$. We have that the Paley graph is a $(q-1)/2$ regular graph on q vertices.

We see that the set $E(P)$ in the definition of Paley graph is well-defined because $x - y \in (\mathbb{F}_{p^n}*)^2$ if and only if $y - x \in (\mathbb{F}_{p^n}*)^2$. Also, because $x - y = -1(y - x)$, showing that $-1 \in (\mathbb{F}_{p^n}*)^2$ will prove this fact.

We can express the graph in terms of the quadratic residue character on $\mathbb{F}_q$, $\chi(x) = x^{(q-1)/2}$ (and for q prime, the Legendre symbol is used, $\chi(x) = \frac{x}{q}$). Then the Paley graph (or quadratic residue graph P_q) has vertex set $\mathbb{F}_q$ and edge set $\{xy : \chi(x - y) = 1\}$.

In other words, the connection set is the set of quadratic residues in $GF(p^n)$. [**5**].

We have inclusions $P(q) \hookrightarrow P(q^n)$ for any n and this allows us to take a chain of Ramanujan graphs

$P(p) \to P(p^3) \to P(p^9) \to ...$

There is an advantage in implementing an odd power degree sequence $P(p) \to P(p^3) \to P(p^9) \to ...$ over an even degree sequence such as $P(p) \to P(p^2) \to P(p^4) \to ...$ which is in even degree sequences, any two vertices in $P(p^{2^i})$ become adjacent in $P(p^{2^{i+1}}$ but in odd power degree sequences, the nonadjacent vertices remain nonadjacent.

Therefore we characterize Paley graphs $P(p^n)$ for $n \in \mathbb{N}$ as a "quasirandom sequence,"

DEFINITION 3.2. A *quasirandom sequence* is a sequence of graphs tending to the identically-p function, with density p.

The Paley graph sequence is quasirandom with density $\frac{1}{2}$. We see that quasirandomness is a property of a sequence of graphs instead of a single graph. So we have that as we take the number of vertices tending to infinity, the graph resembles a random graph. This additionally lets us recommend larger number of vertex parameters in the choice of Paley graph.

In the sequence of odd degree Paley graphs, we have homomorphisms of smaller graphs into larger Paley graphs. (pg. 62) [**11**].

DEFINITION 3.3. A *walk* in G is a homomorphism of a path into G so $\hom(P_k, G)$ counts the number of walks with $k-1$ steps in G.

REMARK 3.4. We can express this as the sum of the entries of A^{k-1} where A is the adjacency matrix of G.

In our cryptographic hash function, we conduct a walk into G, which is an element of $\hom(P_k, G)$ given the message is of length k.

3.1. Explicit Construction of Paley Graphs. For $p = 5$, we may calculate the squares:

$$1^2 = 1, \quad 2^2 = 4 = -1 \tag{3.2}$$

Therefore the Paley graph is $Circ(5:1)$, the 5-cycle.

For $p = 13$, we may calculate the squares:

$$1^2 = 1, \quad 2^2 = 4, \quad 3^2 = -4, \quad 4^2 = 3, \quad 5^2 = -1, \quad 6^2 = -3 \tag{3.3}$$

Seeing that the values are in $\{\pm 1, \pm 3, \pm 4\}$, therefore the Paley graph is $Circ(13 : 1, 3, 4)$. In other words, there are connections at every neighboring vertex, every 3 vertices, and every 4 vertices when the vertices are arranged in a cycle [**5**].

We let $A = (a_{ij}), i, j \in \mathbb{F}_q$ be the adjacency matrix of a directed Paley graph, P_q. Then its entries satisfy $a_{ij} = \chi(j - i), i, j \in \mathbb{F}_q$. We can set $B = A - I_q$ and set C as a $(q+1) \times (q+1)$ matrix found from B by adding a row and column of ones to it on the upper left corner, giving that C is a Hadamard matrix.

For $q = 3$ this allows us to obtain the following three matrices,

$$A = \begin{pmatrix} 0 & + & - \\ - & 0 & + \\ + & - & 0 \end{pmatrix}, B = \begin{pmatrix} - & + & - \\ - & - & + \\ + & - & - \end{pmatrix}, C = \begin{pmatrix} + & + & + & + \\ + & - & + & - \\ + & - & - & + \\ + & + & - & - \end{pmatrix}.$$

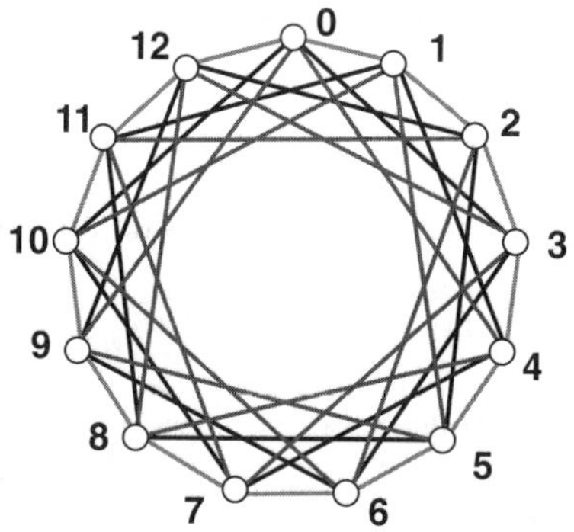

FIGURE 4. Paley Graph of Order 13 [**5**]

DEFINITION 3.5. (Paley Graphs are Strongly regular graphs) A graph is *strongly regular* with parameters (k, a, b) if it is a k-regular incomplete graph. This will give that two adjacent vertices have $a \geq 0$ common neighbors and two nonadjacent vertices will have $b \geq 1$ common neighbors.

We can construct an adjacency matrix from this.

It is possible to show that a Paley graph is a strongly regular graph with q vertices and $((q-1)/2, (q-5)/4, (q-1)/4)$ as its parameters.

Therefore the eigenvalues of the adjacency matrix are $(q-1)/2$ as well as the roots of $x^2 + x - (q-1)/4$, or $(-1 \pm \sqrt{q})/2$. Additionally, these graphs satisfy the bound of $2\sqrt{(q-1)/2 - 1}$ since for $q >> 0$, their ratio is approximately $2\sqrt{2}$.

3.2. Calculation of vertex set and edge set for P_{q^n}. We have seen above some examples of finding vertices and connections in Paley graphs $P(q)$ where q is a prime.

The vertex set of P_9 are the set of elements of the finite field $GF(9)$. In this case, 9 is not a prime number so we must use the following procedure to find the finite field $GF(9) = GF(3^2)$. [**4**]

Step 1: list all monic polynomials of degree $n = 2$ over $GF(3)[x]$:

$$x^2, x^2+x+2, x^2+1, x^2+2x, x^2+2, x^2+2x+1, x^2+x, x^2+2x+2, x^2+x+1. \quad (3.4)$$

Step 2: Find the irreducible polynomials in the list:

$$x^2+1,\ x^2+x+2,\ x^2+2x+2. \quad (3.5)$$

(Above polynomials $p(x)$ are irreducible in $\mathbb{Z}_3$, since $p(0), p(1), p(2) \neq 0 \mod 3$).

Step 3: Choose an irreducible polynomial; say x^2+x+2.

We let a be a zero of x^2+x+2, and because $x^2+x+2 \equiv 0 \mod (x^2+x+2)$ we have the relations,

$$a^2 = -a - 2 = 2a + 1 \mod 3 \quad (3.6)$$

Step 4: Construct elements of $GF(9)$ by writing down linear combinations of the basis $\{1, a\}$ over $GF(3)$:

$$0,\ 1,\ 2,\ a,\ a+1,\ a+2,\ 2a,\ 2a+1,\ 2a+2 \quad (3.7)$$

Note that the powers of a correspond to the above list:

(3.8) $a^1 = a, a^2 = 2a{+}1, a^3 = 2a{+}2, a^4 = 2, a^5 = 2a, a^6 = a{+}2, a^7 = a{+}1, a^8 = 1.$

These elements are the vertex set.

Step 5: Find the quadratic residues by choosing an irreducible polynomial; say $x^2 + 1$.

Let a be a zero of this polynomial. Then by $x^2+1 \equiv 0 \mod (x^2+1)$, we know that $a^2 = -1$.

We find the quadratic residues are:

$$\begin{aligned}(3.9) \qquad 1^2 = 1,\ a^2 = -1 = 2,\ (2a)^2 = 2, 2^2 = 1,\ (a+1)^2 \\ = 2a,\ (2a+1)^2 = a,\ (a+2)^2 = a,\ (2a+2)^2 = 2a.\end{aligned}$$

Step 6: Test vertices of $P(9)$ to see if they are adjacent.

We give a list of calculations for vertices of $P(9)$. When the difference of two vertices is a quadratic residue, they are adjacent.

For example, we list the calculations of some edges in $P(9)$:

TABLE 1. If QR, then there is edge

u - v	difference	QR	edge
(2+2a) - 2	2a	Yes	Yes
1-2	2	Yes	Yes
1- (1+2a)	a	Yes	Yes
(2+2a) - (1+a)	1+a	No	No
(2+2a) - (1+a)	1+a	No	No

4. 2-Lifted Paley Graphs Based Hash Function Scheme

We propose the following three-step process for our expander hash function.

(1) Choose a signing from the collection of length k signings, $\{s\}$

We have that randomness is maximized for there are 2^m possible 2-lifts for m edges. Choosing a signing is a similar process to choosing a random initialization vector to a hash function, as a key to a block cipher function.

(2) Apply the signing to the Payley graph P_q for blocks of the message.

Bilu-Linial's 2-lifts are used to increase the expander properties of the graph. They were created as an operation that doubles the size of the graph without blowing up the eigenvalues. They are able to reduce the degree of the edges of the Paley graph while keeping the eigenvalues within the original bound that implies good spectral expansion.

(3) For each lifted Payley graph, we conduct a random walk on the vertices based on the message.

We have that a Paley graph is a Cayley graph over the group of integers modulo a prime p, denoted as $\mathbb{Z}/p$ [**18**]. Generators are given by quadratic residues modulo p. We use the procedure described in Definition 2.1 to have the final result.

5. Expander Properties of 2-lifted Graphs

We now state the results that show that 2-lifted graphs satisfy properties of expander graphs. The eigenvalues of a 2-lift of a graph G are within the bounds of the base graph's eigenvalues by Bilu-Linial's analysis. Furthermore, they conjecture that every d-regular graph has a signing with spectral radius at most $2\sqrt{d-1}$.

LEMMA 5.1. *Let A be the adjacency matrix of a graph G, and let A_s be the signed adjacency matrix associated with a 2-lift $\hat{G}$. Then every eigenvalue of A and of A_s are eigenvalues of $\hat{G}$. In addition, the multiplicity of each eigenvalue of $\hat{G}$ is the sum of its multiplicities in A and A_s.*

PROOF. We have that the adjacency matrix of $\hat{G}$ is,

$$\hat{A} = \begin{pmatrix} A_1 & A_2 \\ A_2 & A_1 \end{pmatrix} \tag{5.1}$$

where A_1 is the adjacency matrix of $(V, s^{-1}(1))$ and A_2 is the adjacency matrix of $(V, s^{-1}(-1))$, so that $A = A_1 + A_2$ and $A_s = A_1 - A_2$. Let v be the eigenvector of eigenvalue λ_1. Then $\hat{v} = (vv)$ is an eigenvector of $\hat{A}$ with eigenvalue λ_1. Similarly, if u is an eigenvector of A_s with eigenvalue λ_2 then $\hat{u} = (u - u)$ is an eigenvector of $\hat{A}$ with eigenvalue λ_2. Because any $\hat{v}$ is perpendicular with any $\hat{u}$ and they form $2n$ eigenvectors, they span $\hat{A}$ and are its eigenvectors. □

Therefore, we have the following [**1**]:

PROPOSITION 5.2. The new eigenvalues of a 2-lift of G are the eigenvalues of the corresponding signing matrix.

This will lead to their following conjecture [**19**],

CONJECTURE 5.3. Every d-regular adjacency matrix A has a signing A_S with $||A_S|| \leq 2\sqrt{d-1}$.

5.1. Choosing signings. We give the following definitions of good expanders and signings and their composition as given by [**2**]. Then we state the lemma that allows a good signing to be composed with a good expander family to give a good expander family of graphs.

DEFINITION 5.4. Let $f_n : \{0,1\}^t \times [n] \times [n] \to \{0,1\}$, with $t = O(\log n)$. For $r \in \{0,1\}^t, f_n$ defines a graph on n vertices, $G_{f_n}(r)$ where edge i and j are adjacent if and only if $f_n(r,i,j) = 1$. We say that f_n is a *∂-probabilistic strong explicit description of a (n,d)-good expander graph* if given n, f_n can be computed in *poly* $\log(n)$ time and with probability of at least $1 - \partial$ (over a uniform choice of r), $G_{f_n}(r)$ is an (n,d)-good expander graph.

DEFINITION 5.5. Let $h_n : \{0,1\}^t \times [n] \times [n] \to \{-1,1\}$, with $t = O(\log n)$. For $r \in \{0,1\}^t, h_n$ defines a signing s_{h_n} of G on n vertices, given by $s_{h_n}(r)(i,j) = h_n(r,i,j)$. We say that h_n is a *∂-probabilistic strongly explicit description of a (n,d)-good signing* if given n, h_n can be computed in *poly* $\log(n)$ time and for any $\log_2 n, \gamma(d)$-sparse d-regular graph G on n vertices, h_n defines an (n,d)-good signing with probability of at least $1 - \partial$ (over a uniform choice of r).

LEMMA 5.6. *Let f_n be a ∂_1-probabilistic strongly explicit description of an (n,d)-good expander, and h_n a ∂_2-probabilistic strongly explicit description of an*

(n,d)-good signing. Then their composition is a $(\eth_1+\eth_2)$-probabilistic strongly explicit description of an $(2n,d)$-good expander.

Bilu and Linial's discussion [**2**] shows us that such explicitness can be achieved for constructions based on 2-lifts.

An explicit construction of such sample space is suggested by [**13**].

In their construction, the size of the sample space is polynomial in m when $k=O(\log m)$ and $1/\epsilon = poly(m)$.

In their description, they consider a (ϵ,k)-wise independent space Ω_m as a function $w:\{0,1\}^t\to\{-1,1\}^m$ where $|\Omega_m|=2^t$. From [**13**] it follows that w can be computed efficiently and also that given $r\in\{0,1\}^t, p\in[m]$, the pth coordinate of $w(r)$, denoted by $w(r)_p$, can be computed efficiently (in $poly\log(m)$ time).

Take $k=\binom{n}{2}$ and index the elements of $\{-1,1\}^k$ into unordered pairs $(i,j)\in\binom{[n]}{2}$. Let function $h_n(r,i,j)=w(r)_{i,j}$ be a description of a signing. [**2**] gives discussion showing that h_n is a $\frac{1}{n^2}$-probabilitistic strongly explicit description of an (n,d)-good signing for k,ϵ.

For G an (n,d)-good expander, with $n\geq\frac{1}{\eth}$ For $0\leq i\leq\ell=\log_2(N/n)$, we define $n_i:=n(2^i), m_i:=\binom{n_i}{2}$, and $k_i=d\log_2 n_i$. We have a description of an (ϵ_i,k_i)-wise independent space of bit strings of length m_i given by $w_i:\{0,1\}^{t_i}\to\{-1,1\}^{m_i}$, where element ϵ_i is chosen uniformly at random from this space. This gives us an (n_i,d)-good signing with probability at least $1-\frac{1}{n_i^2}$.

So we have a collection of $\frac{1}{n_i^2}$-probabilistic strongly explicit descriptions of (n_i,d)-good signings given by $h_{n_i}(r,p,q)=w_i(r)_{p,q}$.

Now with our description of paley graph $G=P_q$ denoted by f_n, if adjacency in G can be decided in $poly\log(n)$ then f_n is a 0-probabilistic strongly explicit description of an (n,d)-good expander. Then it follows from Lemma 5.6 the composition of these functions gives an $\frac{1}{n}$-probabilistic strongly explicit description of an (N,d)-good expander graph.

6. Implementation Efficiency

We analyze the efficiency of our construction step-by-step.

6.1. Step 1. While constructing a probability space that stores all 2^n possible assignments of $x_1,...,x_n\in\{0,1\}$ Bernouilli random variables would require much storage, we are able to construct a much smaller probability space, given by Naor and Naor [**12**]. The number of random bits needed in order to generate the random variables is $O(\log n+\log\frac{1}{\epsilon})$.

When $k=O(\log m)$ and $1/\epsilon=poly(m)$, the size of the sample space is polynomial in m.

They require a 3-stage process to generate the random variables.

- A polynomial size family $\mathcal{F}$ of $\{0,1\}^n$ vectors is generated with the property that for all subsets $S\subseteq\{1,...,n\}$, for a vector r chosen from $\mathcal{F}$ uniformly at random, $\Pr\chi'_S(r)=1\geq\beta$ for some constant β.

- The vectors $r_1, ..., r_\ell$ are sampled via a Markov Process from $\mathcal{F}$ such that for all subsets $S \subseteq \{1, ..., n\}, \Pr \forall i, 1 \leq i \leq \ell, \chi'_S(r_i) = 0 \leq \epsilon$, and ℓ will turn out to be $O(\log \frac{1}{\epsilon})$.
- Let $\vec{a} = (a_1, ..., a_\ell)$ be chosen uniformly at random from $\{0,1\}^\ell$. The assignment to random variables $x_1, ..., x_n$ is a combination of the vectors sampled at the stage above,

$$x = \sum_{i=1}^{\ell} a_i r_i \tag{6.1}$$

A method is presented that involves the ring $GF[2]$. Assume that n is a power of 2. Here is the algorithm to generate a random vector r. This algorithm uses $O(\log n)$ random bits and generates a distinguisher with probability of at least $\beta = \frac{1}{8}$. The complexity of computing the value of the random variable x_i is $O(1)$ operations on words of size $O(\log n)$ [**12**].

6.2. Step 2. Based on [**13**] (Theorem 10), there exists a deterministically efficiently computable huge graph $G = \{g_n\}_n$ which satisfies the $D(n) - 0/1$ law for $D(n) = \frac{n}{2\log(1/\epsilon)}(1 - o(1))$, and where g_n has size $2^{\theta(n)}$ and edge density $p'(n)$ such that $|p'(n) - p(n)| < \epsilon(n)$.

It is possible to deterministically compute an irreducible polynomial of degree n' in $GF(3)[X]$ in time $poly(n') = poly(n)$ and therefore can efficiently calculate the field $F = GF(3^{n'})$. This follows from Shoup's theorem for computing irreducible polynomials in $GF[q][X]$ in time $poly(q, m)$ [**16**], [**17**]. This is in the case where we need to compute the monic irreducible polynomials for a P_{q^n} graph where q is prime and n is an integer power.

In the case where we have P_q and q is prime, then it is pretty efficient to compute the neighbors of any given vertex. It only requires us to consider $\lceil q/2 \rceil$ integers, square them, then modulo by q and find out if that result or if q minus it is a smaller value to keep.

6.3. Step 3. In our example with the Paley-13 graph, operations cost per bit of message is about 13 additions. As shown in AppendixA, the cost per bit of input to the hash function is approximately q additions for P_q, q prime.

7. Cryptanalysis of our hash function

7.1. Large expansion. Paley graphs are optimal expander graphs, so they have large expansion constants. Applying the 2-lift operation is seen to preserve the value of the second largest eigenvalue by Lemma 5.1 and therefore preserves the expansion property.

By definition of the expansion constant c we know that it is at least $\frac{2(k-\mu_1)}{2k-2\mu_1}$ and therefore depends on the first and second largest eigenvalues.

7.2. Short diameter. Paley graphs have short diameters; for our choice of Paley graphs with recommended q prime and $10 \leq q \leq 200$ it is possible to see that the diameter is bounded by q. For odd degree exponents n for Paley graphs, P_{p^n}, it can be shown that the diameter is equal to 2.

7.3. Large girth (length of the shortest cycle in graph). For Cayley hashes, a large girth is always required.

For recommended parameters, q large, the girth of the graph P_q will have lower bounds of approximately $\lceil q/2 \rceil$.

The girth of a random k-regular graph is known to be $\log_{k-1} n$. Since we are working with 2-lifted Paley graphs, we see that they are random, since we have applied the random 2-lift.

From the parameters of $P(q)$, we know that it is $(q-1)/2$-regular from Definition 3.5. So we know that the eigenvalues of the 2-lifted Paley graphs are the same and therefore they are also $(q-1)/2$-regular. We calculate that the girth is equal to $\log_{(q-1)/2-1} q$ for a Paley graph on q vertices.

Therefore, as we choose q large, we will see that Paley graphs may not have very large cycle lengths, but we argue that the 2-lifts can make cycles in the graph will lead to larger girth.

We know that with high probability that after one edge flip or lift between two copies of the Paley graphs that the girth will have increased by a factor of two. This is possible since we have now connected the shortest cycles in each graph rooted at a certain vertex to each other, doubling the length of the path of the shortest cycle. Now given that a nearby vertex is connected to the vertex where there was an edge lift, we will further increase the chance of making the girth larger.

7.4. Efficiency. Computation of the neighbors of any given vertex should be very efficient. As seen in Appendix A, the computation of neighbors of any vertex is simply addition modulo q by a certain value in the set of quadratic residues of $GF(q)$. It just takes 1 addition modulo p^n for P_{p^n}.

7.5. Collision, preimage, second preimage resistance. Finding collisions in the graph is similar to finding cycles in the graph. While we have large girth for our graph, we can also decrease the existing cycles with our tool of 2-lifts, which will break cycles by crossing edges. If the graph does not have small cycles, then this problem is very hard, and is related to the difficulty of solving the quadratic residuosity problem. Since our attacker is assumed not to know the initialization vector of our 2-lift, or the signing, this makes it increasingly difficult for them to decrypt the hash values.

8. Acknowledgements

We thank Zvi Kedem for insightful discussions and helpful comments. Also, we thank the members of the GC Cryptography Seminar for their talks that clarified some concepts that are used. Lastly, I would like to thank Delaram Kahrobaei for her suggestions in revising the paper and for her encouragement.

Appendix A. Examples and Analysis

A.1. Example with P_{13}. We know that a 2-lift of a graph is given by a ± 1 assignment of edges of the graph, corresponding to which edge is duplicated or to be lifted with a flip.

We choose the signing $s \in \{0,1\}^m$ randomly.

We calculate the edges of P_{13} using the method of squaring numbers from

$$1, ..., \lceil 13/2 \rceil$$

.

So we have edge set given by $Circ\{13 : 1, 3, 4\}$.

The P_{13} graph has 39 edges.

Step 1:

Say we have a signing for a 13-Paley graph given by

$[1, -1, 1, 1, -1, 1, 1, 1, 1, -1, 1, 1, 1, 1, 1, 1, 1, 1, 1, -1, 1, 1, 1, 1, 1, 1, 1, 1, 1, 1, 1, 1, 1, 1, 1, 1, 1, 1, -1]$ (flip the 2nd, 5th, 10th, 20th, and last edges).

Step 2:

From the above calculation, we have that the adjacency matrix is:

Now say we have a message given by m=" A scenic route to Los Angeles is along the coast."

The equivalent ASCII in binary code for this is: 00100000 01000001 00100000 01110011 01100011 01100101 01101110 01101001 01100011 00100000 01110010 01101111 01110101 01110100 01100101 00100000 01110100 01101111 00100000 01001100 01101111 01110011 00100000 01000001 01101110 01100111 01100101 01101100 01100101 01110011 00100000 01101001 01110011 00100000 01100001 01101100 01101111 01101110 01100111 00100000 01110100 01101000 01100101 00100000 01100011 01101111 01100001 01110011 01110100.

Step 3:

For a prime power q, the P_q graph is $Cay(\mathbb{Z}/p\mathbb{Z}, S)$ where S is the set of quadratic residues or generators.

In the base P_{13} graph, $P_{13} = Cay(\mathbb{Z}/13\mathbb{Z}, \{1, 3, 4\})$.

So there are 3 possible generators in this case, given by actions of $\pm 1, \pm 3, \pm 4$ on the set of vertices $\{1, ..., 13\}$.

Call these operations $a_0 = +1, a_1 = +3, a_2 = +4, a_3 = -1, a_4 = -3, a_5 = -4$.

This gives us the mapping

$$\begin{aligned} \pi : \{a_0, a_1, ..., a_5\} &\to \{0, 1, 2, 3, 4, 5\} \\ a_0 &\mapsto 0, \\ a_1 &\mapsto 1, \\ a_2 &\mapsto 2, \\ a_3 &\mapsto 3 \\ a_4 &\mapsto 4 \\ a_5 &\mapsto 5 \end{aligned}$$

We have that on the P_{13} graph that there are 39 edges, clearly seen from the fact that there are 3 sets of edges on the vertex set of size 13.

Since there are 6 maps, we can use the ASCII code in base 6.

In the modified graph, we have a vertex set of $\{1, ..., 13, 1', ..., 13'\}$ so now we modulo out by 26.

In addition, for certain pairs of edges, they no longer connect to the same pairs of vertices but have switched connections.

For this example, the pairs of edges that have switched connections include $(13, 1)$ and $(13', 1') = (39, 14)$, $(4, 5)$ and $(17, 18)$, $(5, 8)$ and $(18, 21)$, $(3, 12)$ and $(16, 25)$, $(9, 13)$ and $(22, 39)$. Choose a starting vertex $v_0 = 1$ in the construction. So we have generating set $\{\pm 1, \pm 3, \pm 4\}$ except at edges $(1, 13), (39, 14), (4, 5), (17, 18)$ which do not connect in terms of $+1, -1$ (respectively), $(5, 8), (18, 21), (3, 12), (16, 25)$ by $+4, -4$, and $(9, 13)$ and $(22, 39)$ by $+3, -3$. In these cases, for example, we see that 1 is connected to 39 and 13 is connected to 14. We want to add in additive maps by an amount of 13 and in subtractive maps an amount of 13 to the previous coordinates.

So in these pairs we must add an extra add 13 map to the maps a_3 when the walk reaches $1, 14, 5, 18$, to map a_0 when walk reaches $13, 39, 4, 17$, a_1 when the walk reaches $5, 18, a_4$ for $8, 21$, and a_2 for $3, 16, 9, 22, a_5$ when the walk encounters $12, 25, 13, 39$.

The algorithm will decompose a message m into μ digits, $m = m_0 ... m_{\mu-1}$ such that $m_i \in \{0, ..., 5\}$ and some size μ blocks. We know that $2^\mu = 12$ for undirected 13-regular graphs, giving $\mu \approx 4$.

2223 5022 3223 3141 1203 3143 4122 4125 1502 3000 5051 5524 1015 2324 2110 4411 5222 4510 0215 3234 5154 2553 3102 0541 3305 1540 3101 5125 2405 3544 2420 3534 3445 0004 5500 3524 4252 500

Now we will let each block of integers dictate the walk with the starting vertex as 1.

For example, for the first block, we have operations $2 = a_2 = +4, 3 = a_3 = -1, 5 = a_5 = -4, 0 = a_0 = +1, 1 = a_1 = +3, 4 = a_4 = -3$.

So 1 is mapped to 5, then to 9, then to 13, then to 12.

We note that the walk is still in $\mod 13$ unless we reach one of the crossed edges, where it will then be in the duplicated graph of vertices labelled $\{(\mod 13) + 13\}$. We know that there must be extra storage that stores the vertices that correspond to crossed edges.

So we can see therefore that computing the walk on each block will take 13 additions in $F_q = F_{13}$. An integer addition in F_q is the same as integer addition plus another addition by $-q$ if the value of the result is over 13 or 26 depending on the copy of the graph P_q. So the cost per bit of message is about 13 additions.

List of Figures

References

[1] A. Amit and N. Linial, *Random lifts of graphs: edge expansion*, Combin. Probab. Comput. **15** (2006), no. 3, 317–332, DOI 10.1017/S0963548305007273. MR2216470

[2] Y. Bilu and N. Linial, *Lifts, discrepancy and nearly optimal spectral gap*, Combinatorica **26** (2006), no. 5, 495–519, DOI 10.1007/s00493-006-0029-7. MR2279667

[3] D. X. Charles, E. Z. Goren, and K. E. Lauter, *Families of Ramanujan graphs and quaternion algebras*, Groups and symmetries, CRM Proc. Lecture Notes, vol. 47, Amer. Math. Soc., Providence, RI, 2009, pp. 53–80. MR2500554

[4] A. Dy, *Constructing the 9-paley graph*, https://sites.google.com/site/mymathclassroom/discrete-mathematics/constructing-the-9-paley-graph.

[5] A.N. Elsawy, *Paley graphs and their generalizations*, http://arxiv.org/pdf/1203.1818v1.pdf, 2012.

[6] F. R. K. Chung, R. L. Graham, and R. M. Wilson, *Quasi-random graphs*, Combinatorica **9** (1989), no. 4, 345–362, DOI 10.1007/BF02125347. MR1054011

[7] R. L. Graham and J. H. Spencer, *A constructive solution to a tournament problem*, Canad. Math. Bull. **14** (1971), 45–48. MR0292715

[8] S. Hoory, N. Linial, and A. Wigderson, *Expander graphs and their applications*, Bull. Amer. Math. Soc. (N.S.) **43** (2006), no. 4, 439–561 (electronic), DOI 10.1090/S0273-0979-06-01126-8. MR2247919

[9] M. Krebs and A. Shaheen, *Expander families and Cayley graphs*, Oxford University Press, Oxford, 2011. A beginner's guide. MR3137611

[10] D. A. Levin, Y. Peres, and E. L. Wilmer, *Markov chains and mixing times*, American Mathematical Society, Providence, RI, 2009. With a chapter by James G. Propp and David B. Wilson. MR2466937

[11] L. Lovász, *Large networks and graph limits*, American Mathematical Society Colloquium Publications, vol. 60, American Mathematical Society, Providence, RI, 2012. MR3012035

[12] J. Naor and M. Naor, *Small-bias probability spaces: efficient constructions and applications*, SIAM J. Comput. **22** (1993), no. 4, 838–856, DOI 10.1137/0222053. MR1227764

[13] M. Naor, A. Nussboim, and E. Tromer, *Efficiently constructible huge graphs that preserve first order properties of random graphs*, Theory of cryptography, Lecture Notes in Comput. Sci., vol. 3378, Springer, Berlin, 2005, pp. 66–85, DOI 10.1007/978-3-540-30576-7_5. MR2168478

[14] A. Nilli, *On the second eigenvalue of a graph*, Discrete Math. **91** (1991), no. 2, 207–210, DOI 10.1016/0012-365X(91)90112-F. MR1124768

[15] C. Petit, K. Lauter, and J.-J. Quisquater, *Cayley Hashes: A Class of Efficient Graph-based Hash Functions*, Preprint (2007).

[16] V. Shoup, *New algorithms for finding irreducible polynomials over finite fields*, Math. Comp. **54** (1990), no. 189, 435–447, DOI 10.2307/2008704. MR993933

[17] V. Shoup, *Searching for primitive roots in finite fields*, Math. Comp. **58** (1992), no. 197, 369–380, DOI 10.2307/2153041. MR1106981

[18] D. A. Spielman, *Rings, Paths, and Paley Graphs*, http://www.cs.yale.edu/homes/spielman/561/lect05-12.pdf, 2012.

[19] N. Srivastava, *Graph sparsification iii: Interlacing polynomials and ramanujan graphs of every degree*, 2014.

[20] S. P. Vadhan, *Pseudorandomness*, Found. Trends Theor. Comput. Sci. **7** (2011), no. 1-3, front matter, 1–336. MR3019182

251 Mercer St., New York, New York 10012
E-mail address: ssy248@nyu.edu

Selected Published Titles in This Series

677 **Delaram Kahrobaei, Bren Cavallo, and David Garber, Editors,** Algebra and Computer Science, 2016

674 **Bogdan D. Suceavă, Alfonso Carriazo, Yun Myung Oh, and Joeri Van der Veken, Editors,** Recent Advances in the Geometry of Submanifolds, 2016

673 **Alex Martsinkovsky, Gordana Todorov, and Kiyoshi Igusa, Editors,** Recent Developments in Representation Theory, 2016

672 **Bernard Russo, Asuman Güven Aksoy, Ravshan Ashurov, and Shavkat Ayupov, Editors,** Topics in Functional Analysis and Algebra, 2016

671 **Robert S. Doran and Efton Park, Editors,** Operator Algebras and Their Applications, 2016

670 **Krishnendu Gongopadhyay and Rama Mishra, Editors,** Knot Theory and Its Applications, 2016

669 **Sergiĭ Kolyada, Martin Möller, Pieter Moree, and Thomas Ward, Editors,** Dynamics and Numbers, 2016

668 **Gregory Budzban, Harry Randolph Hughes, and Henri Schurz, Editors,** Probability on Algebraic and Geometric Structures, 2016

667 **Mark L. Agranovsky, Matania Ben-Artzi, Greg Galloway, Lavi Karp, Dmitry Khavinson, Simeon Reich, Gilbert Weinstein, and Lawrence Zalcman, Editors,** Complex Analysis and Dynamical Systems VI: Part 2: Complex Analysis, Quasiconformal Mappings, Complex Dynamics, 2016

666 **Vicenţiu D. Rădulescu, Adélia Sequeira, and Vsevolod A. Solonnikov, Editors,** Recent Advances in Partial Differential Equations and Applications, 2016

665 **Helge Glöckner, Alain Escassut, and Khodr Shamseddine, Editors,** Advances in Non-Archimedean Analysis, 2016

664 **Dihua Jiang, Freydoon Shahidi, and David Soudry, Editors,** Advances in the Theory of Automorphic Forms and Their L-functions, 2016

663 **David Kohel and Igor Shparlinski, Editors,** Frobenius Distributions: Lang-Trotter and Sato-Tate Conjectures, 2016

662 **Zair Ibragimov, Norman Levenberg, Sergey Pinchuk, and Azimbay Sadullaev, Editors,** Topics in Several Complex Variables, 2016

661 **Douglas P. Hardin, Doron S. Lubinsky, and Brian Z. Simanek, Editors,** Modern Trends in Constructive Function Theory, 2016

660 **Habib Ammari, Yves Capdeboscq, Hyeonbae Kang, and Imbo Sim, Editors,** Imaging, Multi-scale and High Contrast Partial Differential Equations, 2016

659 **Boris S. Mordukhovich, Simeon Reich, and Alexander J. Zaslavski, Editors,** Nonlinear Analysis and Optimization, 2016

658 **Carlos M. da Fonseca, Dinh Van Huynh, Steve Kirkland, and Vu Kim Tuan, Editors,** A Panorama of Mathematics: Pure and Applied, 2016

657 **Noé Bárcenas, Fernando Galaz-García, and Mónica Moreno Rocha, Editors,** Mexican Mathematicians Abroad, 2016

656 **José A. de la Peña, J. Alfredo López-Mimbela, Miguel Nakamura, and Jimmy Petean, Editors,** Mathematical Congress of the Americas, 2016

655 **A. C. Cojocaru, C. David, and F. Pappalardi, Editors,** SCHOLAR—a Scientific Celebration Highlighting Open Lines of Arithmetic Research, 2015

654 **Carlo Gasbarri, Steven Lu, Mike Roth, and Yuri Tschinkel, Editors,** Rational Points, Rational Curves, and Entire Holomorphic Curves on Projective Varieties, 2015

For a complete list of titles in this series, visit the
AMS Bookstore at **www.ams.org/bookstore/conmseries/**.